METHODS IN MOLECULAR BIOLOGY

Series Editor
John M. Walker
School of Life and Medical Sciences
University of Hertfordshire
Hatfield, Hertfordshire, UK

For further volumes:
http://www.springer.com/series/7651

For over 35 years, biological scientists have come to rely on the research protocols and methodologies in the critically acclaimed *Methods in Molecular Biology* series. The series was the first to introduce the step-by-step protocols approach that has become the standard in all biomedical protocol publishing. Each protocol is provided in readily-reproducible step-by-step fashion, opening with an introductory overview, a list of the materials and reagents needed to complete the experiment, and followed by a detailed procedure that is supported with a helpful notes section offering tips and tricks of the trade as well as troubleshooting advice. These hallmark features were introduced by series editor Dr. John Walker and constitute the key ingredient in each and every volume of the *Methods in Molecular Biology* series. Tested and trusted, comprehensive and reliable, all protocols from the series are indexed in PubMed.

Effector-Triggered Immunity

Methods and Protocols

Edited by

Thomas A. Kufer

Department of Immunology, Institute of Nutritional Medicine, University of Hohenheim, Stuttgart, Baden-Württemberg, Germany

Maria Kaparakis-Liaskos

Department of Physiology Anatomy & Microbiology, La Trobe University, Bundoora, VIC, Australia

Humana Press

Editors
Thomas A. Kufer
Department of Immunology, Institute of Nutritional Medicine
University of Hohenheim
Stuttgart, Baden-Württemberg, Germany

Maria Kaparakis-Liaskos
Department of Physiology Anatomy & Microbiology
La Trobe University
Bundoora, VIC, Australia

ISSN 1064-3745 ISSN 1940-6029 (electronic)
Methods in Molecular Biology
ISBN 978-1-0716-2448-7 ISBN 978-1-0716-2449-4 (eBook)
https://doi.org/10.1007/978-1-0716-2449-4

This Humana imprint is published by the registered company Springer Science+Business Media, LLC, part of Springer Nature.
The registered company address is: 1 New York Plaza, New York, NY 10004, U.S.A.

Preface

Effector triggered immunity (ETI), first described in plants, is the induction of defense pathways mediated by recognition of the action of pathogen-derived effector molecules in the host cell. Recent evidence supports that this system also works in mammals to complement the detection of microbes by pattern-recognition receptors (PRR) that recognize conserved microbial structures, commonly referred to as pathogen-associated molecular patterns (MAMP).

The molecular details of this sophisticated immune response are emerging and offer new insights into infection biology and immunity. In the volume *Effector Triggered Immunity in Plants and Animals* of the Methods in Molecular Biology series, we present a collection of 18 protocols to study ETI both in plants and animals from eminent groups in the field. Topics covered in this volume include the genetic manipulation of plant and animal pathogens and host cells and the analysis of key host responses, with a focus on mammalian cells. These include techniques used for the analysis of inflammasome activation, cell death pathways, and mitochondria damage in response to pathogens. Technically these protocols cover a broad spectrum of immunological, biochemical, cell biological, and structural biology approaches to examine ETI. Each chapter discusses recent literature in the field and provides detailed, step-by step protocols with additional tips and tricks to provide a useful companion for expert researchers, as well as great resource for the novice researcher, or researchers new to the ETI field. All these make this book a valuable resource for lab scientists ranging from the student up to the PI level.

Stuttgart, Baden-Württemberg, Germany — ***Thomas A. Kufer***
Bundoora, VIC, Australia — ***Maria Kaparakis-Liaskos***

Contents

Contributors

SYEDA FARHANA AFROZ • *Institute for Molecular Bioscience (IMB), The University of Queensland, Brisbane, QLD, Australia; IMB Centre for Inflammation and Disease Research, The University of Queensland, Brisbane, QLD, Australia; Australian Infectious Diseases Research Centre, The University of Queensland, Brisbane, QLD, Australia*

NATALIE J. BITTO • *Department of Physiology, Anatomy and Microbiology, La Trobe University, Melbourne, VIC, Australia; Research Centre for Extracellular Vesicles, School of Molecular Sciences, La Trobe University, Melbourne, VIC, Australia*

PETR BROZ • *Department of Biochemistry, University of Lausanne, Epalinges, Switzerland*

YUEN-YAN CHANG • *Dynamics of Host-Pathogen Interactions Unit, Institut Pasteur, UMR3691 CNRS, Paris, France; Division of Molecular and Cellular Biology, National Institute of Child Health and Human Development, NIH, Bethesda, MD, USA*

DARIA CHIRITA • *CIRI (Centre International de Recherche en Infectiologie), Univ Lyon, Inserm U1111, Lyon, France; Université Claude Bernard Lyon 1, CNRS, UMR5308, ENS de Lyon, Univ Lyon, Lyon, France*

SEONG H. CHOW • *Infection Program, Biomedicine Discovery Institute and Department of Biochemistry & Molecular Biology, Monash University, Clayton, VIC, Australia*

NICHOLAS D. CONDON • *Institute for Molecular Bioscience (IMB), The University of Queensland, Brisbane, QLD, Australia*

BEN A. CROKER • *Division of Allergy, Immunology & Rheumatology, Department of Pediatrics, University of California San Diego, La Jolla, CA, USA*

AKSHAY A. D'CRUZ • *Centre for Inflammatory Diseases, Department of Medicine, School of Clinical Sciences, Monash University, Clayton, VIC, Australia*

BENJAMIN DEMARCO • *Department of Biochemistry, University of Lausanne, Epalinges, Switzerland; Kennedy Institute of Rheumatology, NDORMS, University of Oxford, Oxford, UK*

KORNELIA ELLWANGER • *Department of Immunology, Institute of Nutritional Medicine, University of Hohenheim, Stuttgart, Baden-Württemberg, Germany*

JOST ENNINGA • *Dynamics of Host-Pathogen Interactions Unit, Institut Pasteur, UMR3691 CNRS, Paris, France*

MARIA ERICSSON • *Department of Cell Biology, Harvard Medical School, Boston, MA, USA*

MATTHIAS GEYER • *Institute of Structural Biology, University Hospital Bonn, University of Bonn, Bonn, Germany*

ELIZABETH L. HARTLAND • *Centre for Innate Immunity and Infectious Diseases, Hudson Institute of Medical Research, Clayton, VIC, Australia; Department of Molecular and Translational Science, Faculty of Medicine, Nursing and Health Sciences, Monash University, Clayton, VIC, Australia*

THOMAS HENRY • *CIRI (Centre International de Recherche en Infectiologie), Univ Lyon, Inserm U1111, Lyon, France; Université Claude Bernard Lyon 1, CNRS, UMR5308, ENS de Lyon, Univ Lyon, Lyon, France*

ADRIAN HEROD • *Department of Microbiology and Immunology, Dalhousie University Halifax, Halifax, NS, Canada*

INGA V. HOCHHEISER • *Institute of Structural Biology, University Hospital Bonn, University of Bonn, Bonn, Germany*

Veit Hornung • *Gene Center and Department of Biochemistry, Ludwig-Maximilians-Universität München, Munich, Germany*
Yvan Jamilloux • *CIRI (Centre International de Recherche en Infectiologie), Univ Lyon, Inserm U1111, Lyon, France; Université Claude Bernard Lyon 1, CNRS, UMR5308, ENS de Lyon, Univ Lyon, Lyon, France; Internal Medicine, University Hospital Croix-Rousse, Hospices Civils de Lyon, Lyon, France*
Kostya Kanyuka • *Plant Pathology and Entomology, NIAB, Cambridge, UK; Biointeractions and Crop Protection, Rothamsted Research, Harpenden, UK*
Maria Kaparakis-Liaskos • *Department of Physiology, Anatomy and Microbiology, La Trobe University, Melbourne, VIC, Australia; Research Centre for Extracellular Vesicles, School of Molecular Sciences, La Trobe University, Melbourne, VIC, Australia*
Ronan Kapetanovic • *Institute for Molecular Bioscience (IMB), The University of Queensland, Brisbane, QLD, Australia; IMB Centre for Inflammation and Disease Research, The University of Queensland, Brisbane, QLD, Australia; Australian Infectious Diseases Research Centre, The University of Queensland, Brisbane, QLD, Australia; Friedrich Miescher Institute for Biomedical Research, Basel, Switzerland*
Thomas A. Kufer • *Department of Immunology, Institute of Nutritional Medicine, University of Hohenheim, Stuttgart, Baden-Württemberg, Germany*
Franz Leissing • *RWTH Aachen University, Institute for Biology I, Unit of Plant Molecular Cell Biology, Aachen, Germany*
Pengfei Li • *Centre for Innate Immunity and Infectious Diseases, Hudson Institute of Medical Research, Clayton, VIC, Australia; Department of Molecular and Translational Science, Faculty of Medicine, Nursing and Health Sciences, Monash University, Clayton, VIC, Australia; National Clinical Research Center for Hematologic Diseases, Soochow University, Suzhou, China*
Andreas Linder • *Gene Center and Department of Biochemistry, Ludwig-Maximilians-Universität München, Munich, Germany; Department of Medicine II, University Hospital, Ludwig-Maximilians-Universität München, Munich, Germany*
Tobias I. Link • *Department of Phytopathology, University of Hohenheim, Stuttgart, Germany*
Flora Magnotti • *CIRI (Centre International de Recherche en Infectiologie), Univ Lyon, Inserm U1111, Lyon, France; Université Claude Bernard Lyon 1, CNRS, UMR5308, ENS de Lyon, Univ Lyon, Lyon, France*
Kathleen McCaffrey • *Centre for Innate Immunity and Infectious Diseases, Hudson Institute of Medical Research, Clayton, VIC, Australia; Department of Molecular and Translational Science, Faculty of Medicine, Nursing and Health Sciences, Monash University, Clayton, VIC, Australia*
Nora Mellouk • *Dynamics of Host-Pathogen Interactions Unit, Institut Pasteur, UMR3691 CNRS, Paris, France*
Etienne Meunier • *Institute of Pharmacology and Structural Biology (IPBS), University of Toulouse, CNRS, Toulouse, France*
Jonas Moecking • *Institute of Structural Biology, University Hospital Bonn, University of Bonn, Bonn, Germany*
Thomas Naderer • *Infection Program, Biomedicine Discovery Institute and Department of Biochemistry & Molecular Biology, Monash University, Clayton, VIC, Australia*
Ralph Panstruga • *RWTH Aachen University, Institute for Biology I, Unit of Plant Molecular Cell Biology, Aachen, Germany*

GANG PEI • *Institute for Immunology, Friedrich-Loeffler-Institut, Greifswald – Insel Riems, Germany*
RÉMI PLANÈS • *Institute of Pharmacology and Structural Biology (IPBS), University of Toulouse, CNRS, Toulouse, France*
SARAY RAMOS • *Department of Biochemistry, University of Lausanne, Epalinges, Switzerland*
ANJA REINSTÄDLER • *RWTH Aachen University, Institute for Biology I, Unit of Plant Molecular Cell Biology, Aachen, Germany*
JOHN ROHDE • *Department of Microbiology and Immunology, Dalhousie University Halifax, Halifax, NS, Canada*
JULIE RYU • *Department of Microbiology and Immunology, Dalhousie University Halifax, Halifax, NS, Canada*
LISA SANCHEZ • *Dynamics of Host-Pathogen Interactions Unit, Institut Pasteur, UMR3691 CNRS, Paris, France; Université de Paris, Sorbonne Paris Cite, Paris, France*
KARIN SANTONI • *Institute of Pharmacology and Structural Biology (IPBS), University of Toulouse, CNRS, Toulouse, France*
HEIDRUN STEINLE • *Department of Immunology, Institute of Nutritional Medicine, University of Hohenheim, Stuttgart, Germany*
MATTHEW J. SWEET • *Institute for Molecular Bioscience (IMB), The University of Queensland, Brisbane, QLD, Australia; IMB Centre for Inflammation and Disease Research, The University of Queensland, Brisbane, QLD, Australia; Australian Infectious Diseases Research Centre, The University of Queensland, Brisbane, QLD, Australia*
HANNAH THIERON • *RWTH Aachen University, Institute for Biology I, Unit of Plant Molecular Cell Biology, Aachen, Germany*
RACHELIA R. WIBAWA • *Centre for Innate Immunity and Infectious Diseases, Hudson Institute of Medical Research, Clayton, VIC, Australia; Department of Molecular and Translational Science, Faculty of Medicine, Nursing and Health Sciences, Monash University, Clayton, VIC, Australia*

Chapter 1

A Brief Introduction to Effector-Triggered Immunity

Thomas A. Kufer and Maria Kaparakis-Liaskos

Abstract

Detection of microbes by the host is essential to restrict microbial colonization, to clear pathogens, and to mount adapted defense reactions, and thus is the key function of the innate immune systems of plants and mammals. Here we provide an introduction into pathogen recognition by the innate immune system of both plants and animals. We will particularly focus on the concept of effector-triggered immunity, and similarities and differences in its function between plants and animals.

Key words Innate immunity, Concepts, Plants, Animals, PRR

1 Introduction

Immunity in mammals is conferred by a complex multilayered response that includes the innate immune response, a quick but rather broad recognition of pathogens, and a more fine-tuned response to defined antigens that provides lifelong memory, being the adaptive immune response. Historically often seen as two independent arms of immunity and the host defense response, we now know that activation of innate immune responses is key to driving adaptive immunity. This concept was first proposed by Charles Janeway, who stated that pattern-recognition receptors (PRR) exist in the host and function to interact with conserved microbial structures – pathogen-associated molecular patterns (PAMP) [1] to induce immunity, a concept that has since been well established by experimental data [2]. Over the period of more than two decades, several PRRs have been identified, and most of these are functionally well characterized, including the membrane bound Toll-like receptors (TLRs), the C-type lectins, and the cytosolic NOD-like receptor (NLR) families. Cognate interactions of these receptors with their respective ligands trigger intracellular signaling pathways leading to the induction of downstream signaling events that ultimately result in the production of pro-inflammatory cytokines and

Thomas A. Kufer and Maria Kaparakis-Liaskos (eds.), *Effector-Triggered Immunity: Methods and Protocols*, Methods in Molecular Biology, vol. 2523, https://doi.org/10.1007/978-1-0716-2449-4_1,

antimicrobial effectors. The paradigm of discrimination of self from non-self by the recognition of pathogen-derived motifs was later expanded to the recognition of cell and tissue damage by Polly Matzinger who introduced the danger model [3]. This model proposed that upon cellular damage, normally compartmentalized host cell proteins are released and can therefore function as danger-associated molecule patterns (DAMPs) that are accessible for detection by their respective receptor of the innate immune system. Examples of DAMPs that are released by host cells and are capable of initiating a PRR-mediated inflammatory response include heat shock proteins, nuclear proteins, and metabolites such as ATP [4].

Plants also have a sophisticated innate immune system that shows similarities to mammalian innate immunity. Similar to TLRs and C-type lectins in mammals, plants evolved membrane bound leucine-rich repeat (LRR) containing receptor-kinases that allow for sensing of PAMPs; hence, this system is referred to in plants as PAMP-triggered immunity (PTI) [5]. Plants however also express cytosolic proteins that confer resistance to pathogens. This cytosolic sensing system was indeed the first immune defense system to be identified in plants and mediates race (i.e., cultivar)-specific resistance to pathogens [6]. Most of these sensor proteins (also termed resistance (R) proteins) of the cytosolic plant immune system belong to a large family of nucleotide-binding, leucine-rich repeat (NLR) proteins and confer resistance to specific pathogen effectors, commonly called avirulence proteins, a concept genetically first revealed by the gene-for-gene hypothesis of Harold Henry Flor [7, 8]. By contrast to the recognition of PAMPs, it has been shown that this plant intracellular surveillance system in most cases detects indirectly the cellular activity of pathogen effector proteins, and not the specific pathogen-derived molecules directly, and thus is conceptionally different from PTI.

In many cases, immune responses can be triggered by sensing the activity or binding of a cellular host protein that is either targeted, degraded, or modified by the pathogen effector. A common theme in this type of immune response seems to be the interaction of the host protein with a NLR that is perturbed by the effector activity, often induced by post-transcriptional modification or competitive binding, leading to liberation and activation of the NLR. This indirect recognition of pathogen effectors was termed "effector-triggered immunity" (ETI) [9]. It should be noted that PTI and ETI are interconnected, and recent work suggests that recognition of PAMPs might be needed to trigger ETI responses (*see* ref. 10). It is now becoming evident that effector-triggered immunity, i.e., the indirect recognition of the activity and consequences of pathogen effectors in host cells, also exists in mammalian innate immunity [11].

Taken together, this provides a conceptual framework for two functionally different but interconnected pathways of how pathogens can be detected by the host, which can occur by (i) direct sensing of pathogen-derived substances (PAMPs) and (ii) indirect sensing of the consequences of pathogen effector protein functions.

2 ETI in Plants

Following the initial genetic evidence for gene-for-gene resistance in plants [8], the first resistance (R) proteins were cloned. It turned out that most of these resistance proteins are members of the nucleotide-binding domain and leucine-rich repeat receptor family (NLR), containing either an N-terminal Toll/interleukin-1 receptor (TIR) or coiled-coil (CC) domain. Different plants have diverging repertories of NLR gene sets ranging from 0 to over 1000 NLRs. For example, the model plant *Arabidopsis thaliana* encodes for over 150 different NLR proteins [12]. By contrast, mammals have a rather restricted set of NLR proteins, consisting of approximately 20 different NLR proteins [13]. Based on studies examining the pathogen effector avirulence protein Pto (AvrPto) from *Pseudomonas syringae* and its interaction with its cellular targets the Pto kinase and the R-protein Prf, Dangel and Jones proposed the "guard" hypothesis [14]. In this model, which is widely accepted, the pathogen effector AvrPto is sensed by its action on the cellular target Pto that is under surveillance by a R protein (Prf). This model was expanded by the "decoy" model, where the pathogen effector targets a cellular protein that has no essential function in immunity, which is guarded by an R protein. Pathogen-mediated degradation of the decoy then leads to liberation and thus activation of the R protein [15]. The results of R protein activation are the induction of a localized hypersensitive cell death response (HR), to control the spreading of the pathogen and the induction of transcriptional responses that confer widespread immune responses [9].

We know many examples of pathogen recognition by this system, albeit the molecular details of R protein and NLR activation are only beginning to emerge. Research over the last 20 years revealed a complex interplay of R and NLR proteins. This includes formation of heterodimers of NLRs, the use of "helper" NLRs, and different complex formation of activated NLRs. Effector mechanisms of R proteins are not always understood in detail at present, but activation of MAP kinase signaling, transcriptional reprogramming by a set of transcription factors involved in stress responses, the WRKY family [16], and the formation of aggregates termed resistosome which form at membranes and can act as pores that are linked with the induction of cell death [17] might be common mechanisms.

3 ETI in Mammals

In mammals, ETI was only recently recognized as being part of the innate immune response [12, 18], and the stimuli of this pathway are only emerging. As many ETI triggers confer perturbation of the homeostasis of cellular events, they were also termed homeostasis-altering molecular processes (HAMPs) [19].

Many adapted pathogens express virulence factors, also called effector proteins, mostly of enzymatic activity, that are essential for pathogen invasion and replication. This is well exemplified by invasive bacteria such as *Salmonella* that use a set of secreted effector proteins to manipulate host cell pathways in order to gain access to cells, to provide a niche for their replication and to subvert immune responses [20]. A common activity of secreted bacterial effectors is the use of post-translational modification of host proteins to change their activity by ADP-ribosylation [21], ubiquitination [22], phosphorylation, and dephosphorylation [23], among other effects. Key targets of these pathogen effectors are small GTPases of the host that regulate F-actin dynamics and signal transduction [21], and components of the NF-κB and MAPK pathway of innate immune signaling [24]. Several sensors for such modification have been identified and are discussed below. As pathogen effectors also use chemical reactions not found in the host cell [25], one could speculate that the host might have also evolved mechanisms to detect these, a hypothesis worth addressing in future work.

One of the first sensors associated with ETI in mammalian cells was the NLR protein NLRP3, which forms a high molecular weight cytosolic complex with the apoptosis-associated speck-like protein containing a CARD (ASC) that is termed inflammasome [26]. Inflammasomes can also be formed by other NLR proteins and act as platforms that activate caspase-1 which subsequently processes the key pro-inflammatory cytokines IL-1β and IL-18 from their progenitors. At the same time, active caspase-1 also cleaves gasdermin D that in the processed from forms pores in the cell membrane that are needed to release IL-1β, and ultimately can also lead to a form of programmed cell death called pyroptosis, if not healed by cellular machinery [26]. The NLRP3 inflammasome responds to a large variety of different stimuli, including crystalline urea, bacterial pore-forming toxins, and ATP. The exact mechanism of inflammasome activation is still enigmatic, but disintegration of the host cell membrane and disintegration of cellular compartments by pathogen activity and toxins are emerging as key mechanisms of NLRP3 activation. The sensing of membrane damage that can be induced by bacterial toxins fits well with the concept of ETI. Another inflammasome-forming NLR protein, NLRP1B, is activated by proteolytic cleavage of its N-terminus. Of note, NLRP1B is cleaved by *Bacillus anthracis* lethal toxin, and NLRP1B ubiquitination by the *Shigella* effector IpaH7.8 leads to proteasomal degradation [27–30]. Both events liberate the C-terminal fragment of NLRP1B that induces inflammasome formation. Compared with the plant immune system, this corresponds to the integrated-decoy model that was proposed by Peter Dodds for NLR pairs that harbor atypical domains [31], such as is the case for the function to find (FIIND) domain in NLRP1B.

More recently, pyrin (also known as marenostrin or TRIM20) that also forms an inflammasome was shown to react toward bacterial GTPases that modify host proteins [32]. Pyrin is activated in response to modifications of RhoA [33, 34]; functionally this is mediated by phosphorylation of pyrin by the RhoA-activated kinases PKN1 and PKN2, which allow binding of pyrin to 14-3-3 proteins which inhibits the pyrin inflammasome formation [35]. This scenario very closely mimics the "guard" mechanism of ETI in plants.

NOD1, an NLR protein that acts as a PRR for bacterial peptidoglycan [13], was shown to respond to changes in cellular RhoGTPase activity [36] and F-actin remodeling induced upon bacterial invasion [37, 38]. Induction of inflammatory pathways upon F-actin depolymerization was also reported for the NOD1 related protein NOD2 [39]. How NOD1/2 sense GTPase and actin perturbation is not clear at present. However, all these events induce cellular stress, accumulating in activation of the integrated stress response (ISR), which is a hallmark upon infection with many pathogens [40]. Recent work suggests that cellular stress is a unifying concept and can well explain how specific PRRs are wired to ETI recognition. This is exemplified by recent work that showed that NOD1 and NOD2 are also activated by the endogenous lipid shingosine-1 phosphate (S1P) that acts as a ligand for these NLRs [31]. Changes in mitochondrial activity are also commonly observed upon infection with various pathogens, and mitochondria play central roles in orchestrating a cellular response upon pathogen detection, and in particular upon viral infections [41]. Notably, mitochondria are also involved in the regulation of the NLRP3 inflammasome [33, 34].

4 Conclusion

Since the first discovery of race-specific resistance in plants and innate immune detection of microbes by the innate immune system in mammals, we have come a long way in understanding the molecular facets of the immune system. In both plants and mammals, a tailored immune response involves both sensing of PAMPs and activation of ETI responses [11]. Both systems use similar sensors of the NLR protein family that show remarkable homologies also in their wiring to cellular signaling [42]. Even though plant and animal NLR proteins seem to have arisen from convergent evolution [43], taking an integrative approach might help to provide a better understanding of both pathways. Indeed, recent data shows that NLRs from plants and animals both form complexes of similar structures, as exemplified by the NLRC4 inflammasome of mammalian cells [44] and the ZAR1 resistosome of plant cells [45], respectively.

Understanding in detail how a cell detects the activity of pathogen effectors and checks homeostasis of cellular processes is suited to provide novel insights into the basis of immunity and into challenging the arms race between pathogens and their hosts. In this book, we provide a set of protocols from leading experts in the field to study specific pathogen effectors and their action in host cells for the analysis of both plant and animal cell-mediated ETI responses.

References

1. Janeway CA Jr (1989) Approaching the asymptote? Evolution and revolution in immunology. Cold Spring Harb Symp Quant Biol 54(Pt 1): 1–13
2. Medzhitov R, Janeway C Jr (2000) Innate immune recognition: mechanisms and pathways. Immunol Rev 173:89–97. https://doi.org/10.1034/j.1600-065x.2000.917309.x
3. Matzinger P (1994) Tolerance, danger, and the extended family. Annu Rev Immunol 12: 991–1045. https://doi.org/10.1146/annurev.iy.12.040194.005015
4. Patel S (2018) Danger-associated molecular patterns (DAMPs): the derivatives and triggers of inflammation. Curr Allergy Asthma Rep 18(11):63. https://doi.org/10.1007/s11882-018-0817-3
5. Couto D, Zipfel C (2016) Regulation of pattern recognition receptor signalling in plants. Nat Rev Immunol 16(9):537–552. https://doi.org/10.1038/nri.2016.77
6. Jones JD, Vance RE, Dangl JL (2016) Intracellular innate immune surveillance devices in plants and animals. Science 354(6316): aaf6395. https://doi.org/10.1126/science.aaf6395
7. van Wersch S, Tian L, Hoy R, Li X (2020) Plant NLRs: the whistleblowers of plant immunity. Plant Commun 1(1):100016. https://doi.org/10.1016/j.xplc.2019.100016
8. Staskawicz BJ, Ausubel FM, Baker BJ, Ellis JG, Jones JD (1995) Molecular genetics of plant disease resistance. Science 268(5211): 661–667. https://doi.org/10.1126/science.7732374
9. Jones JD, Dangl JL (2006) The plant immune system. Nature 444(7117):323–329. https://doi.org/10.1038/nature05286
10. Yuan M, Ngou BPM, Ding P, Xin XF (2021) PTI-ETI crosstalk: an integrative view of plant immunity. Curr Opin Plant Biol 62:102030. https://doi.org/10.1016/j.pbi.2021.102030
11. Kufer TA, Creagh EM, Bryant CE (2019) Guardians of the cell: effector-triggered immunity steers mammalian immune defense. Trends Immunol 40(10):939–951. https://doi.org/10.1016/j.it.2019.08.001
12. Lopes Fischer N, Naseer N, Shin S, Brodsky IE (2020) Effector-triggered immunity and pathogen sensing in metazoans. Nat Microbiol 5(1):14–26. https://doi.org/10.1038/s41564-019-0623-2
13. Kufer TA, Fritz JH, Philpott DJ (2005) NACHT-LRR proteins (NLRs) in bacterial infection and immunity. Trends Microbiol 13(8):381–388. https://doi.org/10.1016/j.tim.2005.06.004
14. Dangl JL, Jones JD (2001) Plant pathogens and integrated defence responses to infection. Nature 411(6839):826–833. https://doi.org/10.1038/35081161
15. van der Hoorn RA, Kamoun S (2008) From guard to decoy: a new model for perception of plant pathogen effectors. Plant Cell 20(8): 2009–2017. https://doi.org/10.1105/tpc.108.060194
16. Wani SH, Anand S, Singh B, Bohra A, Joshi R (2021) WRKY transcription factors and plant defense responses: latest discoveries and future prospects. Plant Cell Rep 40(7):1071–1085.

https://doi.org/10.1007/s00299-021-02691-8
17. Saur IML, Panstruga R, Schulze-Lefert P (2021) NOD-like receptor-mediated plant immunity: from structure to cell death. Nat Rev Immunol 21(5):305–318. https://doi.org/10.1038/s41577-020-00473-z
18. Stuart LM, Paquette N, Boyer L (2013) Effector-triggered versus pattern-triggered immunity: how animals sense pathogens. Nat Rev Immunol 13(3):199–206. https://doi.org/10.1038/nri3398
19. Liston A, Masters SL (2017) Homeostasis-altering molecular processes as mechanisms of inflammasome activation. Nat Rev Immunol 17(3):208–214. https://doi.org/10.1038/nri.2016.151
20. Azimi T, Zamirnasta M, Sani MA, Soltan Dallal MM, Nasser A (2020) Molecular mechanisms of salmonella effector proteins: a comprehensive review. Infect Drug Resist 13:11–26. https://doi.org/10.2147/IDR.S230604
21. Aktories K (2011) Bacterial protein toxins that modify host regulatory GTPases. Nat Rev Microbiol 9(7):487–498. https://doi.org/10.1038/nrmicro2592
22. Berglund J, Gjondrekaj R, Verney E, Maupin-Furlow JA, Edelmann MJ (2020) Modification of the host ubiquitome by bacterial enzymes. Microbiol Res 235:126429. https://doi.org/10.1016/j.micres.2020.126429
23. Grishin AM, Beyrakhova KA, Cygler M (2015) Structural insight into effector proteins of Gram-negative bacterial pathogens that modulate the phosphoproteome of their host. Protein Sci 24(5):604–620. https://doi.org/10.1002/pro.2636
24. Brodsky IE, Medzhitov R (2009) Targeting of immune signalling networks by bacterial pathogens. Nat Cell Biol 11(5):521–526. https://doi.org/10.1038/ncb0509-521
25. Chambers KA, Scheck RA (2020) Bacterial virulence mediated by orthogonal post-translational modification. Nat Chem Biol 16(10):1043–1051. https://doi.org/10.1038/s41589-020-0638-2
26. Schroder K, Tschopp J (2010) The inflammasomes. Cell 140(6):821–832. https://doi.org/10.1016/j.cell.2010.01.040
27. Levinsohn JL, Newman ZL, Hellmich KA, Fattah R, Getz MA, Liu S, Sastalla I, Leppla SH, Moayeri M (2012) Anthrax lethal factor cleavage of Nlrp1 is required for activation of the inflammasome. PLoS Pathog 8(3):e1002638. https://doi.org/10.1371/journal.ppat.1002638
28. Chavarria-Smith J, Vance RE (2013) Direct proteolytic cleavage of NLRP1B is necessary and sufficient for inflammasome activation by anthrax lethal factor. PLoS Pathog 9(6):e1003452. https://doi.org/10.1371/journal.ppat.1003452
29. Chui AJ, Okondo MC, Rao SD, Gai K, Griswold AR, Johnson DC, Ball DP, Taabazuing CY, Orth EL, Vittimberga BA, Bachovchin DA (2019) N-terminal degradation activates the NLRP1B inflammasome. Science 364(6435):82–85. https://doi.org/10.1126/science.aau1208
30. Sandstrom A, Mitchell PS, Goers L, Mu EW, Lesser CF, Vance RE (2019) Functional degradation: a mechanism of NLRP1 inflammasome activation by diverse pathogen enzymes. Science 364(6435):eaau1330. https://doi.org/10.1126/science.aau1330
31. Pei G, Zyla J, He L, Moura-Alves P, Steinle H, Saikali P, Lozza L, Nieuwenhuizen N, Weiner J, Mollenkopf HJ, Ellwanger K, Arnold C, Duan M, Dagil Y, Pashenkov M, Boneca IG, Kufer TA, Dorhoi A, Kaufmann SH (2021) Cellular stress promotes NOD1/2-dependent inflammation via the endogenous metabolite sphingosine-1-phosphate. EMBO J 40(13):e106272. https://doi.org/10.15252/embj.2020106272
32. Jamilloux Y, Magnotti F, Belot A, Henry T (2018) The pyrin inflammasome: from sensing RhoA GTPases-inhibiting toxins to triggering autoinflammatory syndromes. Pathog Dis 76(3):fty020. https://doi.org/10.1093/femspd/fty020
33. Zhong Z, Liang S, Sanchez-Lopez E, He F, Shalapour S, Lin XJ, Wong J, Ding S, Seki E, Schnabl B, Hevener AL, Greenberg HB, Kisseleva T, Karin M (2018) New mitochondrial DNA synthesis enables NLRP3 inflammasome activation. Nature 560(7717):198–203. https://doi.org/10.1038/s41586-018-0372-z
34. Zhou R, Yazdi AS, Menu P, Tschopp J (2011) A role for mitochondria in NLRP3 inflammasome activation. Nature 469(7329):221–225. https://doi.org/10.1038/nature09663
35. Park YH, Wood G, Kastner DL, Chae JJ (2016) Pyrin inflammasome activation and RhoA signaling in the autoinflammatory diseases FMF and HIDS. Nat Immunol 17(8):914–921. https://doi.org/10.1038/ni.3457
36. Keestra AM, Winter MG, Auburger JJ, Frassle SP, Xavier MN, Winter SE, Kim A, Poon V, Ravesloot MM, Waldenmaier JF, Tsolis RM, Eigenheer RA, Baumler AJ (2013) Manipulation of small Rho GTPases is a pathogen-induced process detected by NOD1. Nature

496(7444):233–237. https://doi.org/10.1038/nature12025

37. Kufer TA, Kremmer E, Adam AC, Philpott DJ, Sansonetti PJ (2008) The pattern-recognition molecule Nod1 is localized at the plasma membrane at sites of bacterial interaction. Cell Microbiol 10(2):477–486. https://doi.org/10.1111/j.1462-5822.2007.01062.x

38. Bielig H, Lautz K, Braun PR, Menning M, Machuy N, Bruegmann C, Barisic S, Eisler SA, Andreel M, Zurek B, Kashkar H, Sansonetti PJ, Hausser A, Meyer TF, Kufer TA (2014) The cofilin phosphatase slingshot homolog 1 (SSH1) links NOD1 signaling to actin remodeling. PLoS Pathog 10(9):e1004351. https://doi.org/10.1371/journal.ppat.1004351

39. Legrand-Poels S, Kustermans G, Bex F, Kremmer E, Kufer TA, Piette J (2007) Modulation of Nod2-dependent NF-kappaB signaling by the actin cytoskeleton. J Cell Sci 120 (Pt 7):1299–1310. https://doi.org/10.1242/jcs.03424

40. Rodrigues L, Graca RSF, Carneiro LAM (2018) Integrated stress responses to bacterial pathogenesis patterns. Front Immunol 9:1306. https://doi.org/10.3389/fimmu.2018.01306

41. West AP, Shadel GS, Ghosh S (2011) Mitochondria in innate immune responses. Nat Rev Immunol 11(6):389–402. https://doi.org/10.1038/nri2975

42. Maekawa T, Kufer TA, Schulze-Lefert P (2011) NLR functions in plant and animal immune systems: so far and yet so close. Nat Immunol 12(9):818–826. https://doi.org/10.1038/ni.2083

43. Urbach JM, Ausubel FM (2017) The NBS-LRR architectures of plant R-proteins and metazoan NLRs evolved in independent events. Proc Natl Acad Sci U S A 114(5):1063–1068. https://doi.org/10.1073/pnas.1619730114

44. Zhang L, Chen S, Ruan J, Wu J, Tong AB, Yin Q, Li Y, David L, Lu A, Wang WL, Marks C, Ouyang Q, Zhang X, Mao Y, Wu H (2015) Cryo-EM structure of the activated NAIP2-NLRC4 inflammasome reveals nucleated polymerization. Science 350(6259):404–409. https://doi.org/10.1126/science.aac5789

45. Wang J, Hu M, Wang J, Qi J, Han Z, Wang G, Qi Y, Wang HW, Zhou JM, Chai J (2019) Reconstitution and structure of a plant NLR resistosome conferring immunity. Science 364(6435):eaav5870. https://doi.org/10.1126/science.aav5870

Chapter 2

Lambda Red Recombineering in *Shigella flexneri*

Adrian Herod, Julie Ryu, and John Rohde

Abstract

Shigellosis remains a major cause of severe diarrheal disease and death throughout the world. Vaccine development against shigellosis has been hampered by an incomplete understanding of the molecular mechanisms by which *Shigella* spp. causes disease and difficulties in manipulating *Shigella* spp. genomes. While homologous recombination protocols for the construction of precise gene deletions exist, construction of mutants in *S. flexneri* has not become commonplace. We describe the steps for construction of gene deletions using λ-red recombination using tools that we have developed in our laboratory.

Key words *Shigella*, λ-Red recombination, Conjugation, Congo red, Vaccine development

1 Introduction

Shigellosis continues to wreak a heavy burden on humanity across the globe. In recent decades, the mortality due to shigellosis has decreased dramatically; however, molecular diagnostic techniques suggest that the incidence has remained largely unchanged from previous decades [1]. The increasing frequency of *Shigella* spp. clinical isolates that display resistance to one or more frontline antibiotics are worrying [1]. An effective vaccine against shigellosis would need to provide protection against all four *Shigella* species: *dysenteriae*, *flexneri*, *bongori*, and *sonnei* as well as the multiple serotypes within these species (reviewed in 2). Multiple avenues of vaccine development are being pursued including recombinant subunits, LPS, and the formulation of live-attenuated vaccines [3–5]. The development of live-attenuated vaccines relies on the routine manipulation of *Shigella* spp. genomes to inactivate gene products and test their fitness for ability to cause disease and elicit immune responses in cell culture, the available animal models that exist, and finally human trials.

The past decade has seen an advance in the knowledge of the strategies used by the cyto-invasive pathogen *S. flexneri*. These

Thomas A. Kufer and Maria Kaparakis-Liaskos (eds.), *Effector-Triggered Immunity: Methods and Protocols*, Methods in Molecular Biology, vol. 2523, https://doi.org/10.1007/978-1-0716-2449-4_2,

include the role of complement, the identification of a novel intracellular LPS-sensing pathway, and the role of interferon during infection [6–9]. These studies have all relied on specific mutants of *S. flexneri* that helped to make these discoveries using both host and pathogen tools. While huge advancements in the areas of innate immunology and imaging of host-pathogen interactions have been made, advancement in the genetic manipulation of *S. flexneri* genomes has been slower, and the process of creating precise mutations in *S. flexneri* genomes is not routine in most laboratories. The advent of lambda red recombination for gene replacement technologies (often referred to as "recombineering") in *Escherichia coli* and other Gram-negative bacteria was developed in 2000 by Datsenko and Wanner [10]. However, for reasons yet to be elucidated, these same mutagenesis platforms are less efficient in *S. flexneri* by orders of magnitude. Our laboratory has developed λ-red protocols and reagents that have been optimized for use in *S. flexneri* [11] (Fig. 1).

Our approach for creating *S. flexneri* mutants with the λ-red system was informed by two important publications. The first was the description of the "Keio collection" by Baba and coworkers explaining an approach for creating a large collection of mutants for phenotypic screens in *E. coli* [12]. The second was a step-by-step λ-red mutagenesis strategy in *Salmonella enterica* by Joyce Karlinsey [13]. These articles provided the framework for our own protocol and reagents. We have constructed plasmids that encode the *sacB* gene from *Bacillus subtilis*, the gene product of *sacB* (levansucrase), which is toxic to Gram-negative bacteria in the presence of sucrose. Therefore, counterselections against the plasmids pRR008 and pRR009 can be carried out at 30 °C or 37 °C instead of the high temperature selection that was described for the original pKD46 plasmid [10]. These plasmids can be introduced to *S. flexneri* by conjugation using *E. coli* strains that supply the transfer functions for RK2 plasmids. This protocol was written to enable laboratories to manipulate *Shigella* genomes easily with equipment that is likely to exist in their own laboratories or within their department (*see* **Note 1**).

Laboratories new to genetic manipulation should first construct deletions in *ipaD* and *mxiD* to serve as positive and negative controls for the Congo red phenotype (*see* **Note 2**). The *Shigella* spp. Type 3 Secretion Apparatus (T3SA) is encoded on the virulence plasmid and becomes active in the presence of the clothing dye Congo red [14, 15]. Strains that contain a functional T3SA grow as red colonies on media containing Congo red, and strains that lack the virulence plasmid or that harbor mutations that render the T3SA inactive are pink. This phenotype can vary somewhat depending on the stock of Congo red used; therefore, it is useful

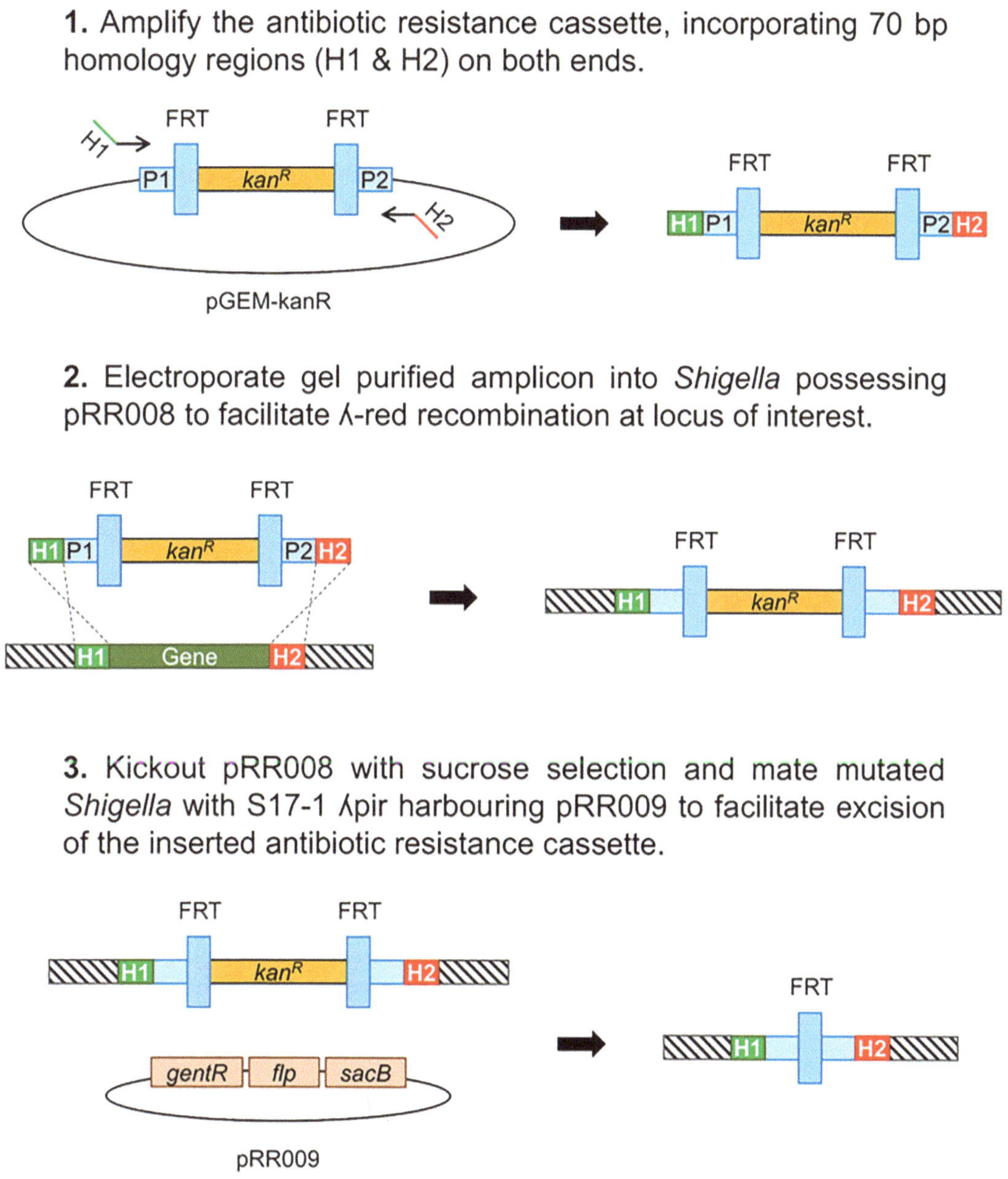

Fig. 1 Outline of lambda red recombination strategy to delete a gene of interest in *Shigella* using the FRT-flanked kanamycin resistance cassette from pGEM-kanR, followed by excision of the cassette facilitated by pRR009

to include *ipaD* and *mxiD* mutants as controls. The *ipaD* mutation is deregulated for T3SA activity resulting in a colony that is darker than the wild-type parent [16]. By contrast, the *mxiD* mutation results in a nonfunctional T3SA and grows as a pink colony on Congo red containing media [17].

We have constructed mutants with cassettes that encode resistance to tetracycline or kanamycin [10]. While both selections are effective, we find that the kanamycin cassette results in fewer cases of polar mutations. The kanamycin cassette that we constructed is flanked by flippase recognition target (FRT) sites. Expression of the "flippase" recombinase (encoded by pRR009) results in the excision of the selection cassette leaving in place an 18-nucleotide scar sequence [11, 12]. In most cases, removal of the antibiotic resistance marker results in a nonpolar mutation. However, this cannot be assumed, and all mutants must be complemented in order to establish causality of the phenotype being studied.

2 Materials

2.1 Oligo Design, Amplicon Construction, and Purification

1. Oligonucleotides containing sequence to be targeted within *Shigella* genome and sequence for disruption cassette. Oligonucleotides for diagnostic PCR (Fig. 1a).
2. Design the forward primer to include 50 base pairs of the sequence directly upstream from the gene or region to be deleted, referred to as homology region 1 (H1), followed by the 20-base pair priming site located upstream from the kanamycin resistance cassette, referred to as priming site 1 (P1). The forward primer should consist of a total of 70 base pairs (H1 + P1) and is referred to as λ-red forward primer 1 (LR1F).
3. Design the reverse primer to include 50 base pairs of the sequence directly downstream from the gene or region to be deleted, referred to as homology region 2 (H2), followed by the 20-base pair priming site located downstream from the kanamycin/tetracycline resistance cassette, referred to as priming site 2 (P2). The reverse primer should consist of a total of 70 base pairs (H2 + P2) and is referred to as λ-red reverse primer 1 (LR1R).
4. Design a primer that will anneal 100–300 base pairs upstream from H1 as a forward primer. This primer should be between 20 and 25 base pairs and is referred to as LR2F. LR2F is used in combination with a 20-base pair reverse primer that anneals to P2 to confirm the integration of the antibiotic resistance cassette into the gene or region of interest.
5. Design a primer that will anneal 100–300 base pairs downstream from H2 as reverse primer. This primer should be between 20 and 25 base pairs and is referred to as LR2R. LR2R is used in combination with a 20-base pair forward primer that anneals to P1 to confirm the integration of the antibiotic resistance cassette into the gene or region of interest.

Table 1
List of λ-red recombination primers

Prime name	Prime sequence (5′–3′)
ipaD LR1F	ATCCAAGAGCCATAATAATATATGGCTCTTCCTGTAAGGAAATAACCATGA TTCCGGGGATCCGTCGACC
ipaD LR1R	AGTATTATTTACATTATGCATGGCGCACCTCAGAAATGGAGAAAAAG TTTTGTAGGCTGGAGCTGCTTCG
ipaD LR2R	GAGAAGTTTGCGAACTATGG
mxiD LR1F	TTCTGCAAGAGAATTATCGAATATTACTTTTTTAGTGAAAAAAATAAATGA TTCCGGGGATCCGTCGACC
mxiD LR1R	ACATCAAGCATACATCACCTACTTTATAGTTAGTAATTTAAGTATGAAACTG TAGGCTGGAGCTGCTTCG
mxiD LR2F	ATGTATTTTTGCACAATCAAAC
P1	ATTCCGGGGATCCGTCGACC
P2	TGTAGGCTGGAGCTGCTTCG

6. When the oligos arrive, dilute them in water into 100 μM stocks for long-term storage at −20 °C and also dilute them to 10 μM working stocks. Primers for ʎ-red recombination of *ipaD* and *mxiD* are listed in Table 1 and described in Fig. 2.
7. PCR reagents:10× reaction buffer, 10 mM dNTPs, 10 μM forward primer, 10 μM reverse primer, Taq polymerase (*see* **Note 3**), template DNA, nuclease-free H_2O (Table 2).
8. Thermocycler, DNA polymerase.
9. Agarose gel electrophoresis reagents and equipment.
10. *Shigella* spp. such as *S. flexneri* M90T Sm [18] available from ATCC.
11. An *E. coli* strain that supplies R2K transfer functions in trans (e.g., S17-1 λpir).
12. Antibiotics (e.g., kanamycin and gentamicin).
13. 20% L-arabinose in water, sterilized by autoclaving.
14. Template plasmid for kanamycin resistance cassette (pGEM-kanR) (Table 3).
15. Plasmid encoding L-arabinose-inducible λ-red recombination machinery, pRR008 [8] (Table 3).
16. Trypticase soy broth (TSB).
17. Trypticase soy agar (TSA) plates containing relevant antibiotics (25-μg/mL kanamycin, 15-μg/mL gentamicin).
18. 5% Congo red (CR) solution in water.

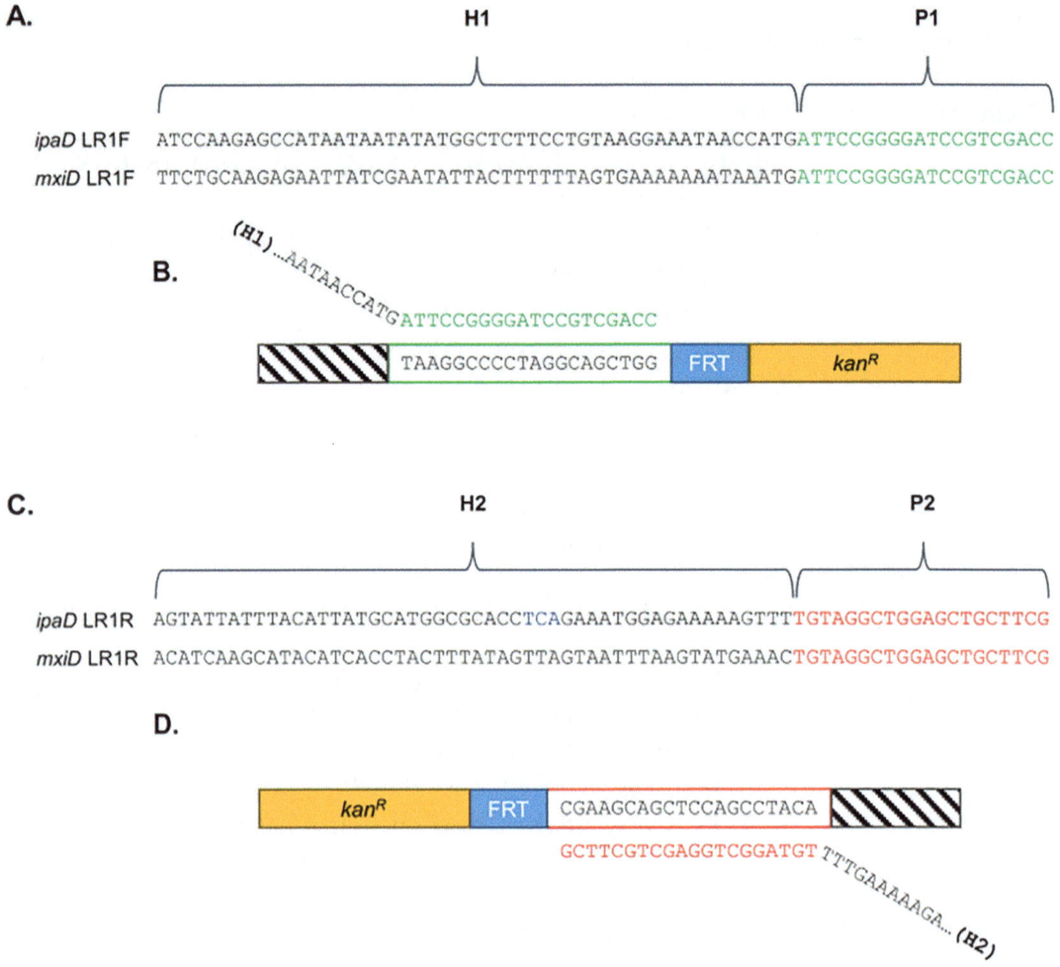

Fig. 2 Primer design for lambda red recombination in *Shigella*. (**a** and **c**) λ-red forward (LR1F) and reverse (LR1R) primers for deletion of *ipaD* and *mxiD*. Primers are separated into the 70-bp homology regions upstream (H1) and downstream (H2) of the targeted gene and the 20-bp priming sites upstream (P1) and downstream (P2) of the antibiotic resistance cassette. (**b** and **d**) Priming of *ipaD* LR1F and LR1R to P1 and P2, respectively, on pGEM-kanR. The 70-bp H1 and H2 sequences are incorporated into the amplicon in subsequent rounds of amplification

19. TSA Congo red plates (CR-TSA): Add 2 mL/L of Congo red to TSA before autoclaving, and add antibiotics after autoclaving if required.
20. Reagents to purify amplicons from agarose gels.
21. 37 °C incubator.
22. 30 °C incubator.
23. Centrifuge and sterile tubes to spin 50 mL at 4000 × *g* at 4 °C (e.g., Eppendorf 5810R or similar).
24. Ice-cold sterile deionized water.

Table 2
PCR setup for λ-red recombination

Reagent	Volume
10× reaction buffer	2.5 μL
10-mM dNTPs	0.5 μL
10-μM forward primer	0.5 μL
10-μM reverse primer	0.5 μL
Taq polymerase	0.125 μL
Template DNA	Variable
Nuclease-free H_2O	Up to 25 μL

Table 3
List of plasmids

Plasmid	Selection	Purpose
pGEM-kanR	Kanamycin	Kanamycin resistance cassette template with FRT and priming sites
pRR008	Gentamicin, sucrose	λ-red recombination machinery
pRR009	Gentamicin, sucrose	Flippase

2.2 Counterselection of Plasmids and Excision of Antibiotic Cassettes

1. Trypticase soy broth (TSB) and trypticase soy agar (TSA) plates containing relevant antibiotics (25-μg/mL kanamycin, 15-μg/mL gentamicin). When used, sucrose is added to 5% before autoclaving.
2. MacConkey agar plates containing relevant antibiotics (25-μg/mL kanamycin, 15-μg/mL tetracycline, 15-μg/mL gentamicin).
3. Plasmid encoding flippase, plasmid pRR009 is identical to pRR003 described in [8] except that the antibiotic selection is gentamicin instead of tetracycline (Table 3).

2.3 Recharging of Electroporation Cuvettes

1. 10% bleach.
2. 0.2 M HCl.
3. 70% ethanol.

3 Methods

3.1 PCR

1. Use 5–10 ng of template DNA, as template in a 25-uL reaction. PCR setup and thermocycling conditions are found in Tables 2 and 4.

Table 4
Thermocycling conditions for λ-red recombination-related PCRs

Step	Cycle	Temp	Time
Denaturation	1	95 °C	2 min
Denaturation	35	95 °C	30 s
Annealing	35	54 °C	30 s
Extension	35	68 °C	2 min
Extension	1	68 °C	5 min
Hold	–	4 °C	–

2. Run PCR mixtures on a 0.8% agarose gel and observe the DNA fragment (~1.5 kb) under long wave UV light. Excise gel slice and purify using a commercially available gel purification column (*see* **Note 5**). Resuspend purified fragment in dH_2O.

3.2 Mating

1. Streak out M90T-Sm and S17-1 λpir with pRR008 onto CR-TSA and TSA with gentamicin, respectively, and incubate overnight in an incubator at 37 °C.
2. For dry mating, select a single colony from the TSA-gentamicin plate containing S17-1 λpir with pRR008 and streak a single line on a fresh TSA plate. Next, select a single colony from the CR-TSA plate containing M90T-Sm and streak directly over the previous streak to mix the two colonies. Place the TSA plate in an incubator overnight at 30 °C.
3. For wet mating, grow overnight cultures of M90T-Sm and S17-1 λpir with pRR008 in TSB (include gentamicin for the S17-1 λpir strain harboring pRR008). The following day, aliquot 100 μL of each culture into separate 1.5-mL tubes and centrifuge at 13,000 × *g* for 1 min. Remove the supernatant from each and resuspend each pellet in 100 μL fresh TSB. Pipette 10 μL of S17-1 λpir with pRR008 and 90 μL of M90T-Sm to the center of a new TSA plate. Place the TSA plate in an incubator overnight at 30 °C.
4. Use an inoculation loop to collect a small amount of bacterial growth from the mating plate where growth from both strains overlapped and streak onto a MacConkey agar with gentamicin plate. Place the MacConkey-gentamicin plate in an incubator overnight at 30 °C.
5. Observe for growth of yellow colonies (lactose negative *S. flexneri*), which will indicate successful mating of pRR008 into M90T-Sm. Streak a lactose negative colony on a CR-TSA-gentamicin plate and incubate at 37 °C.

3.3 Electroporation

1. Pick a single CR positive colony of M90T-Sm harboring pRR008 from the CR-TSA-gentamicin plate and inoculate 2 mL of TSB, and incubate overnight shaking at 30 °C.
2. Inoculate 50 mL of TSB in a 250-mL flask with 0.5 mL of overnight culture. Add 5 mL of 20% L-arabinose (sterilized by autoclaving). Grow shaking at 37 °C to an OD_{600} between 0.4 and 0.6 (usually about 3 h).
3. Transfer the culture to a 50-mL sterile plastic conical flask and load into a centrifuge pre-cooled to 4 °C. Centrifuge at 3220 × *g* for 10 min.
4. Resuspend the bacterial pellet in 50 mL of ice-cold sterile dH_20. Centrifuge at 3220 × *g* at 4 °C for 10 min.
5. Repeat **step 4** again twice.
6. Resuspend the bacteria in 500 μL of water (usually the residual liquid). Transfer 50 μL of the resuspended bacteria to a 2-mM electroporation cuvette along with 5–10 μL of purified knockout cassette.
7. Electroporate bacteria at 2.5 kV. Add 1 mL of TSB and incubate at 37 °C with shaking at 200 rpm.
8. Transfer the bacteria to a sterile 1.5-mL tube and centrifuge at 13,000 × *g* for 1 min at room temperature. Remove 700 μL, resuspend bacteria in remaining liquid and plate on a CR-TSA plate containing kanamycin, and grow overnight at 37 °C to select for integration of the knockout cassette.

3.4 Counterselection and Diagnostic PCR

1. Use sterile toothpicks to patch colonies onto newly labeled CR TSA-kanamycin plates containing 37 °C overnight.
2. Resuspend colonies from the plate into separate tubes with 100 μL of ddH2O and heat tubes at 95 °C for 5 min. Resuspension should appear slightly cloudy, but not turbid. Use 2 μL of colony resuspensions as PCR template to confirm the insertion of the kanamycin resistance cassette into the gene of interest using LR2F and P2 or LR2R and P1 primers.
3. Run PCR mixtures on a 1% agarose gel to determine if knockout has been attained. Examine gel for presence of ~1.5–1.7 kb DNA fragment (length will depend on sequence location of the LR2F or LR2R primer).
4. Repatch colonies with successful recombination from the CR kanamycin plate onto a CR sucrose plate and incubate at 37 °C overnight (*see* **Note 5**).
5. Patch desired knockouts from CR-TSA containing 5% sucrose onto both CR-TSA (no antibiotic) and CR-TSA-gentamicin plates. Incubate at 37 °C. Gentamicin sensitivity indicates loss of pRR008.

3.5 Removal of Cassette

1. Streak out desired knockout (e.g., *ipaD::KAN*) and S17-1 λpir with pRR009 onto CR-TSA and TSA with gentamicin plates, respectively, and incubate overnight in an incubator at 37 °C.
2. For dry mating, select a single colony from the TSA-gentamicin plate containing S17-1 λpir with pRR009 and streak a single line on a fresh TSA plate. Next, select a single colony from the CR-TSA plate containing *S. flexneri* knockout of interest and streak directly over the previous streak to mix the two colonies. Place the TSA plate in an incubator overnight at 30 °C.
3. For wet mating, grow overnight cultures of desired *S. flexneri* knockout and S17-1 λpir with pRR008 in TSB (include gentamicin for the S17-1 λpir strain harboring pRR008). The following day, aliquot 100 μL of each culture into separate 1.5-mL tubes and centrifuge at 13,000 × *g* for 1 min. Remove the supernatant from each tube and resuspend each pellet in 100-μL fresh TSB. Pipette 10 μL of S17-1 λpir with pRR009 and 90 μL of M90T-Sm to the center of a new TSA plate. Place the TSA plate in an incubator overnight at 30 °C.
4. Use an inoculation loop to collect a small amount of bacterial growth from the mating plate where growth from both strains overlapped and streak onto a MacConkey agar with gentamicin plate. Place the MacConkey-gentamicin plate in an incubator overnight at 30 °C.
5. Observe for growth of yellow colonies (lactose negative *S. flexneri*), which will indicate successful mating of pRR008 into the *S. flexneri* knockout of interest. Streak a lactose negative colony on a CR-TSA-gentamicin plate and incubate at 37 °C.
6. Use an inoculation loop to streak a CR-TSA-gentamicin plate and incubate at 37 °C.
7. Use sterile toothpicks to patch several (20–30) colonies onto one CR-TSA-gentamicin plate and one CR-TSA-kanamycin plate and incubate overnight at 37 °C. Kanamycin sensitivity indicates the loss of the disruption cassette. If none of the bacteria display sensitivity to kanamycin, use an inoculation loop to streak another CR-TSA-gentamicin plate (*see* **Note 5**).
8. Follow steps for diagnostic PCR and loss of the pRR009 plasmid using the same steps as in Heading 3.4 **step 2** through to **step 5**.

3.6 Recharging of Electroporation Cuvettes (See Note 6)

1. Fill each used electroporation cuvette with 10% bleach immediately after removing the transformed *Shigella* to kill any remaining bacteria. Let the 10% bleach stand in the cuvettes for no longer than 5 min as additional waiting can corrode the cuvettes. Rinse cuvettes three times with distilled water.

2. Fill each electroporation cuvette with 0.2-M HCl to eliminate any residual DNA from the previous transformation. Let the HCl stand in the cuvettes for no longer than 10 min as additional waiting can corrode the cuvettes. Rinse cuvettes three times with distilled water.
3. Rinse each cuvette three times with 70% ethanol. For the final rinse, decant the 70% ethanol in a sterile hood and remove residual ethanol within the cuvette using a pipette or needle. Invert the cuvettes in the hood and leave to air-dry for 1 h.
4. Add electroporation cuvettes caps into a small beaker with 10% bleach and leave for 5 min to kill any residual bacteria. Rinse three times with distilled water.
5. Rinse electroporation cuvette caps three times with 70% ethanol. For the final rinse, remove the caps from the 70% ethanol filled beaker in a sterile hood and leave to air-dry for 1 h.
6. Once dry, assemble the electroporation cuvettes and caps together. Store in a dry environment to prevent future cuvette corrosion.

4 Notes

1. The infectious dose for *S. flexneri* is very low and can be a cause for laboratory-acquired infection. In cases where cultures of *S. flexneri* may be transported from one laboratory to another (e.g., to use a centrifuge or electroporation apparatus), standard operating procedures for safe transport need to be established.
2. Stocks of *S. flexneri* should be maintained on CR-TSA to ensure that the T3SA is functioning properly. Constructions of *mxiD* (CR negative) and *ipaD* (CR very positive) are excellent choices for the first mutants to construct for two reasons. First, the in vitro phenotype associated with CR allows for ease of screening (fewer false positives). Second, these mutants serve as controls for the CR phenotype during routine manipulation of *S. flexneri*.
3. We use economical DNA polymerase for generation of the knockout cassettes and for diagnostic PCR.
4. We routinely gel-purify our knockout cassettes. The presence of even small amounts of contaminating template DNA results in high background of false positives to be screened for successful knockouts.
5. The sucrose counterselection may not be successful after a single passage on TSA sucrose plates. Time is saved by patching a second time on TSA sucrose plates before selecting for sensitivity to gentamicin to ensure that the plasmid has been lost. In some cases, removal of the knockout cassette may take as many

as three cycles of (i) streaking on CR-TSA-gentamicin plates and (ii) patching individual colonies on CR-TSA and CR-TSA-kanamycin plates to identify kanamycin-sensitive bacteria.

6. Electroporation cuvettes are surprisingly costly; reuse of electroporation cuvettes is straightforward and results in less waste. We have observed no cases of cross-contamination of knockout cassette DNA or bacterial cultures following the protocol described for recharging of electroporation cuvettes.

References

1. von Seidlein L, Kim DR, Ali M, Lee H, Wang X, Thiem VD, Canh DG, Chaicumpa W, Agtini MD, Hossain A, Bhutta ZA, Mason C, Sethabutr O, Talukder K, Nair GB, Deen JL, Kotloff K, Clemens J (2006) A multicentre study of *Shigella* diarrhoea in six Asian countries: disease burden, clinical manifestations, and microbiology. PLoS Med 3(9): e353
2. Barry EM, Pasetti MF, Sztein MB, Fasano A, Kotloff KL, Levine MM (2013) Progress and pitfalls in *Shigella* vaccine research. Nat Rev Gastroenterol Hepatol 10(4):245–255
3. Turbyfill KR, Clarkson KA, Vortherms AR, Oaks EV, Kaminski RW (2018) Assembly, biochemical characterization, immunogenicity, Adjuvanticity, and efficacy of *Shigella* artificial Invaplex. mSphere 3(2):e00583–e00517
4. Cohen D, Atsmon J, Artaud C, Meron-Sudai S, Gougeon ML, Bialik A, Goren S, Asato V, Ariel-Cohen O, Reizis A, Dorman A, Hoitink CWG, Westdijk J, Ashkenazi S, Sansonetti P, Mulard LA, Phalipon A (2021) Safety and immunogenicity of a synthetic carbohydrate conjugate vaccine against *Shigella flexneri* 2a in healthy adult volunteers: a phase 1, dose-escalating, single-blind, randomised, placebo-controlled study. Lancet Infect Dis 4:546–558
5. Ranallo RT, Barnoy S, Thakkar S, Urick T, Venkatesan MM (2006) Developing live *Shigella* vaccines using lambda Red recombineering. FEMS Immunol Med Microbiol 47(3): 462–469
6. Milivojevic M, Dangeard AS, Kasper CA, Tschon T, Emmenlauer M, Pique C, Schnupf P, Guignot J, Arrieumerlou C (2017) ALPK1 controls TIFA/TRAF6-dependent innate immunity against heptose-1,7-bisphosphate of gram-negative bacteria. PLoS Pathog 13(2):e1006224
7. Gaudet RG, Guo CX, Molinaro R, Kottwitz H, Rohde JR, Dangeard AS, Arrieumerlou C, Girardin SE, Gray-Owen SD (2017) Innate recognition of intracellular bacterial growth is driven by the TIFA-dependent cytosolic surveillance pathway. Cell Rep 19(7):1418–1430
8. Li P, Jiang W, Yu Q, Liu W, Zhou P, Li J, Xu J, Xu B, Wang F, Shao F (2017) Ubiquitination and degradation of GBPs by a *Shigella* effector to suppress host defence. Nature 551(7680): 378–383
9. Wandel MP, Pathe C, Werner EI, Ellison CJ, Boyle KB, von der Malsburg A, Rohde J, Randow F (2017) GBPs inhibit motility of *Shigella flexneri* but are targeted for degradation by the bacterial ubiquitin ligase IpaH9.8. Cell Host Microbe 22(4):507–518
10. Datsenko KA, Wanner BL (2000) One-step inactivation of chromosomal genes in *Escherichia coli* K-12 using PCR products. Proc Natl Acad Sci U S A 97(12):6640–6645
11. Sidik S, Kottwitz H, Benjamin J, Ryu J, Jarrar A, Garduno R, Rohde JR (2014) A *Shigella flexneri* virulence plasmid encoded factor controls production of outer membrane vesicles. G3 (Bethesda) 4(12):2493–2503
12. Baba T, Ara T, Hasegawa M, Takai Y, Okumura Y, Baba M, Datsenko KA, Tomita M, Wanner BL, Mori H (2006) Construction of *Escherichia coli* K-12 in-frame, single-gene knockout mutants: the Keio collection. Mol Syst Biol 2:2006
13. Karlinsey JE (2007) lambda-Red genetic engineering in *Salmonella enterica* serovar Typhimurium. Methods Enzymol 421:199–209
14. Sakai T, Sasakawa C, Makino S, Kamata K, Yoshikawa M (1986) Molecular cloning of a genetic determinant for Congo red binding ability which is essential for the virulence of *Shigella flexneri*. Infect Immun 51(2):476–482
15. Parsot C, Ménard R, Gounon P, Sansonetti PJ (1995) Enhanced secretion through the *Shigella flexneri* Mxi-Spa translocon leads to assembly of extracellular proteins into

macromolecular structures. Mol Microbiol 16(2):291–300

16. Demers B, Sansonetti PJ, Parsot C (1998) Induction of type III secretion in *Shigella flexneri* is associated with differential control of transcription of genes encoding secreted proteins. EMBO J 17(10):2894–2903
17. Allaoui A, Menard R, Sansonetti PJ, Parsot C (1993) Characterization of the *Shigella flexneri* ipgD and ipgF genes, which are located in the proximal part of the mxi locus. Infect Immun 61:1707–1714
18. Onodera NT, Ryu J, Durbic T, Nislow C, Archibald JM, Rohde JR (2012) Genome sequence of *Shigella flexneri* serotype 5a strain M90T Sm. J Bacteriol 194(11):3022

Chapter 3

Using Genomic Deletion Mutants to Investigate Effector-Triggered Immunity During *Legionella pneumophila* Infection

Rachelia R. Wibawa, Pengfei Li, Kathleen McCaffrey, and Elizabeth L. Hartland

Abstract

Legionella pneumophila is an intracellular bacterial pathogen that uses a type IV secretion system (T4SS), termed Dot/Icm, to secrete more than 330 virulence effector proteins into the infected host cell. Many Dot/Icm effectors are involved in biogenesis of the *Legionella*-containing vacuole (LCV), which allows intracellular bacterial replication in environmental amoebae and alveolar macrophages. Through their activity, some effectors trigger the mammalian host immune response in a phenomenon termed effector-triggered immunity (ETI). Here, we describe a protocol to create and use *L. pneumophila* genome deletion mutants to identify effector(s) that alter pro-inflammatory cytokine production and bacterial clearance in the lungs of mice.

Key words *Legionella pneumophila*, Mutagenesis, Effector-triggered immunity, In vivo infection, T4SS

1 Introduction

Legionella pneumophila is an accidental human pathogen that causes a severe pneumonia termed Legionnaires' disease [1]. *L. pneumophila* is found in both man-made and natural aquatic environments, often coexisting within a diverse range of free-living amoebae, including *Acanthamoeba* spp., *Vermamoeba* spp., and *Vahlkampfia* spp. [2, 3]. Upon accidental inhalation, *L. pneumophila* primarily replicates in alveolar macrophages in susceptible people [4, 5]. After phagocytosis into either mammalian macrophages or amoebae, *L. pneumophila* evades phagosome maturation by remodeling the vacuole into a replicative niche termed the *Legionella*-containing vacuole (LCV) [6, 7]. Establishment of the LCV and subsequent bacterial replication requires the Dot/Icm type IV secretion system (T4SS), which secretes bacterial

Thomas A. Kufer and Maria Kaparakis-Liaskos (eds.), *Effector-Triggered Immunity: Methods and Protocols*, Methods in Molecular Biology, vol. 2523, https://doi.org/10.1007/978-1-0716-2449-4_3,

proteins termed effectors into the infected host cell [8–12]. To facilitate bacterial virulence in a diverse host range, *L. pneumophila* has amassed a large number of Dot/Icm effectors. To date, approximately 330 proteins are predicted to be effectors of the Dot/Icm T4SS [13]. While the Dot/Icm effector proteins are important to allow the survival and replication of *L. pneumophila* in various protozoan species, some effector proteins are detrimental to bacterial persistence in mammalian host [14, 15]. Deletion of the genes encoding such effectors results in higher bacterial load of the mutants in mice. It is postulated that this phenotype may arise from recognition of the effectors or their activities by the host immune response.

The mechanism of effector recognition by the immune system is termed effector-triggered immunity (ETI). ETI was originally described in plants, but recently, some *L. pneumophila* effectors have been reported to trigger ETI in mammalian macrophages [16]. *L. pneumophila* possesses at least five effectors (Lgt1-3, SidI, and SidL) that dampen host translation elongation and reduce the unfolded protein response (UPR) [17]. However, this activity also leads to an increase in the transcription of the pro-inflammatory cytokines IL1α and IL1β [18, 19], which then activate bystander cells to secrete TNF and type I IFN, leading to a robust pro-inflammatory response in the lung [20].

Dot/Icm effector activities that benefit *L. pneumophila* in the protozoan host may also trigger an unfavorable outcome during infection of mammalian hosts. The Dot/Icm effector protein LamA is an amylase that reduces host glycogenolysis, leading to an elevated level of glucose in the host cytosol [21]. LamA activity is required to prevent encystation of amoebae, a process in which protozoa become dormant and which is unfavorable for *L. pneumophila* replication [21, 22]. However, the high level of glucose in macrophages triggers a shift into the pro-inflammatory M1 macrophage phenotype, which leads to pro-inflammatory cytokine secretion and antimicrobial activities [23]. Similar to LamA, another Dot/Icm effector protein LegC4 was reported to dampen *L. pneumophila* replication in mice by potentiating host pro-inflammatory cytokine production, although the significance of LegC4 during infection of amoebae is unknown [15, 24].

The mouse is a well-established in vivo model for the study of the host immune response against *L. pneumophila*. However, of all in-bred mice, only A-strain mice are permissive to *L. pneumophila* replication. This susceptibility is due to a polymorphism in the gene encoding the NLR family, apoptosis inhibitor protein 5 (Naip5), leading to defective flagellin recognition [25–28]. Flagellin recognition by Naip5 is followed by the formation of an inflammasome complex with the NOD-like receptor CARD domain containing 4 (NLRC4), triggering an inflammatory cell death called pyroptosis, the release of IL1β and IL-18, and rapid clearance of

L. pneumophila [27, 28]. Therefore, to accommodate the study of flagellated *L. pneumophila* in vivo, A-strain mice are typically used.

The identification of Dot/Icm effectors that contribute to ETI in *L. pneumophila* remains a challenge due to the large Dot/Icm effector repertoire, their functional redundancy, host target redundancy, and the phenomenon of metaeffectors that antagonize the activities of other effectors [29]. To approach this question systematically, here we describe methods to identify effectors that may trigger a host immune response using mutants lacking genomic regions enriched for effector genes. First, we describe the mutagenesis process to create *L. pneumophila* genomic deletion mutants (Fig. 1). We used the suicide vector pSR47s to create deletion mutants. This vector comprises (i) an origin of replication (*ori*) R6K which only replicates in the presence of *pir* gene, (ii) kanamycin resistance cassette (kan^R) to allow selection, and (iii) *sacB* which encodes for an enzyme that converts sucrose into a toxic product to provide selection pressure for the second homologous recombination event (Fig. 1). To enable homologous recombination, sequences flanking the genetic region to be deleted are cloned into the multi-cloning site (MCS) of this vector.

After mutagenesis, the growth rate of the mutants is then validated in vitro as bacterial growth rate directly influences the host-pathogen interaction. The mutants are then used to infect A-strain mice, and the bacterial CFU is enumerated by serial dilution (Fig. 2). Finally, the level of pulmonary cytokines after infection with wild-type or mutant *L. pneumophila* derivatives is measured using cytometric bead array (CBA) (Fig. 3). An example of a genomic deletion mutant, ΔE, that exhibits reduced CFU and a reduced IFNγ response is shown, as well as a single gene deletion, Δ*legC4*, from the ΔE region that exhibits wild-type CFU but reduced IFNγ response, suggesting ETI by LegC4 [24].

2 Materials

2.1 Bioinformatics to Design the Genomic Mutants

1. Genome sequence of *L. pneumophila* strain 130b, or any other strain of interest.
2. List of putative Dot/Icm effector-encoding genes.
3. List of secretion system-encoding genes.
4. Artemis genome browser and annotation tool (Wellcome Sanger).

2.2 Cloning the Flanking Sequence Into Suicide Vector

1. Suicide vector pSR47s.
2. Proofreading DNA polymerase, such as Phusion (NEB) or PfuUltra II Fusion (Agilent), and the appropriate buffer as supplied by the manufacturer.

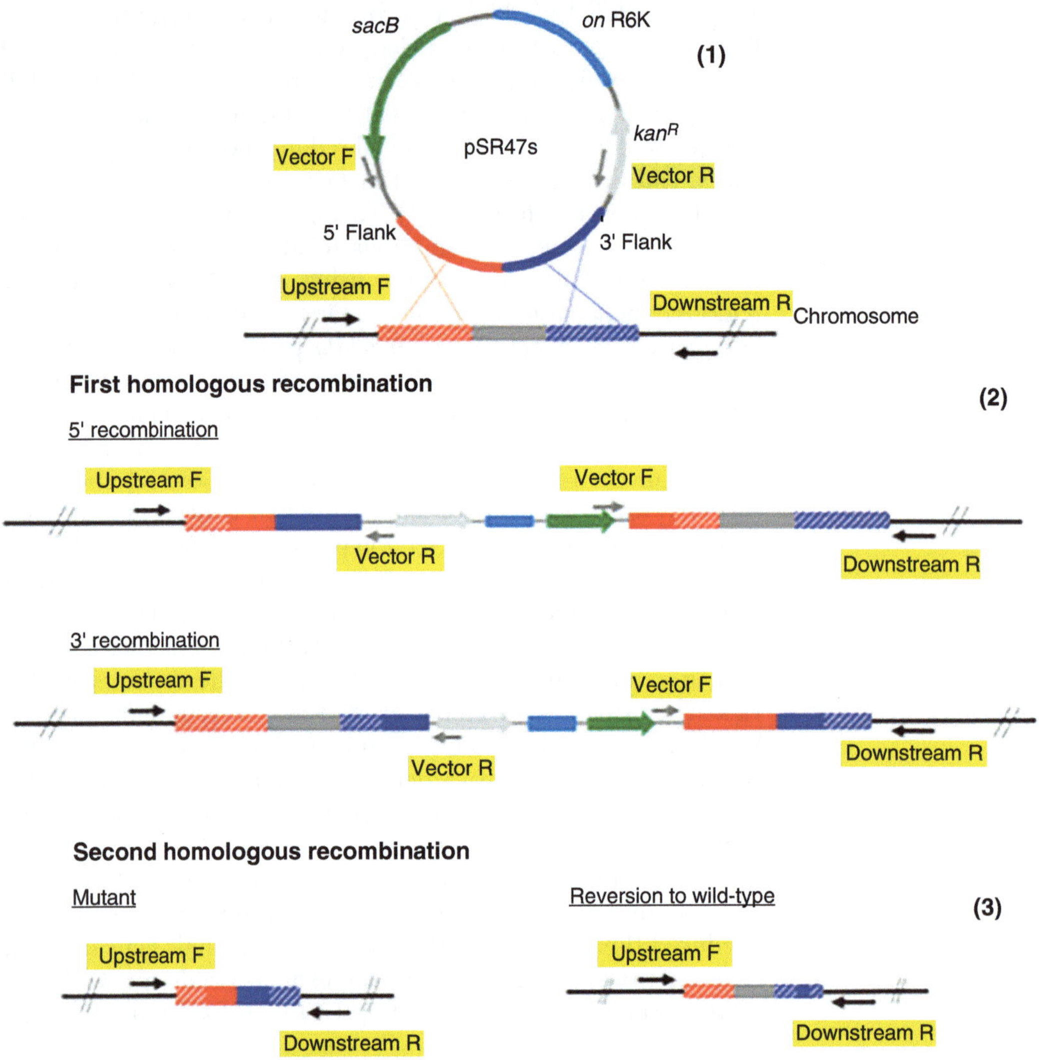

Fig. 1 *Legionella* mutagenesis using suicide vector pSR47s. The suicide vector pSR47s (1) is transformed into *Legionella* using electroporation. The first homologous recombination then occurs (2), such that the vector is integrated into bacterial chromosome to create merodiploid species. The addition of sucrose to growth media then puts a selection pressure for the second homologous recombination to occur (3), yielding either a deletion mutant strain or reversion to the wild-type genotype primers required to screen each stage of mutagenesis which is highlighted in yellow

3. Polymerase chain reaction (PCR) machine.
4. PCR tubes.
5. 1.5-mL microfuge tubes.
6. 10-mM dNTP mix.
7. Dimethyl sulfoxide (DMSO).
8. Agarose.

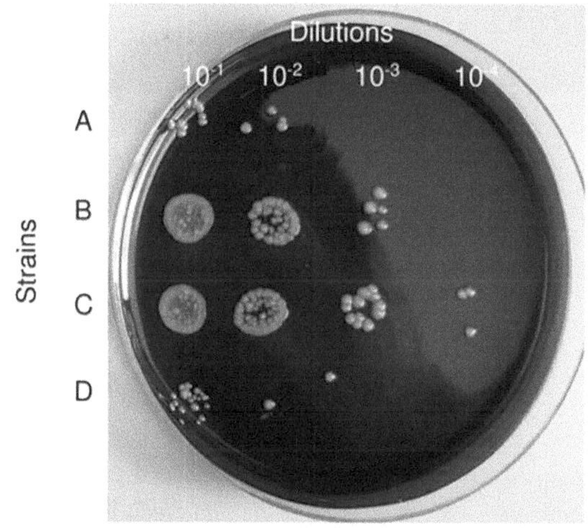

Fig. 2 Example of serial plating on BCYE plates for CFU enumeration. Each drop corresponds to 25 μL of culture plated. From top to bottom: different strains being plated. From left to right: increasing dilutions of culture

9. 1× Tris-Acetate-EDTA (TAE) buffer (40-mM Tris, 1-mM EDTA disodium salt, 0.114% glacial acetic acid (v/v), pH 8.3).
10. Restriction enzyme.
11. Calf intestinal phosphatase (CIP).
12. T4 ligase.
13. Competent *pir*-positive *Escherichia coli* strain.
14. Luria broth (LB) and/or Luria agar (LA) plus 100 μg/mL kanamycin.
15. Super-optimal broth with catabolite repression (SOC) broth.
16. Taq polymerase master mix such as GoTaq Green 2× Master-Mix (Promega).
17. UV trans-illuminator.
18. ZymoClean™ DNA purification kit (Zymo Research).
19. ZymoClean™ gel DNA extraction kit (Zymo Research).
20. Primers flanking the multi-cloning site (MCS) of pSR47s.
    Vector F: 5′ GTGAACGGCAGGTATATGT 3′
    Vector R: 3′ AGCTATGACCATGATTACGC 5′

2.3 *Legionella Mutagenesis*

1. *L. pneumophila* strain 130b (ATCC BAA-744), or any other strain of interest.
2. Black charcoal yeast extract (BCYE) agar plus 25 μg/mL kanamycin (BCYK).
3. BCYE plus 5% sucrose (BCYS).

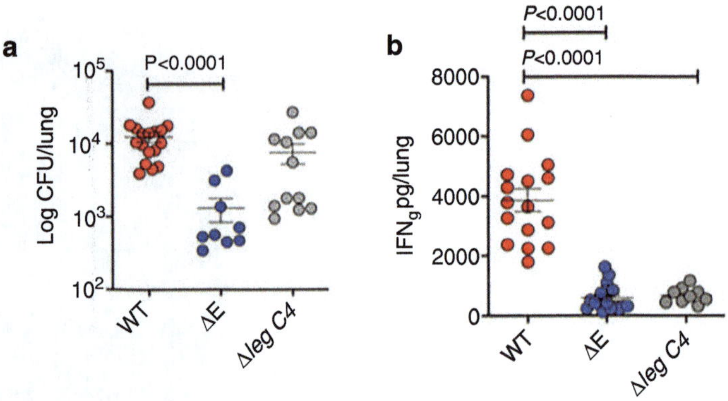

Fig. 3 Bacterial load and pulmonary IFNγ response after infection with either wild-type *L. pneumophila*; the genomic deletion mutant, ΔE; or single effector gene deletion, Δ*legC4*. A-strain mice were infected with wild-type (WT) or mutant derivatives (ΔE or Δ*legC4*) of *L. pneumophila* 130b. (**a**) At day 3 after infection, lungs were harvested and homogenized. After permeabilization to release intracellular bacteria, CFU was enumerated by plating onto BCYE plates (**b**) The level of cytokines from the lung homogenate was then quantified using CBA Mouse Flex set kit. Error bars represent standard error of the mean (SEM). Each data point is one mouse. Statistical analysis was done using Mann-Whitney test. Primers for flanking regions to construct ΔE mutant in 130b 5′ Flank: Forward 5′-gagggatccgaccctgttggcagctatc-3′ Reverse 5′-tatccctacctcttgtggag-3′; 3′ Flank: Forward 5′-ctccacaagaggtagggataatggcagacaaatgttaaggaa-3′ Reverse 5′-agcggccgccggttggagtagtttggaca-3. Primers outside of region E used to confirm deletion: Upstream 5′-gaacacgcgcttgtaagaaaa-3′, Downstream: 5′-aaattgcaaagataagattgtact-3′. Primers for flanking regions to construct Δ*legC4* in 130b (*LPW_20041*) 5′ Flank: Forward 5-′-tcctgcagcccggggatccacccggatataaccaccac-3′, Reverse 5-′-ttacaggcttctgacaagcgaaatacaatg-3′. 3′ Flank: Forward 5′-ttgtcagaagcctgtaagtactgcatcg-3′, Reverse: 5-′-ctccaccgcggtggcggccgccctcaagtgatttccgtg-3′. Primers outside the *legC4* coding region used to confirm deletion: Forward 5′-aacgcttactgttcccggta-3′, Reverse 5′-gcgtgcttcccagttcaaaa-3′

4. ACES-buffered yeast extract (AYE) broth (1% (w/v) ACES buffer, 1% (w/v) yeast extract, pH 6.9) supplemented with 1-mM iron (III) nitrate ($Fe(NO_3)_3$) and 1-mM L-cysteine hydrochloride.
5. Electroporator and 2-mm cuvettes.
6. Sterile 1.5-mL microfuge tubes.
7. Sterile 30-mL bottles.
8. Sterile 1× PBS.
9. Sterile dH_2O.
10. 10% glycerol (v/v) in dH_2O, filter sterilized with 0.45-μm filter.

11. Taq polymerase master mix such as GoTaq Green 2× Master-Mix (Promega).
12. Primers upstream and downstream of the target genomic region (Fig. 1)

2.4 Measuring the Growth of Legionella mutants In Vitro

1. Sterile 30-mL bottles.
2. Sterile 96-well plate.
3. Microplate reader that detects absorbance, has temperature control, and has shaking option.
4. Spectrophotometer and cuvettes.

2.5 Infection of A-Strain Mice

1. A-strain mice aged 7–12 weeks old, male and/or female.
2. *L. pneumophila* wild-type 130b and mutant strains.
3. BCYE agar plates.
4. Spectrophotometer and cuvettes.
5. Sterile 1.5-mL microfuge tubes.
6. Anesthetic machine comprising of an induction chamber, an oxygen cylinder, an isoflurane evaporator, and a carbon chamber to capture isoflurane waste.
7. Sterile 1× PBS.
8. FACS buffer: 1× PBS with 0.1% BSA and 2-mM EDTA.

2.6 Enumeration of CFU and Cytokine Bead Array (CBA)

1. Sterile 10-mL conical tubes.
2. Small dissection kit.
3. CO_2 euthanasia chamber comprising a CO_2 cylinder with adjustable flow rate connected to a sealed chamber.
4. Standard benchtop homogenizer with probe.
5. Sterile 1× PBS.
6. 80% ethanol.
7. 10% (w/v) saponin in dH_2O, filter sterilized with 0.2-μm filter.
8. BCYE agar plates.
9. Standard benchtop centrifuge.
10. Clear 96-well plates.
11. FACS buffer.
12. CBA mouse Flex set (BD).
13. Bovine serum albumin (BSA).

3 Methods

3.1 Creation of *L. pneumophila* Genomic Deletion Mutants

3.1.1 Design of Genomic Regions to Be Deleted

1. Using Artemis, annotate the effector-encoding genes and secretion system component-encoding genes onto the genome of *L. pneumophila*. This will allow the visual identification of effectors that are located in close proximity.
2. Group nearby effector-encoding genes, excluding all secretion system components and genes known to be important for bacterial viability (*see* **Note 1**).

3.1.2 Clone the Flanking Sequence of Designed Genomic Regions into Suicide Vector pSR47s

Amplification and Ligation of the Flanking Regions

1. Design primers to amplify the upstream (5′) and downstream (3′) regions flanking the designed genomic regions (*see* **Note 2**). These flanking regions will enable homologous recombination into the bacterial genome (Fig. 1). The primers used to amplify these regions should also add an 18–20-bp overhang sequence to the end of the 5′ region that complements the start of the 3′ region. Alternatively, an overhang sequence complementary to the end of the 5′ region can be added to the start of the 3′ region and vice versa. The primers should also add restriction enzyme sites to enable ligation into the suicide vector.
2. Amplify the 5′ and 3′ flanking regions using PCR (*see* **Note 3**). Run a fraction of the PCR product on agarose gel to check the success of amplification (*see* **Note 4**). Clean up the PCR reaction to get rid of the PCR by-products.
3. Combine the two flanking regions using overlapping PCR. The overlapping PCR consists of two steps: (i) annealing the two flanking regions through the complementary overhang sequence to create a combined DNA fragment and (ii) amplifying the combined DNA fragment (*see* **Note 5**).
4. Run all of the PCR product on agarose gel. Visualize the DNA using UV transilluminator and cut the band with the right size. Purify DNA from the gel.

Restriction Digest and Ligation

1. Perform restriction digest on the purified overlap PCR product and vector using the enzymes of choice according to the manufacturer's protocol (*see* **Note 6**).
2. After digestion, it is also advisable to treat the cut vector using calf intestinal alkaline phosphatase (CIP) as per manufacturer's protocol to reduce self-ligation. Do not treat the cut insert with CIP, as this will prevent ligation completely (*see* **Note 7**).
3. Purify the digested insert and vector, and perform ligation using T4 ligase according to the manufacturer's protocol (*see* **Notes 8** and **9**).

Transformation into Chemically Competent *pir*-Positive *Escherichia coli*

1. Add all of the ligation products to 50–100 μL of chemically competent *pir*-positive *E. coli* strain (*see* **Notes 10** and **11**) and incubate on ice for 20–30 min to allow the plasmid to come into contact with the bacteria.
2. Heat shock at 42 °C for 90 s and immediately cool the tube on ice for 2–3 min.
3. Recover bacteria at 37 °C in Super Optimal Broth with catabolite repression (SOC) broth for 1.5–2 h (*see* **Note 12**).
4. Plate the recovered bacteria onto Luria broth (LB) agar plate with the appropriate antibiotic. Incubate plate at 37 °C overnight.
5. Screen for the success of transformation using colony PCR with Vector F and Vector R primers (*see* **Note 13**, Fig. 1, see Subheading 2.2).

   For each colony to be tested, set up 20 μL GoTaq Green reaction (as per manufacturer's protocol). Pick a colony from the transformation plate, patch on a new LB agar plate with antibiotic, and inoculate into each PCR reaction. PCR should then be set up as per the GoTaq manufacturer's protocol.
6. Run the colony PCR product on agarose gel and identify positive colonies (colonies that have taken up the correct construct).
7. Extract plasmid from positive colonies and confirm with Sanger sequencing.

3.1.3 Legionella mutagenesis

Legionella Transformation

1. Streak out *Legionella* on black charcoal yeast extract (BCYE) agar plate. Incubate at 37 °C for 3 days.
2. Inoculate a loopful of *Legionella* into 1 mL PBS. Spin down at top speed (≥10,000 rcf) for 1 min at room temperature. Remove supernatant and resuspend in 1-mL ice-cold sterile dH_2O. From this point on, keep the bacteria on ice (*see* **Note 14**).
3. Spin at top speed, 1 min, at 4 °C. Remove supernatant and resuspend in 1-mL ice-cold sterile dH_2O.
4. Spin at top speed, 1 min, 4 °C. Remove supernatant and resuspend in 1-mL ice-cold sterile 10% (v/v) glycerol.
5. Spin at top speed, 1 min, 4 °C. Remove supernatant and resuspend in 100-μL ice-cold sterile 10% (v/v) glycerol.
6. Add 1 μg of the suicide vector into the bacteria suspension (*see* **Note 15**). Transfer the suspension into a clean 2-mm electroporation cuvette.
7. Electroporate at 2.4 kV, 200 Ω, 0.25 μF, and immediately recover in 0.9-mL ACES broth with supplement in a 30-mL tube (*see* **Note 14**). Incubate the culture at 37 °C for 5–6 h with agitation.

8. Plate culture on BCYE plus 25 μg/mL kanamycin (BCYK) agar plate. Incubate at 37 °C for 3 days.
9. After 3 days, screen for positive transformants using colony PCR. In the first recombination, we are looking for colonies where the suicide vector has undergone homologous recombination into the *Legionella* genome. Use primer set Vector F + downstream R primer and Vector R + upstream F primer to screen for both possibilities of recombination (Fig. 1, *see* **Note 16**). Pick a single colony from the transformation plate, patch on fresh BCYK plate, and inoculate into each PCR reaction (*see* **Note 17**).
10. Run PCR products on agarose gel to identify positive colonies.

Sucrose Selection

1. Using streak dilution, streak positive transformants obtained in "*Legionella* Transformation" for single colonies onto BCYE plus 5% sucrose (BCYS) plate. Incubate at 37 °C for 3 days.
2. Screen for colonies that have undergone the second homologous recombination using colony PCR. For the colony PCR, use the upstream F and downstream R primers (Fig. 1, *see* **Note 18**). As in **step 9** of "*Legionella* Transformation", pick a single colony from the BCYS plate, patch on fresh BCYE agar, and inoculate into each PCR reaction.
3. Run PCR products on agarose gel to identify positive colonies. Streak out positive colonies and stock at −80 °C for long-term storage (*see* **Note 19**). The resulting colonies will contain either the genomic deletion or wild-type genetic sequence (Fig. 1).

3.2 Validate the Growth of Legionella mutants In Vitro

The deletion of putative effector-encoding genes or genomic regions may influence the bacterial growth rate, which in turn can confound the results of infection. Therefore, we suggest ensuring that the mutants have a similar growth rate to wild-type *Legionella* before commencing other experiments.

1. Streak out *Legionella* wild-type and mutant strains on BCYE plates. Incubate at 37 °C for 3 days.
2. Inoculate ACES broth containing supplements with one to two single colonies from the BCYE plate. Incubate at 37 °C for 18 h with shaking (200 rpm).
3. We typically expect an OD_{600} of 3–4 after 18 h of incubation. Given an OD_{600} measurement above 1.0 is often inaccurate, we recommend making a 1:10 dilution of the culture to be measured by adding 100-μL culture to 900-μL media in the spectrophotometer cuvette. Measure the OD_{600} of this culture and adjust the starting culture to an OD_{600} of 1. Afterward, serially dilute this culture to OD_{600} by doing 1:10 dilution twice. Add 200 μL of culture into each well in a clear sterile

96-well plate in triplicate (*see* **Note 20**). To control for the background absorbance of the media, include media only control in triplicate.

4. Load plate into microplate reader. Set the plate reader to read at least every 2 h for 40–50 h, keeping the temperature at 37 °C and shaking at 200 rpm (*see* **Note 21**).
5. Data analysis:
   (a) Average the triplicate absorbance reading at each desired time point.
   (b) Subtract the average background absorbance (media only control) from the average absorbance at each desired time point.
   (c) Plot the absorbance against time to generate a growth curve. A comparison between the wild-type and mutant strains can then be done to determine whether there is any alteration to the mutant growth rate. We typically use unpaired t-test or one-way ANOVA to identify whether there is any differential absorbance value between wild-type and mutant strains at any given time points.

3.3 Infection of A-Strain Mice and Harvesting Lungs

The procedure involving animals outlined in this paper was performed under the approval from the University of Melbourne Animal Ethics Committee. An animal ethics approval needs to be obtained from the relevant animal ethics committee prior to the commencement of any work involving animals and 3-R guidelines must be strictly applied.

1. Resuspend wild-type and mutant *L. pneumophila* 130b in 1-mL sterile PBS. Adjust the OD_{600} to 1.0 (*see* Subheading 3.2). Perform 1 in 20 dilutions to get a final concentration of 5.0×10^7 bacteria/mL.
2. Confirm the accuracy of inoculum by plating the bacteria and counting the CFU. Serially dilute the inoculum up to 10^{-5} dilution (i.e., five times 1:10 dilutions) and plate 50–100 μL of this dilution onto BCYE plates in duplicate. Incubate plates for 3 days at 37 °C and enumerate the CFU.
3. Prior to infection, weigh mice to determine the starting weight.

   Perform **steps 4** and **5** in a BSC-II biosafety cabinet, as there is a high risk of aerosol generation.
4. Anesthetize mice by placing them into a sealed induction chamber with 2–4% isoflurane and 1–2-L/min oxygen flow. Initially, isoflurane should be administered at 4%, and once the animal is unconscious (*see* **Note 22**), it should ideally be reduced to 2% to prevent them from gaining consciousness.

5. Grab mouse by the loose skin on the neck and administer 50 μL of the inoculum (2.5×10^6 cfu) into one nostril. If the mouse regains consciousness during this process, return the mouse into the chamber, placing it on its back (*see* **Note 23**). Ensure that the breathing of the mouse is calm before resuming infection. After the mouse has received all of the inoculum dose, return mouse to the chamber on its back for a minute to allow it to breathe in any remaining dose.
6. Return mouse to the cage in a supine position. The mouse should wake up within a minute.
7. Monitor the weight and disease progression daily. A-strain mice typically start losing weight at day 1 after infection and continue to lose weight at day 2. However, at day 3 after infection, the weight should start to return to original weight. Other symptoms to expect include reduced activity, slightly hunched/ruffled appearance, and slight alteration to the breathing rate.
8. At each time point, euthanize mice based on the approved procedure. We typically use CO_2 asphyxiation delivered at 20% chamber volume/min followed by cervical dislocation to ensure death.
9. Harvest the right lung into 2.5-mL cold sterile PBS in a 10-mL conical tube to be used for CFU enumeration and cytokine assay. Keep all samples on ice from this point onward.

3.4 Measure the Bacterial Numbers After In Vivo Infection

The number of bacteria infecting the lung directly influences the level of inflammation, immune cell recruitment, and cytokine production. Deletion of some genes/genetic regions may lead to a reduction in bacterial virulence, which in turn will reduce the magnitude of host immune response.

1. Homogenize the right lung in the 10-mL conical tube. Perform this procedure a BSC-II as it generates aerosols. Between samples, wash the probe twice with 80% ethanol and once with sterile PBS. The homogenate can then be used for both CFU quantification and cytokine measurement (*see* Subheading 3.5).
2. Aseptically take 250 μL of lung homogenate into an 1.5-mL tube and add 25 μL of 1% saponin (final concentration of 0.1%). Incubate at 37 °C for 30 min.
3. Serially dilute the homogenate (*see* **Note 24**). Drop 25 μL of each dilution in quadruplicate onto BCYE plates (Fig. 2, *see* **Note 25**). Incubate plate at 37 °C for 3 days and enumerate the number of CFU from each mouse (*see* **Note 26**).

3.5 Measure the Levels of Cytokines in Lung Homogenate

1. Take 500 μL of lung homogenate from **step 1**, Subheading 3.4 (*see* **Note 27**). Spin down at ≥10,000 rcf for 5 min to pellet cells. Sometimes after centrifuging, the supernatant is also split into two phases. Invert once to mix the two phases, but not the cells. Transfer 5-μL supernatant into clean 96-well plate (*see* **Note 28**).
2. Prepare standard curve for the following cytokines: IL-1α, IL-12, IL-6, TNFα, GM-CSF, IL-10, IFNγ, IL-17A, and MCP1. Resuspend all standard cytokines in 2-mL PBS plus 3% BSA, to result in 5000-pg/mL final concentration. Serially dilute eight times using 1:2 dilution. The concentration of the standard curve should then be 5000 pg/mL, 2500 pg/mL, 1250 pg/mL, 625 pg/mL, 312.5 pg/mL, 156.25 pg/mL, 80 pg/mL, 40 pg/mL, and 20 pg/mL. For the 0 pg/mL, use PBS only. Aliquot 5 μL of each standard to a well in a 96-well plate.
3. Prepare capture beads from the CBA Flex kit to detect cytokines. For each bead, we typically use 0.1 μL of beads/sample. For each sample, use 5 μL of beads mix in PBS (*see* **Note 29**).
4. Prepare PE detection reagent from the CBA Flex kit in the same manner as the capture beads (*see* **Note 29**).
5. Add the capture beads and PE detection reagent to the samples and standard. Incubate the mix in the dark at room temperature for 2 h.
6. Add 200 μL of CBA wash buffer or FACS buffer. Spin down cells at 600 rcf, 5 min, 4 °C. Using multichannel pipette, take out 175 μL of supernatant, and add 60 μL of fresh CBA wash buffer or FACS buffer. The samples are now ready for FACS analysis.
7. Perform flow cytometry analysis and data analysis as per manufacturer's instructions.

3.6 Identification of Dot/Icm Effectors that Trigger the Mammalian Immune Response

If a particular genomic deletion mutant strain displays reduced pro-inflammatory cytokine response compared to the wild-type parent strain, it may be inferred that this mutant lacks an effector that triggers the immune response or that the mutant exhibits decreased bacterial load in the lung. Deletion of the relevant single effector-encoding genes can then be performed according to methods outlined in Subheading 3.1 to identify the genetic basis of the phenotype. Afterward, the replication of the mutant in the murine lung, as well as the level of pulmonary cytokine, can be measured as outlined in Subheadings 3.3 and 3.4 (Fig. 3).

4 Notes

1. A recent study has utilized transposon mutagenesis to identify genes important for *L. pneumophila* viability [30]. The results from this study can be used as a guideline. However, different strains of *L. pneumophila* may have different requirements for viability. Therefore, it is possible that some mutants will fail the mutagenesis process because the deletion of certain genes renders the bacteria nonviable. We have also suggested validating the growth of *Legionella* mutants (*see* Subheading 3.2) to ensure that the mutant's in vitro growth rate is not affected by deletion.
2. We have successfully deleted genomic regions sized over 50 kb using 3-kb-sized flanking regions. Smaller-sized flanking regions can be used when deleting smaller regions to reduce PCR time. We typically use 500–600-bp flanking regions to delete single genes sized up to 2 kb.
3. We recommend the use of proofreading enzymes for amplification as these regions may be recombined into the bacterial genome, and it is advisable to minimize second-site mutation during mutagenesis. We use either PfuUltra II (Agilent) or Phusion (NEB), but other proofreading enzymes may be used too.
4. If the correct PCR product is not obtained, try to optimize the annealing temperature using either gradient or touchdown PCR.
5. The typical reaction for our overlapping PCR is as follows:

Reaction name	Reaction setup	Thermocycler setup
Overlap PCR step 1	50-ng 5′ flank 50-ng 3′ flank 1-mM dNTP mix 0.5 μM each primer 1.5-μL DMSO 10-μL Phusion-HF buffer 0.5-μL Phusion-HF polymerase Nuclease-free dH2O to 50 μL	98 °C, 1 min initial denaturation *Start cycle 10×* 98 °C, 10s denaturation Tann*, 30s annealing 72 °C, 30s elongation *End cycle*
Overlap PCR step 2	50-μL step 1 product 1-mM dNTP mix 0.5 μM each primer 1.5-μL DMSO 10-μL Phusion-HF buffer 0.5-μL Phusion-HF polymerase Nuclease-free dH2O to 50 μL	98 °C, 1-min initial denaturation *Start cycle 35×* 98 °C, 10-s denaturation Tann*, 20-s annealing 72 °C, 30-s/kb elongation *End cycle* 72 °C, 5-min final elongation

6. We usually do the restriction digest for 3 h to ensure that the digestion is complete. If desired, a small fraction of the cut vector can be run on agarose gel side to side with the uncut vector. If the digest is complete, the cut vector should be completely linearized, unlike the circular uncut vector.
7. We use Quick CIP alkaline phosphatase (NEB), which is compatible with CutSmart buffer (NEB). Therefore, if digest is performed in CutSmart buffer, 1-μL alkaline phosphatase can directly be added into each 20-μL digest reaction after digest is completed.
8. We typically do either a 1-h ligation at room temperature or overnight at 4 °C. For large inserts that are difficult to clone, try ligating at 16 °C overnight.
9. It is recommended to add a vector-only ligation control during ligation reaction. This will give an indication on whether the vector has self-ligated.
10. When using the suicide vector pSR47s, it is important to use *pir*-positive strain, as its origin of replication (*ori*) requires the *pir* gene to replicate. We typically use either PIR1 or PIR2 *E. coli* strains. PIR1 strain carries higher copy number than PIR2; however, PIR2 is better at tolerating toxic inserts.
11. Here a protocol for transformation into chemically competent strains is provided. However, transformation can also be done using electroporation with electro-competent strains.
12. While this step only involves the use of 1-mL culture, we recommend performing the recovery in 30-mL bottles/tubes to provide good aeration to culture.
13. If desired, positive and negative controls can also be included in the colony PCR. As negative controls, colonies from the vector-only ligation control (*see* **Note 10**) can be used. Alternatively, the uncut vector can be used as well. In addition, a no template control can also be included to ensure that the PCR reaction was not contaminated. A pSR47s construct containing any insert can be used as positive control.
14. Performing the transformation at cold temperature may improve the efficiency of transformation. It is also recommended to chill the electroporation cuvette in the freezer for at least 30 min before use.
15. The quality of the suicide vector prep is essential to a successful transformation. Additionally, a concentration of 100 ng/μL or above is desired. Lower prep concentration may also be used, although this decreases the transformation efficiency.

16. Guidelines for expected size in the first mutagenesis step:
    (a) *If 5′ recombination had occurred:*
    Vector F + downstream primer → expect a product sized as big as the 5′ flanking region + genetic region to be deleted + 3′ flanking region. Often this is too big to amplify and as such PCR will not yield any product.
    Vector R + upstream primer → expect a product sized as big as the 5′ flanking region + 3′ flanking region.
    (b) *If 3′ recombination had occurred:*
    Vector F + downstream primer → expect a product sized as big as the 5′ flanking region + 3′ flanking region.
    Vector R + upstream primer → expect a product sized as big as the 5′ flanking region + genetic region to be deleted + 3′ flanking region. Often this is too big to amplify and as such PCR will not yield any product.
17. Colonies can also be patched first onto BCYE plate containing the appropriate antibiotic. Incubate the plate at 37 °C for overnight and perform colony PCR on the next day.
18. Expected size for the second mutagenesis screen:
    (a) If colony has reverted to wild-type, expect a product sized as big as the 5′ flanking region + genetic region to be deleted + 3′ flanking region. Often this is too big to amplify and as such PCR will not yield any product.
    (b) If colony has lost the genetic region of interest: expect a product sized as big as the 5′ flanking region + 3′ flanking region.
19. If desired, genomic DNA can be extracted from the mutant culture and either Sanger sequencing using the upstream/downstream primer can be performed to ensure that the desired region has been deleted. Whole genome sequencing can also be done to ensure that there is no second-site mutation.
20. As the plate will be incubated at 37 °C for 40–50 h, it is advisable to aliquot 200-μL sterile PBS into the surrounding wells to reduce evaporation of culture during experiment.
21. Setting the plate reader can be tricky, as not all plate readers can do 2-h intervals. We use 5-min intervals for reading, but only plot the data for every 2 h to ease analysis.
22. The mouse breathing rate should be slow and calm, and mouse should not attempt to wake up when the chamber is tilted or move when one of the paws is gently pinched.
23. It is important to prevent the nose from touching anything as this may lead to a loss of inoculum dose.

24. We typically plate 10^{-1} to 10^{-5} dilutions for wild-type *L. pneumophila* infection. However, some mutants may display either increased or decreased CFU; thus, the dilution should be adjusted accordingly.
25. As with all bacterial plates, the BCYE plates should be incubated in the correct orientation (i.e., inverted, lid on the bottom), as the opposite orientation may lead to the formation of condensation on the lid. This may fall onto the agar plate and interfere with the plating layout. Also ensure that the droplet has been absorbed into the agar before incubating plate. We find that drying the BCYE plates before use is very helpful in reducing the time required for the droplet to be absorbed and reduces the risk of the droplet dispersing.
26. Example calculation to enumerate the CFU per lung:

    Strain A: count 30 colonies at dilution (10^{-2})

    Strain B: count 25 colonies at dilution (10^{-3})

    (a) Multiply CFU with 10 to the power of dilution:

    Strain A → $30 \times 10^2 = 3 \times 10^3$ CFU

    Strain B → $25 \times 10^3 = 2.5 \times 10^4$ CFU

    (b) Divide this number by the amount plated (in mL) to get CFU/mL:

    Strain A → 3×10^3 CFU/25×10^{-3} mL $= 1.2 \times 10^5$ CFU/mL

    Strain B → 2.5×10^4 CFU/25×10^{-3} mL $= 1 \times 10^6$ CFU/mL

    (c) Multiply the result from (b) with the volume of homogenate (in mL) to get CFU/lungs:

    Strain A → 1.2×10^5 CFU/mL $\times$ 2.5 mL $= 3 \times 10^5$ CFU/lungs

    Strain B → 1×10^6 CFU/mL $\times$ 2.5 mL $= 2.5 \times 10^6$ CFU/lungs
27. Lung homogenate may also be stored at −80 °C to be used at a later time for cytokine analysis.
28. CBA can also be done according to the manufacturer's protocol. Here we have outlined the procedure commonly used in our lab to reduce the volume of reagents required for this analysis.
29. Example calculation to prepare bead mix in PBS:

    Cytokines to measure: IL1α, IL6, TNF, and IFNγ (a total of four cytokines)

    Number of samples: 90

    Number of standards: 10

Total number of samples = 90 + 10 = 100 → prepare 10% extra mix to account for pipetting inaccuracy.

(a) Calculate total volume of beads mix required:
5-μL bead/sample × (110% × 100 samples) = 550 μL

(b) Calculate the volume of each bead to be added to the mix:
0.1 μL each bead/sample × (110% × 10 samples) = 11 μL

(c) Calculate the volume of PBS to be added to the mix:
550 μL − (11 μL × 4 cytokines) = 506 μL

References

1. Phin N, Parry-Ford F, Harrison T, Stagg HR, Zhang N, Kumar K, Lortholary O, Zumla A, Abubakar I (2014) Epidemiology and clinical management of Legionnaires' disease. Lancet Infect Dis 14(10):1011–1021. https://doi.org/10.1016/S1473-3099(14)70713-3
2. Conza L, Pagani SC, Gaia V (2013) Presence of *Legionella* and free-living amoebae in composts and bioaerosols from composting facilities. PLoS One 8(7):e68244. https://doi.org/10.1371/journal.pone.0068244
3. Valcina O, Pule D, Malisevs A, Trofimova J, Makarova S, Konvisers G, Berzins A, Krumina A (2019) Co-occurrence of free-living amoeba and *Legionella* in drinking water supply systems. Medicina (Kaunas) 55(8):492. https://doi.org/10.3390/medicina55080492
4. Nash TW, Libby DM, Horwitz MA (1984) Interaction between the legionnaires' disease bacterium (*Legionella pneumophila*) and human alveolar macrophages. Influence of antibody, lymphokines, and hydrocortisone. J Clin Invest 74(3):771–782. https://doi.org/10.1172/JCI111493
5. Copenhaver AM, Casson CN, Nguyen HT, Fung TC, Duda MM, Roy CR, Shin S (2014) Alveolar macrophages and neutrophils are the primary reservoirs for *Legionella pneumophila* and mediate cytosolic surveillance of type IV secretion. Infect Immun 82(10):4325–4336. https://doi.org/10.1128/IAI.01891-14
6. Horwitz MA (1983) The Legionnaires' disease bacterium (*Legionella pneumophila*) inhibits phagosome-lysosome fusion in human monocytes. J Exp Med 158(6):2108–2126. https://doi.org/10.1084/jem.158.6.2108
7. Horwitz MA, Maxfield FR (1984) *Legionella pneumophila* inhibits acidification of its phagosome in human monocytes. J Cell Biol 99(6): 1936–1943. https://doi.org/10.1083/jcb.99.6.1936
8. Berger KH, Isberg RR (1993) Two distinct defects in intracellular growth complemented by a single genetic locus in *Legionella pneumophila*. Mol Microbiol 7(1):7–19. https://doi.org/10.1111/j.1365-2958.1993.tb01092.x
9. Horwitz MA (1987) Characterization of avirulent mutant *Legionella pneumophila* that survive but do not multiply within human monocytes. J Exp Med 166(5):1310–1328. https://doi.org/10.1084/jem.166.5.1310
10. Kwak MJ, Kim JD, Kim H, Kim C, Bowman JW, Kim S, Joo K, Lee J, Jin KS, Kim YG, Lee NK, Jung JU, Oh BH (2017) Architecture of the type IV coupling protein complex of *Legionella pneumophila*. Nat Microbiol 2:17114. https://doi.org/10.1038/nmicrobiol.2017.114
11. Marra A, Blander SJ, Horwitz MA, Shuman HA (1992) Identification of a *Legionella pneumophila* locus required for intracellular multiplication in human macrophages. Proc Natl Acad Sci U S A 89(20):9607–9611. https://doi.org/10.1073/pnas.89.20.9607
12. Swanson MS, Isberg RR (1996) Identification of *Legionella pneumophila* mutants that have aberrant intracellular fates. Infect Immun 64(7):2585–2594
13. Gomez-Valero L, Rusniok C, Carson D, Mondino S, Perez-Cobas AE, Rolando M, Pasricha S, Reuter S, Demirtas J, Crumbach J, Descorps-Declere S, Hartland EL, Jarraud S, Dougan G, Schroeder GN, Frankel G, Buchrieser C (2019) More than 18,000 effectors in the Legionella genus genome provide multiple, independent combinations for replication in human cells. Proc Natl Acad Sci U S A 116(6):2265–2273. https://doi.org/10.1073/pnas.1808016116
14. Park JM, Ghosh S, O'Connor TJ (2020) Combinatorial selection in amoebal hosts drives the evolution of the human pathogen *Legionella pneumophila*. Nat Microbiol 5(4):599–609.

https://doi.org/10.1038/s41564-019-0663-7

15. Shames SR, Liu L, Havey JC, Schofield WB, Goodman AL, Roy CR (2017) Multiple *Legionella pneumophila* effector virulence phenotypes revealed through high-throughput analysis of targeted mutant libraries. Proc Natl Acad Sci U S A 114(48):E10446–E10454. https://doi.org/10.1073/pnas.1708553114
16. Fischer NL, Naseer N, Shin S, Brodsky IE (2020) Effector-triggered immunity and pathogen sensing in metazoans. Nat Microbiol 5(1):14–26
17. Hempstead AD, Isberg RR (2015) Inhibition of host cell translation elongation by *Legionella pneumophila* blocks the host cell unfolded protein response. Proc Natl Acad Sci U S A 112(49):E6790–E6797. https://doi.org/10.1073/pnas.1508716112
18. Barry KC, Ingolia NT, Vance RE (2017) Global analysis of gene expression reveals mRNA superinduction is required for the inducible immune response to a bacterial pathogen. Elife 6:e22707. https://doi.org/10.7554/eLife.22707
19. Fontana MF, Banga S, Barry KC, Shen X, Tan Y, Luo ZQ, Vance RE (2011) Secreted bacterial effectors that inhibit host protein synthesis are critical for induction of the innate immune response to virulent *Legionella pneumophila*. PLoS Pathog 7(2):e1001289. https://doi.org/10.1371/journal.ppat.1001289
20. Copenhaver AM, Casson CN, Nguyen HT, Duda MM, Shin S (2015) IL-1R signaling enables bystander cells to overcome bacterial blockade of host protein synthesis. Proc Natl Acad Sci U S A 112(24):7557–7562. https://doi.org/10.1073/pnas.1501289112
21. Price C, Jones S, Mihelcic M, Santic M, Abu Kwaik Y (2020) Paradoxical pro-inflammatory responses by human macrophages to an amoebae host-adapted legionella effector. Cell Host Microbe 27(4):571–584, e577. https://doi.org/10.1016/j.chom.2020.03.003
22. Aguilar-Diaz H, Carrero JC, Arguello-Garcia-R, Laclette JP, Morales-Montor J (2011) Cyst and encystment in protozoan parasites: optimal targets for new life-cycle interrupting strategies? Trends Parasitol 27(10):450–458. https://doi.org/10.1016/j.pt.2011.06.003
23. Rodriguez-Prados JC, Traves PG, Cuenca J, Rico D, Aragones J, Martin-Sanz P, Cascante M, Bosca L (2010) Substrate fate in activated macrophages: a comparison between innate, classic, and alternative activation. J Immunol 185(1):605–614. https://doi.org/10.4049/jimmunol.0901698
24. Ngwaga T, Hydock AJ, Ganesan S, Shames SR (2019) Potentiation of cytokine-mediated restriction of *Legionella* intracellular replication by a dot/Icm-translocated effector. J Bacteriol 201(14):e00755–e00718. https://doi.org/10.1128/JB.00755-18
25. Brieland J, Freeman P, Kunkel R, Chrisp C, Hurley M, Fantone J, Engleberg C (1994) Replicative *Legionella pneumophila* lung infection in intratracheally inoculated A/J mice. A murine model of human Legionnaires' disease. Am J Pathol 145(6):1537–1546
26. Fitzgeorge RB, Baskerville A, Broster M, Hambleton P, Dennis PJ (1983) Aerosol infection of animals with strains of *Legionella pneumophila* of different virulence: comparison with intraperitoneal and intranasal routes of infection. J Hyg (Lond) 90(1):81–89. https://doi.org/10.1017/s0022172400063877
27. Diez E, Lee SH, Gauthier S, Yaraghi Z, Tremblay M, Vidal S, Gros P (2003) Birc1e is the gene within the Lgn1 locus associated with resistance to *Legionella pneumophila*. Nat Genet 33(1):55–60. https://doi.org/10.1038/ng1065
28. Wright EK, Goodart SA, Growney JD, Hadinoto V, Endrizzi MG, Long EM, Sadigh K, Abney AL, Bernstein-Hanley I, Dietrich WF (2003) Naip5 affects host susceptibility to the intracellular pathogen *Legionella pneumophila*. Curr Biol 13(1):27–36
29. Ensminger AW (2016) *Legionella pneumophila*, armed to the hilt: justifying the largest arsenal of effectors in the bacterial world. Curr Opin Microbiol 29:74–80. https://doi.org/10.1016/j.mib.2015.11.002
30. O'Connor TJ, Adepoju Y, Boyd D, Isberg RR (2011) Minimization of the *Legionella pneumophila* genome reveals chromosomal regions involved in host range expansion. Proc Natl Acad Sci U S A 108(36):14733–14740. https://doi.org/10.1073/pnas.1111678108

Chapter 4

Methods of Bacterial Membrane Vesicle Production, Purification, Quantification, and Examination of Their Immunogenic Functions

Natalie J. Bitto and Maria Kaparakis-Liaskos

Abstract

Bacterial membrane vesicles (BMVs) released by Gram-negative and Gram-positive bacteria are a *bona fide* secretion system that enable the dissemination of bacterial effector molecules, and can trigger a range of responses in the host. The study of BMV production, composition, and functions can give insights into their roles in mediating bacterial survival, pathogenesis, and disease. Furthermore, BMVs can be harnessed to develop cutting-edge nano-therapeutics including targeted chemotherapy delivery, antimicrobials, and novel vaccines. Here we describe routine methods that can be used for small- or large-scale production, isolation, and purification of outer membrane vesicles produced by Gram-negative bacteria, and membrane vesicles produced by Gram-positive bacteria, which we collectively refer to as BMVs. We discuss methods that can be used to visualize BMVs by electron microscopy, and to quantify their DNA, RNA, and protein cargo. We outline a method for the fluorescent labeling of BMVs that can be applied to examine their ability to interact with and enter host cells using a range of in vitro and in vivo biological assays. Finally, we provide a cell culture-based method that can be used to examine a range of immunogenic properties of BMVs, including their cytotoxicity, ability to activate pathogen-recognition receptors (PRRs), induce autophagy and cytokine responses, and modulate cellular pathways.

Key words Outer membrane vesicles (OMVs), Membrane Vesicles (MVs), Bacterial membrane vesicles (BMVs), Production, Purification, Quantification, Composition, Immunogenicity, Cell culture assays, Cytokine production, Nanoparticle analysis, Electron microscopy

1 Introduction

Bacteria ubiquitously secrete nano-sized membrane vesicles that contain a diverse array of bacterial cargo (reviewed in [1]). Gram-negative bacteria release vesicles from their outer membrane, termed outer membrane vesicles (OMVs), while Gram-positive bacteria, which lack an outer membrane, release vesicles from their cytoplasmic membrane, termed membrane vesicles (MVs) (reviewed in [2]). Both OMVs and MVs, collectively referred to as bacterial membrane vesicles (BMVs), range from approximately

Thomas A. Kufer and Maria Kaparakis-Liaskos (eds.), *Effector-Triggered Immunity: Methods and Protocols*,
Methods in Molecular Biology, vol. 2523, https://doi.org/10.1007/978-1-0716-2449-4_4,

50 to 500 nm in size which may vary based on their mechanism of biogenesis. BMVs carry a range of bacterial cargo on their surface and within their lumen, referred to as microbe-associated molecular patterns (MAMPs) that can be delivered to host cells to trigger pattern-recognition receptors (PRRs) and modulate host immune responses (reviewed in [3]). For example, BMVs produced by pathogenic bacteria can deliver cytotoxic proteins to host epithelial cells to induce cell death [4–6], transport DNA and RNA into host cells to activate endosomal Toll-like receptors [7, 8], or deliver bacterial peptidoglycan to the cytoplasmic receptor nucleotide-binding oligomerization domain containing protein 1 (NOD1) and NOD2 to induce autophagy [8–10]. In contrast, BMVs produced by commensal bacteria can promote anti-inflammatory responses in host cells that have protective effects in the host [11, 12] and may promote mutualism between the host and symbiotic bacteria [13]. The production and composition of BMVs vary significantly with bacterial species, strains, and conditions of bacterial growth [7, 14, 15]. Therefore, understanding the production and contents of BMVs is critical to elucidating their immunogenic properties, which can in turn shed light on their roles in host-pathogen interactions and their therapeutic potential. In addition, the biological functions of BMVs can be harnessed to develop cutting-edge nano-therapeutics (reviewed in [16]). For example, their ability to enter target host cells can be utilized for cell-specific drug delivery [17], their anti-inflammatory functions may be harnessed to restore epithelial barrier function in gastrointestinal inflammatory disease [18], and their immune-stimulating and self-adjuvanting properties make them ideal vaccine candidates [19] as demonstrated by the licensed BMV-based meningococcal B vaccine [20].

The general procedures outlined below are broadly applicable to facilitate the production, purification, quantification, and examination of BMVs and their contents, in addition to the analysis of the immunogenic properties of BMVs, with the aim of making the study of BMVs accessible to all researchers. This will ultimately increase our understanding of the multitude of functions of BMVs in the host and will propel their development for use in a range of therapeutic applications.

2 Materials

2.1 Production of BMVs from Planktonic Cultures

1. Glycerol stock of bacteria of interest.
2. Agar plates required to grow bacteria of interest.
3. Liquid culture media required to grow bacteria of interest.
4. Sterile 50 mL tubes or high-speed centrifuge tubes.

5. Benchtop centrifuge or high-speed centrifuge.
6. 0.22 μm filtration unit with sterile bottle.

2.2 Concentration of Bacterial-Free Supernatant Using a Tangential Flow Unit

1. Class II Biosafety cabinet.
2. 80% ethanol.
3. 10 kDa molecular weight cutoff (MWCO) tangential flow unit and tubing.
4. Peristaltic pump.
5. Sterile milliQ water.
6. Flask for collection of filtrate.
7. Ice.
8. Sterile cleaning and storage solutions as recommended by the manufacturer of the tangential flow filter unit, such as sterile milliQ, 10% ethanol, and 0.2N NaOH.

2.3 Isolation of BMVs by Ultracentrifugation

1. Ultracentrifuge.
2. Swing-out rotor and buckets (e.g., Hitachi P28S, Beckman SW 28 Ti).
3. Ultracentrifuge tubes appropriate for the rotor used (e.g., 25 × 89-mm polypropylene tubes).
4. 2-decimal place balance.
5. Sterile phosphate-buffered saline (PBS).
6. Sterile 1.5 mL tubes.

2.4 Purification of BMVs by Density Gradient Ultracentrifugation

1. OptiPrep medium.
2. Cold sterile PBS.
3. Vortex.
4. Ultracentrifuge.
5. Swing-out rotor and buckets (e.g., Hitachi P40ST, Beckman SW 40 Ti).
6. Ultracentrifuge tubes appropriate for the rotor used (14 × 95-mm polypropylene tubes).
7. 10 kDa MWCO centrifugal filtration units.
8. Benchtop centrifuge.
9. Sterile 1.5 mL tubes.

2.5 Verification of BMV-Containing Fractions and Examination of Their Purity by Transmission Electron Microscopy

1. Parafilm.
2. Carbon-coated grids.
3. Poly-L-lysine.
4. PBS.
5. milliQ water.

6. 1% (v/v) glutaraldehyde diluted in PBS.
7. Foil.
8. 2% (v/v) uranyl acetate pH 7.0 diluted in milliQ water.
9. Filter paper.
10. Grid box.
11. Transmission electron microscope.

2.6 Quantification of BMVs

2.6.1 Nanoparticle Counting

1. 1.5 mL tubes.
2. Sample diluent (e.g., phosphate-buffered saline (PBS) pH 7.4 or 50 mM 4-(2-hydroxyethyl)-1-piperazineethanesulfonic acid (HEPES) pH 7.4).
3. Nanoparticle counting instrument (e.g., NanoSight or ZetaView).

2.6.2 Protein Assay

1. Protein assay quantification kit (e.g., Bradford, Lowry, bicinchoninic acid (BCA) assay).
2. Bovine serum albumin (BSA).
3. PBS.
4. Vortex.
5. Cuvettes.
6. Spectrophotometer.

2.7 Quantification of BMV Protein, DNA, and RNA Contents Using Qubit Assays

1. Qubit protein assay kit.
2. Qubit dsDNA assay kit.
3. Qubit RNA assay kit.
4. Qubit fluorometer.
5. Qubit assay tubes.
6. Vortex.

2.8 Fluorescent Labeling of BMVs for Confocal Microscopy

1. Microfuge.
2. Dulbecco's PBS.
3. Sterile 1.5 mL tubes.
4. Vybrant Di (DiO/DiI/DiD) lipophilic fluorescent stain.
5. Vortex.
6. 10 kDa MWCO centrifugal filtration units.
7. Benchtop centrifuge.

2.9 Stimulation of Eukaryotic Cells with BMVs

1. Class II Biosafety cabinet.
2. 80% ethanol.
3. Cell line of interest (e.g., epithelial, endothelial, macrophage).

4. Complete tissue culture media (appropriate media supplemented with fetal calf serum (FCS), L-glutamine, penicillin, and streptomycin, as required for cell line of interest).
5. Plain tissue culture media (without FCS, L-glutamine, penicillin, and streptomycin).
6. Hemocytometer, or other cell counting apparatus.
7. 24-well tissue culture plate.
8. 37 °C, 5% CO_2 incubator.

3 Methods

All procedures should be performed aseptically, using a Bunsen flame or Class II Biosafety cabinet, unless otherwise stated.

3.1 Production of BMVs from Planktonic Cultures

1. Using a bacterial culture that has been passaged at least once since thawing from glycerol, prepare a broth culture by inoculating the appropriate culture media. Incubate the inoculated broth culture at the relevant conditions and for the required length of time to achieve a bacterial culture at mid-log to stationary phase. This is referred to as the seed culture.
2. Using the seed culture, inoculate a larger volume of liquid media so that the final optical density (O.D. 600-nm) is approximately 0.01–0.05. Inoculated media may be divided over multiple bacterial culture flasks if required (*see* **Note 1**).
3. Incubate the inoculated liquid cultures at the relevant conditions and for the appropriate length of time to allow the bacterial culture to reach mid-log phase or another desired stage of bacterial growth, from which to isolate BMVs (*see* Fig. 1a) (*see* **Note 2**).
4. Once the bacterial culture(s) has grown to the desired stage of bacterial growth, pour into sterile, nuclease-free centrifuge tubes and pellet bacteria using a benchtop centrifuge or high-speed centrifuge at 4 °C at 4000 × *g* for 20–30 min, or until bacteria are completely pelleted (*see* Fig. 1b) (*see* **Note 3**).
5. Collect the supernatant and remove any remaining bacteria by filter-sterilizing the supernatant using a 0.22-μm vacuum filtration unit and a sterile flask (*see* Fig. 1c). The resulting bacterial-free supernatant should be kept on ice from herein as it contains BMVs.
6. For the isolation of BMVs from large culture volumes, the culture supernatant may be concentrated using a 10-kDa MWCO tangential flow unit prior to ultracentrifugation (*see* Fig. 1d), to reduce the number of ultracentrifugations being performed (proceed to Subheading 3.2). Alternatively, BMVs may be pelleted without concentration of the supernatant (proceed to Subheading 3.3).

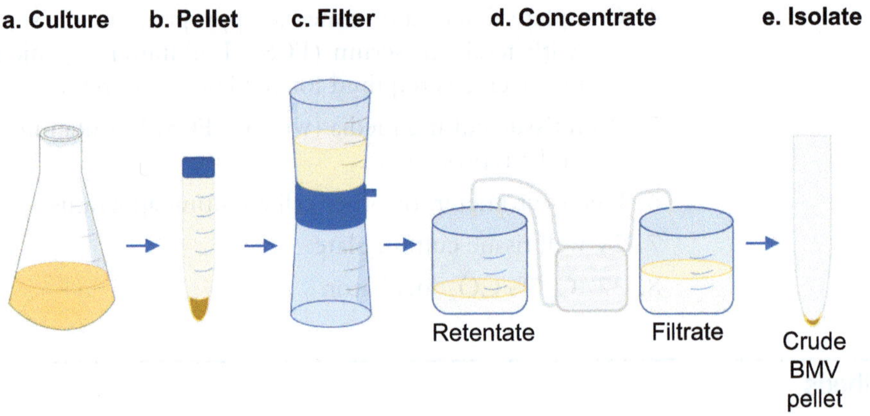

Fig. 1 Workflow for the production and isolation of crude BMVs. Bacteria are cultured in liquid broth (**a**) and then the bacteria are pelleted by centrifugation (**b**). The supernatant is then filtered through a 0.22 μm filter (**c**). For large-scale cultures, the filtered supernatant can be concentrated using a tangential flow filter unit (**d**). The concentrated supernatant can then be subjected to ultracentrifugation, resulting in a crude pellet of BMVs (**e**). Alternatively, for the isolation of BMVs from smaller volumes of bacterial cultures, bacterial-free supernatant (**c**) can next be subjected to ultracentrifugation to obtain a crude pellet of BMVs (**e**)

3.2 Concentration of Bacterial-Free Supernatant Using a Tangential Flow Unit

1. Set up the tangential flow unit and peristaltic pump in a Class II Biosafety cabinet, according to the manufacturer's instructions (*see* **Note 4**), ensuring that all surfaces have been sterilized with 80% ethanol prior to use.
2. Flush the tangential flow unit with 500 mL sterile milliQ water by placing the retentate and filtrate lines in the filtrate flask.
3. Purge the unit and tubing by slowly pumping air until all the liquid is removed from the unit.
4. Place the flask containing bacterial-free supernatant on ice and insert the feed and retentate lines.
5. Slowly increase the pump speed until a pressure of 3 bar or 45 psi is displayed on the pressure gauge.
6. Allow the sample to concentrate to the desired final volume (*see* **Note 5**).
7. Once concentration of the supernatant is complete, reduce the pump speed to slow for 5 min, and then purge the unit with air to recover all residual supernatant. Clean the unit as per the manufacturer's instructions (*see* **Note 4**).

3.3 Isolation of BMVs by Ultracentrifugation

1. Place clean 25 × 89 mm ultracentrifuge tubes into pre-chilled ultracentrifuge buckets. Pipette the bacterial-free supernatant into ultracentrifuge tubes, ensuring all tubes are filled to approximately 1 cm from the top. Place the caps onto the ultracentrifuge buckets and balance to within ±0.02-g using sterile broth media. Once balanced, screw the caps onto the buckets.

2. Carefully place the buckets onto the relevant pre-chilled rotor and lower into the ultracentrifuge.
3. Pellet BMVs at 100,000 × *g* for 2 h at 4 °C (*see* **Note 6**).
4. Without dislodging the BMV pellet (*see* **Note** 7), carefully remove the supernatant using a sterile pipette. Resuspend the BMV pellet in the remaining 100–200-μL of supernatant. Repeat this process to remove the supernatant and to resuspend BMV pellets in each ultracentrifuge tube.
5. Collect the resuspended BMVs using a pipette, and pool them into a sterile 1.5 mL tube on ice. The resulting crude suspension of BMVs may be stored at −20 °C (short term) or −80 °C (long term) prior to purification (*see* Fig. 1e) (*see* **Note 8**).

3.4 Purification of BMVs by Density Gradient Ultracentrifugation

1. Resuspend BMVs in OptiPrep medium (60% w/v iodixanol; *see* **Note 9**) to achieve a final concentration of 45% w/v OptiPrep using ice-cold PBS or relevant buffer (*see* **Note 10**), in a final volume of 2-mL. Place on ice.
2. Using cold sterile PBS, prepare OptiPrep fractions by diluting OptiPrep medium to a final concentration of 40%, 35%, 30%, 25%, and 20% w/v. Vortex all OptiPrep fractions thoroughly and place on ice.
3. In a 14 × 95 mm ultracentrifuge tube (or appropriately sized tube for rotor), generate a continuous gradient by first placing the 45% w/v OptiPrep containing BMVs at the bottom of the tube. Next, sequentially overlay the BMV-containing fraction with 2-mL of 40%, 35%, 30%, 25%, and 20% w/v OptiPrep fractions (*see* **Note 11**), vortexing each fraction thoroughly immediately prior to layering into the ultracentrifuge tube (*see* Fig. 2).
4. Carefully place ultracentrifuge tubes into ultracentrifuge buckets and replace cap. Balance tubes to within ±0.02-g using sterile PBS.
5. Place buckets onto the rotor and fractionate BMVs by ultracentrifugation at 100,000 × *g* for 16 h at 4 °C (*see* **Note 12**).
6. Following ultracentrifugation, bands containing BMVs may be visible in the gradient (*see* Fig. 3a). Carefully remove 1-mL fractions, starting from the top low-density fractions through to the bottom fractions, placing each fraction into a separate sterile 1.5 mL tube. Number these fractions in order of collection. Effective fractionation can be verified by calculating the density of each fraction (*see* **Note 13**).
7. Subsequent removal of residual OptiPrep medium from fractions may be performed using ultracentrifugation (*see* Subheading 3.4.1) or 10-kDa MWCO centrifugal filtration units (*see* Subheading 3.4.2).

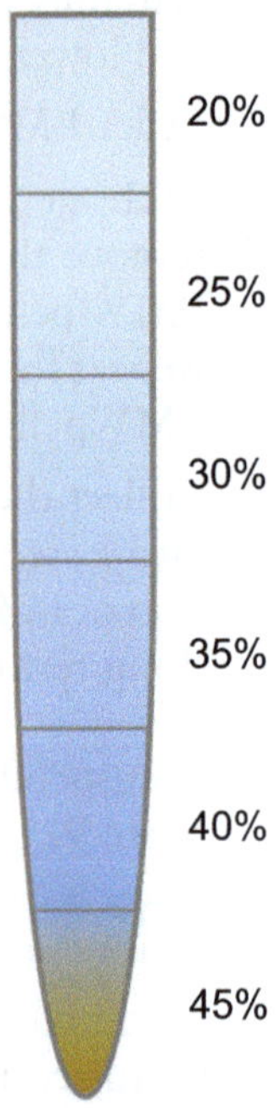

Fig. 2 Preparation of an OptiPrep density gradient. Crude BMVs resuspended in 45% OptiPrep medium are added to the bottom of the ultracentrifuge tube. Sequential layers of 40%, 35%, 30%, 25%, and 20% OptiPrep are then layered to achieve a continuous gradient

3.4.1 Washing BMVs to Remove OptiPrep Medium Using Ultracentrifugation

1. To remove OptiPrep medium from BMV-containing fractions, wash each 1 mL fraction by resuspending in a final volume of 14 mL cold PBS and placing into a 14 × 95 mm ultracentrifuge tube. Pellet the BMVs by ultracentrifugation at 100,000 × *g* for 2 h at 4 °C.
2. Using a sterile pipette, carefully remove all but 100–200-μL of supernatant. Gently pipette the remaining supernatant up and down to resuspend any BMVs pelleted at the bottom of the tube.
3. Perform a second wash by filling the tube with 14 mL fresh PBS, gently pipetting up and down to mix. Pellet BMVs by ultracentrifugation at 100,000 × *g* for 2 h at 4 °C.
4. Carefully using a pipette, remove all but 100–200-μL supernatant and gently pipette up and down the residual supernatant, flushing the bottom of the tube to resuspend pelleted BMVs. Transfer each fraction into a separate sterile 1.5 mL tube and store at −20 °C (short term) or −80 °C (long term).

3.4.2 Washing BMVs Using Centrifugal Filtration Units

1. Place each BMV-containing OptiPrep fraction into separate 4 mL 10-kDa MWCO centrifugal filtration units containing 3 mL cold PBS. Gently pipette up and down to mix well.

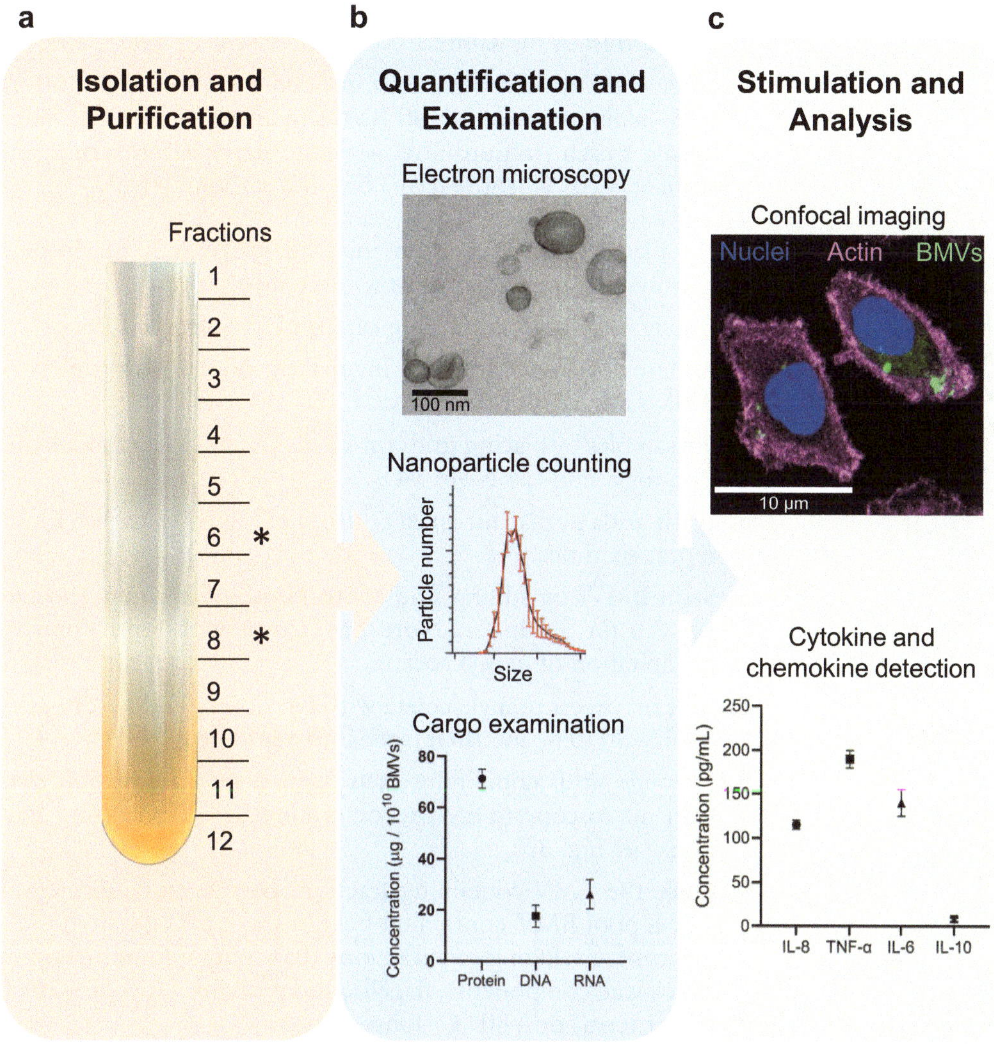

Fig. 3 Overview of the isolation, quantification, examination and analysis of purified BMVs. BMV-containing fractions (indicated by *) obtained from an OptiPrep density gradient (**a**) are examined by transmission electron microscopy to confirm the presence of BMVs (b). The amount of BMVs present within each OptiPrep fraction and purified BMV sample are quantified by nanoparticle counting, and their DNA, RNA, and protein contents can be quantified using a range of assays, including Qubit fluorometric assays (**b**). Purified BMVs can then be used in functional assays to examine their entry into host cells by confocal microscopy or their ability to stimulate cytokine and chemokine production by stimulation of host cells (**c**). BMVs can also be fluorescently labeled if required to facilitate their intracellular detection within host cells by microscopy **(c)**

2. Centrifuge at 4 °C at 4000 × *g* for 15 min or until the majority of liquid has passed through the filter.
3. Empty the filtrate and refill the filtration unit with fresh PBS, gently mixing well by pipetting up and down, taking care to flush off BMVs that have stuck to the filter.

4. Repeat **steps 2–4** until all traces of OptiPrep medium are removed from the sample.
5. After washing is complete, resuspend BMVs in 100–500 μL PBS, taking care to flush off BMVs that have stuck to the filter. Transfer each fraction into a separate sterile 1.5 mL tube and store at −20 °C (short term) or −80 °C (long term).

3.5 Verification of BMV-Containing Fractions and Examination of Their Purity by Transmission Electron Microscopy

1. On a large sheet of parafilm, incubate grids on 20 μL droplets of poly-L-lysine for 1 min at room temperature.
2. Briefly wash grids by placing on 20 μL droplets of PBS.
3. Adhere BMVs to grids by incubating on 10 μL droplets of BMV samples for 10 min.
4. Fix samples by placing grids on 10 μL 1% (v/v) glutaraldehyde-PBS for 5 min (*see* **Note 14**).
5. Wash grids by placing on large drops of milliQ water for 1 min, repeat six times.
6. Stain BMVs by adding grids onto 10-μL of 2% uranyl acetate pH 7.0 for 5 min (*see* **Note 15**). Cover with foil to prevent precipitation of uranyl acetate.
7. Blot off excess uranyl acetate with filter paper and air-dry grids for 15–30 min, and then place into a grid box.
8. Examine grids containing each fraction by transmission electron microscopy (TEM) to determine BMV-containing fractions (*see* Fig. 3b).
9. Once the BMV-containing fractions have been confirmed by TEM, pool BMV-containing fractions into a single sterile 1.5-mL tube, excluding any fractions that contain impurities such as cell wall components, flagella, and proteins. Store at −20 °C (short term) or −80 °C (long term).

3.6 Quantification of BMVs

BMV quantification is a critical step in the analysis of BMV-mediated immunity [21]. BMVs may be quantified by nanoparticle counting, or indirectly based on the amount of protein in the BMV preparation [21]. However, it should be noted that the quantification method selected can influence the outcomes of downstream immunological assays [21]. Standardization of BMV amount by particle number has been shown to enable better differentiation of the immunostimulatory properties of BMVs produced by different species and strains [7, 21].

3.6.1 Nanoparticle Counting

This is a general protocol for the quantification of BMVs using a nanoparticle counting instrument, such as NanoSight nanoparticle tracking analysis (NTA) or ZetaView NTA (*see* **Note 16**).

1. Thaw BMVs on ice. Aseptically remove an aliquot of BMVs into a 1.5-mL tube for quantification. The remainder of the

quantification assay can then be performed using non-aseptic conditions.

2. Perform required washes and calibration of the nanoparticle counting instrument as per the manufacturer's instructions.
3. Immediately prior to use, dilute BMVs in the appropriate assay diluent (e.g., PBS or 50 mM HEPES buffer, pH 7.4) to a range of dilutions (e.g., 1:500, 1:5000, 1:50,000) and vortex thoroughly to dissociate any BMV aggregates (*see* **Note 17**).
4. Starting from the highest dilution, measure each BMV sample, thoroughly flushing the instrument with diluent between samples.
5. When BMVs are diluted appropriately to achieve a particle density within the instrument's minimum and maximum threshold, the output data shows the concentration and size distribution of BMVs (*see* Fig. 3b). If not already factored in, multiply the concentration by the dilution factor to determine the concentration of BMVs in the original BMV preparation (*see* **Note 18**).

3.6.2 Protein Assay

This is a general protocol for the quantification of BMVs using a range of protein assays, including Bradford, Lowry, or BCA assays (*see* **Note 19**).

1. Thaw BMVs on ice. Aseptically remove an aliquot of BMVs and place into a 1.5 mL tube, to be used for protein quantification. The protein assay can be performed using non-aseptic conditions.
2. Prepare serially diluted BSA standards in PBS within the minimum and maximum detection limit of the protein assay of choice.
3. Prepare a range of dilutions of BMVs in PBS (e.g., 1:50, 1:100, 1:500), and vortex thoroughly to mix.
4. Pipette standards and BMV samples into 1 mL cuvettes, in duplicate.
5. Add the required volume of detection reagent as per the protein assay kit instructions, mix well by pipetting, and incubate for the appropriate amount of time at room temperature, protected from light.
6. Using a spectrophotometer, measure the O.D. of the standards and BMV samples at the appropriate wavelength, as per the manufacturer's instructions for the protein assay of choice.
7. Plot the standard curve (O.D. versus protein concentration in μg) and determine the formula of the standard curve.
8. Calculate the protein concentration of the BMV sample within the range of the standard curve (*see* **Note 20**).

9. Multiply the protein concentration by the dilution factor to obtain the protein concentration in the original BMV preparation (*see* **Note 18**).

3.7 Quantification of BMV Protein, DNA, and RNA Contents Using Qubit Fluorometric Quantitation

BMVs contain a range of bacterial effector molecules, or MAMPs, including biologically active proteins [8, 22, 23], DNA [8, 24, 25], and RNA [8, 26, 27]. The amount and type of MAMPs carried by BMVs can influence their function [21]. Therefore, here we describe the quantification of protein, DNA, and RNA associated with BMVs (*see* Fig. 3b). These methods are to be performed as per the manufacturer's guidelines.

1. Thaw BMVs on ice. Aseptically remove an aliquot of BMVs into a 1.5 mL tube for quantification of their contents. The remainder of the assay can be performed using non-aseptic conditions.
2. Prepare the required volume of Qubit protein, DNA, and RNA assay master mixes by adding n(1 μL Qubit reagent) + n(199 μL Qubit buffer), where n = number of samples and standards.
3. Add the Qubit protein, DNA, and RNA standards into Qubit assay tubes using a pipette.
4. Pipette a volume of 1–10-μL of BMVs into Qubit tubes.
5. Add the appropriate Qubit master mix, i.e., Qubit protein, dsDNA, or RNA assay master mix, to each sample and standard to achieve a final volume of 200-μL. Vortex briefly to mix.
6. Incubate samples and standards for 15 min (protein assay) or 2 min (DNA and RNA assays), protected from light.
7. Measure the amount of protein, DNA, and RNA, starting with the standards and then the samples, using a Qubit fluorometer, ensuring the correct sample volume is entered in the input screen.
8. Read the concentration of protein, DNA, or RNA from the display screen. If the correct sample volume was entered into the input screen in **step 7**, the readout will show the final concentration of the BMV sample accounting for the dilution factor.
9. If the particle number of BMVs is known, the amount of protein, DNA, and RNA may be normalized to BMV number to allow comparison of BMV contents produced by different bacterial species or different culture conditions [8].

3.8 Fluorescent Labeling of BMVs for Use in Biological Assays

BMVs can be fluorescently labeled for use in a range of in vitro and in vivo studies, for example, observing the entry of BMVs into host cells [25] (*see* Fig. 3c) and their translocation through the cytoplasm [8, 9], or tracking the biodistribution of BMVs using animal models [28]. This method describes the labeling of BMVs with the lipophilic Vybrant Di fluorescent dye (Molecular Probes); however, it may be applied to any lipophilic stain (e.g., FM4 dyes) by adjusting the final concentration of dye according to the manufacturer's specifications.

1. Thaw BMVs on ice.
2. In a 1.5 mL tube, dilute the desired amount of BMVs in 99 μL DPBS.
3. Briefly centrifuge Vybrant Di fluorescent dye in a microfuge to pellet any dye aggregates.
4. Add 1-μL of Vybrant Di stain to the BMVs to achieve a final dilution of 1:100. Vortex well to mix.
5. Incubate at 37 °C for 20 min on a platform shaker at low speed, protected from light.
6. To remove unbound dye from the BMV sample, add BMV-dye mixture into a 4 mL 10-kDa MWCO centrifugal filtration unit filled with cold DPBS.
7. Centrifuge at 4 °C and 4000 × *g* for 15 min or until the majority of liquid has passed through the filter.
8. Empty the filtrate and refill the filtration unit with fresh DPBS, gently mixing well by pipetting up and down, taking care to flush off any BMVs that have stuck to the filter.
9. Repeat **steps** 7 and **8** three times to remove all unbound Vybrant Di stain.
10. After washing is complete, collect and resuspend the labeled BMVs in a final volume of DPBS to achieve the desired BMV concentration.
11. Transfer labeled BMVs to a sterile 1.5 mL tube wrapped with foil, and either use immediately or store at −20 °C (short term) or −80 °C (long term).

3.9 Stimulation of Cells with BMVs

This general method may be used to examine a range of cellular and immunological processes mediated by BMVs, including, but not limited to, their entry into host cells [9, 25], cytotoxic effects [29], activation of pattern recognition receptors [7, 10, 30], and the cytokine and chemokine responses they induce [31, 32] (*see* Fig. 3c). This method describes the stimulation of host cells seeded in a 24-well tissue culture plate; however, it may be scaled accordingly for stimulation of cells in 96-, 12-, or 6-well tissue culture plates.

1. Prepare all materials including pipettes, tips, tissue culture plates, and pre-warmed media by spraying their surface with 80% ethanol and placing into a Class II Biosafety cabinet.

2. For the examination of cellular entry of BMVs by confocal microscopy, place a sterile glass coverslip in each well of a 24-well plate. Alternatively, cells can be seeded in cell culture chamber slides.
3. Using a relevant cell line, detach cells from the surface of tissue culture plates or flasks as per established methods for specific cell lines, wash cells and resuspend them in the relevant cell culture media.
4. Count cells using a hemocytometer and seed in a 24-well plate at a density of approximately 1×10^5 cells per well.
5. Incubate cells at 37 °C, 5% CO_2 for 24 h or until they are approximately 70–80% confluent.
6. Wash cells once with pre-warmed plain and replace with 200–500-μL of plain media.
7. Stimulate cells with BMVs (fluorescently labeled BMVs if performing confocal microscopy, live cell imaging, or flow cytometry studies) starting at a multiplicity of infection (MOI) of 5×10^4 BMVs per cell [8] or a BMV protein concentration of 50-μg/mL [33] (*see* **Note 21**), as well as appropriate controls (*see* **Note 22**).
8. Incubate cells in a humidified incubator at 37 °C, 5% CO_2 for the appropriate length of time (i.e., 3–4 h for microscopy-based cell entry experiments, including examining the induction of autophagy [8, 9] (*see* **Note 23**); 3–12 h for detection of cytokine expression by qRT-PCR [31]; 16–24 h for the detection of cytokine and chemokine production by enzyme-linked immunosorbent assay (ELISA), cytometric bead array (CBA) [8], or Western immunoblot [9]) (*see* **Note 24**).
9. Following incubation with BMVs, collect the culture supernatant, cells, or coverslips as required and analyze using the appropriate assay.

4 Notes

1. It is recommended that a minimum culture volume of 200-mL is used to produce a sufficient amount of BMVs that can be purified and used in subsequent studies.
2. Bacterial growth stage regulates the size and composition of BMVs [15, 21]; therefore, it is important to carefully select the appropriate growth stage from which to isolate BMVs, and to report the stage of growth at which BMVs are isolated for all studies [21].
3. The duration of centrifugation required to pellet the culture may vary depending on the bacteria of interest.

4. Recommended cleaning solutions, pump pressure, and storage conditions vary with different tangential flow units. Therefore, refer to the manufacturer's guidelines for instructions that are specific to your tangential flow unit. Appropriate cleaning and storage are essential to preserve the sterility of the unit. Once clean, the unit and tubing should be stored aseptically at 4 °C and as per the manufacturer's guidelines.
5. To avoid excessive aggregation of BMVs, it is preferable to concentrate supernatant to a maximum of one-tenth of the initial volume.
6. Caution: ultracentrifuges operate at extreme *g*-force and failure to operate correctly can result in serious injury. Appropriate training is required before use.
7. After ultracentrifugation using a swing-out rotor, the BMV pellet, located in the center of the base of the ultracentrifuge tube, may be very fine and difficult to observe by eye.
8. Crude BMVs contain contaminants such as flagella, membrane fragments, and free proteins; therefore, further purification is required prior to their use in biological and immunological assays, outlined in Subheading 3.4.
9. This protocol describes the purification of BMVs using iodixanol (OptiPrep) density gradient media, as this is most widely used in the BMV field. Alternative density gradient media, such as Percoll [34] or sucrose [33], may be used; however, the gradient should be adjusted to account for differences in density between different density gradient media.
10. Other buffers such as 50 mM HEPES may be used for the resuspension of BMVs, depending on the downstream analyses [35].
11. The percentage of iodixanol used in each layer of the OptiPrep gradient will depend on the bacteria from which BMVs are isolated. As a guide, 20–45% is used in this protocol as it has been verified for both OMV [27] and MV purification [8].
12. To avoid disruption of fractions, adjust the acceleration and deceleration speed of the ultracentrifuge to the appropriate setting for density gradient ultracentrifugation. This will vary depending on the ultracentrifuge used; therefore, refer to the manufacturer's guidelines for optimal settings.
13. To calculate the density of each fraction, number empty 1.5 mL tubes with each fraction number and weigh them to three decimal places. Then, place a 1 mL aliquot of each fraction into the respective tube, and weigh again to three decimal places. To calculate density, subtract the weight of the empty tube from the total weight of each aliquot and divide by the volume (in this case, volume = 1).

14. Caution: glutaraldehyde is a toxic chemical that releases harmful vapors. Use in a fume cabinet while wearing appropriate personal protective equipment (PPE).
15. Caution: uranyl acetate is a radioactive and toxic chemical and should be used in a fume cabinet while wearing appropriate PPE.
16. BMV quantification by particle enumeration may also be performed using nanoparticle flow cytometry [7]; however, we do not provide instructions for their quantification using this method.
17. BMVs may be prone to forming aggregates, which can interfere with their quantification by nanoparticle counting. For instruments that measure particle size, the size distribution of particles can be examined to ensure they are within the expected size range of BMVs (i.e., approximately 500-nm or less). Limiting the freeze-thaw cycles of BMV samples will limit the formation of aggregates, and vortexing thoroughly immediately prior to quantification can dissociate aggregates that may have formed.
18. The quantity of BMVs (particle number or protein concentration) in a given preparation may be normalized to culture volume or bacterial viable count to allow comparison of BMV production by different bacterial species or different culture conditions [8, 21].
19. Different protein assays have been shown to vary in their quantification of BMVs [21]; therefore, careful consideration is required when selecting an appropriate protein assay for BMV quantification. Furthermore, it is important to note that quantification of BMVs by protein assay does not factor in changes in BMV protein cargo that results from different bacterial species and strains [21], growth stages [15, 21], culture conditions [36], or vesicle size [33]. Therefore, protein concentration may not directly correlate with BMV particle number, but rather gives an indication of overall BMV quantity from a single bacterial species with specified growth conditions. In addition, BMV standardization by protein amount for the purpose of immunological assays can mask the immunological effects of other BMV cargoes, such as DNA and RNA [21].
20. Determine the protein concentration of the BMV sample as recommended by the manufacturer for your selected protein assay.
21. Initial experiments may require a dose response to determine the optimal concentration of BMVs required to induce a response of interest. Variables in BMV production can alter

the immunogenic properties of BMVs; therefore, it is essential to clearly define the growth stage and conditions used to generate BMVs as this will alter biological outcomes [21].

22. Positive controls may include agonists or chemical controls. Furthermore, cells may be stimulated with the parent bacteria to enable comparison of cellular responses of bacteria compared to their BMVs [8].
23. Visualization of autophagosome formation for the examination of autophagy requires cells to be transfected with the light chain 3 (LC3)-GFP fluorescent construct. Transfect cells 24 h after seeding with 60 ng LC3-GFP and 170 ng empty vector (e.g., pcDNA3) and incubate overnight before stimulating with BMVs or controls (e.g., 100 nM rapamycin). To assess the degradation of BMVs by autophagy, cells can be pre-treated with bafilomycin A1 to prevent their degradation at a concentration of 10-nM for 30 min prior to stimulation [8, 9].
24. These times are a guide and optimization may be required for the specific cell line and BMVs used. For example, time points may differ when examining immune responses in phagocytic cells such as macrophages, as the rate of BMV entry will differ to that of epithelial cells, thereby affecting the time required to generate a cellular response.

Acknowledgments

This work was supported by the Australian Research Council under Grant DP190101655. Maria Kaparakis-Liaskos is supported by a veski Inspiring Women Fellowship.

References

1. Toyofuku M, Nomura N, Eberl L (2019) Types and origins of bacterial membrane vesicles. Nat Rev Microbiol 17(1):13–24
2. Pathirana RD, Kaparakis-Liaskos M (2016) Bacterial membrane vesicles: biogenesis, immune regulation and pathogenesis. Cell Microbiol 18(11):1518–1524
3. Kaparakis-Liaskos M, Ferrero RL (2015) Immune modulation by bacterial outer membrane vesicles. Nat Rev Immunol 15(6): 375–387
4. Askarian F, Lapek JD Jr, Dongre M et al (2018) *Staphylococcus aureus* membrane-derived vesicles promote bacterial virulence and confer protective immunity in murine infection models. Front Microbiol 9:262–278
5. Yun SH, Park EC, Lee S-Y et al (2018) Antibiotic treatment modulates protein components of cytotoxic outer membrane vesicles of multidrug-resistant clinical strain, *Acinetobacter baumannii* DU202. Clin Proteomics 15(1):28–37
6. Woo JH, Kim S, Lee T et al (2021) Production of membrane vesicles in *Listeria monocytogenes* cultured with or without sub-inhibitory concentrations of antibiotics and their innate immune responses in vitro. Genes (Basel) 12(3):415–424
7. Cecil JD, O'Brien-Simpson NM, Lenzo JC et al (2016) Differential responses of pattern recognition receptors to outer membrane vesicles of three periodontal pathogens. PLoS One 11(4):e0151967–e87

8. Bitto NJ, Cheng L, Johnston EL et al (2021) *Staphylococcus aureus* membrane vesicles contain immunostimulatory DNA, RNA and peptidoglycan that activate innate immune signaling pathways and induce autophagy. J Extracell Vesicles 10(6):e12080–e98
9. Irving AT, Mimuro H, Kufer TA et al (2014) The immune receptor NOD1 and kinase RIP2 interact with bacterial peptidoglycan on early endosomes to promote autophagy and inflammatory signaling. Cell Host Microbe 15(5): 623–635
10. Kaparakis M, Turnbull L, Carneiro L et al (2010) Bacterial membrane vesicles deliver peptidoglycan to NOD1 in epithelial cells. Cell Microbiol 12(3):372–385
11. Kim W, Lee EJ, Bae I-H et al (2020) *Lactobacillus plantarum*-derived extracellular vesicles induce anti-inflammatory M2 macrophage polarization in vitro. J Extracell Vesicles 9(1): 1793514
12. Fábrega M-J, Rodríguez-Nogales A, Garrido-Mesa J et al (2017) Intestinal anti-inflammatory effects of outer membrane vesicles from *Escherichia coli* Nissle 1917 in DSS-experimental colitis in mice. Front Microbiol 8:1274–1286
13. Kim SY, Yi DY (2020) Analysis of the human breast milk microbiome and bacterial extracellular vesicles in healthy mothers. Exp Mol Med 52(8):1288–1297
14. Fábrega M, Aguilera L, Giménez R et al (2016) Activation of immune and defense responses in the intestinal mucosa by outer membrane vesicles of commensal and probiotic *Escherichia coli* strains. Front Microbiol 7:705–718
15. Zavan L, Bitto NJ, Johnston EL et al (2019) *Helicobacter pylori* growth stage determines the size, protein composition, and preferential cargo packaging of outer membrane vesicles. Proteomics 19:e1800209–e23
16. Bitto NJ, Kaparakis-Liaskos M (2017) The therapeutic benefit of bacterial membrane vesicles. Int J Mol Sci 18(6):e1287–e301
17. Li Y, Zhao R, Cheng K et al (2020) Bacterial outer membrane vesicles presenting programmed death 1 for improved cancer immunotherapy via immune activation and checkpoint inhibition. ACS Nano 14(12): 16698–16711
18. Kuhn T, Koch M, Fuhrmann G (2020) Probiomimetics—novel *Lactobacillus*-mimicking microparticles show anti-inflammatory and barrier-protecting effects in gastrointestinal models. Small 16(40):2003158–2003166
19. Wang X, Thompson CD, Weidenmaier C et al (2018) Release of *Staphylococcus aureus* extracellular vesicles and their application as a vaccine platform. Nat Commun 9(1):1379
20. Carter NJ (2013) Multicomponent meningococcal serogroup B vaccine (4CMenB; Bexsero (®)): a review of its use in primary and booster vaccination. BioDrugs 27(3):263–274
21. Bitto NJ, Zavan L, Johnston EL et al (2021) Considerations for the analysis of bacterial membrane vesicles: methods of vesicle production and quantification can influence biological and experimental outcomes. Microbiol Spectr 9(3):e0127321
22. Coelho C, Brown LC, Maryam M et al (2018) *Listeria monocytogenes* virulence factors, including listeriolysin O, are secreted in biologically active extracellular vesicles. J Biol Chem 294(4):1202–1217
23. Zakharzhevskaya NB, Tsvetkov VB, Vanyushkina AA et al (2017) Interaction of *Bacteroides fragilis* toxin with outer membrane vesicles reveals new mechanism of its secretion and delivery. Front Cell Infect Microbiol 7(2): 308–323
24. Biller SJ, McDaniel LD, Breitbart M et al (2017) Membrane vesicles in sea water: heterogeneous DNA content and implications for viral abundance estimates. ISME J 11(2): 394–404
25. Bitto NJ, Chapman R, Pidot S et al (2017) Bacterial membrane vesicles transport their DNA cargo into host cells. Sci Rep 7(1): 7072–7086
26. Choi J-W, Kwon T-Y, Hong S-H et al (2018) Isolation and characterization of a microRNA-size secretable small RNA in *Streptococcus sanguinis*. Cell Biochem Biophys 76(1):293–301
27. Blenkiron C, Simonov D, Muthukaruppan A et al (2016) Uropathogenic *Escherichia coli* releases extracellular vesicles that are associated with RNA. PLoS One 11(8):e0160440–e56
28. Jones EJ, Booth C, Fonseca S et al (2020) The uptake, trafficking, and biodistribution of *Bacteroides thetaiotaomicron* generated outer membrane vesicles. Front Microbiol 11:57
29. Gurung M, Moon DC, Choi CW et al (2011) *Staphylococcus aureus* produces membrane-derived vesicles that induce host cell death. PLoS One 6(11):e27958
30. Wang X, Eagen WJ, Lee JC (2020) Orchestration of human macrophage NLRP3 inflammasome activation by *Staphylococcus aureus* extracellular vesicles. Proc Natl Acad Sci U S A 117(6):3174–3184
31. Tartaglia NR, Breyne K, Meyer E et al (2018) *Staphylococcus aureus* extracellular vesicles elicit an immunostimulatory response in vivo on the

murine mammary gland. Front Cell Infect Microbiol 8:277–294

32. Han E-C, Choi S-Y, Lee Y et al (2019) Extracellular RNAs in periodontopathogenic outer membrane vesicles promote TNF-α production in human macrophages and cross the blood–brain barrier in mice. FASEB J 33(12): 13412–13422
33. Turner L, Bitto NJ, Steer DL et al (2018) *Helicobacter pylori* outer membrane vesicle size determines their mechanisms of host cell entry and protein content. Front Immunol 9: 1466–1476
34. Morein S, Henricson D, Rilfors L (1994) Separation of inner and outer membrane vesicles from *Escherichia coli* in self-generating Percoll gradients. Anal Biochem 216(1):47–51
35. Bonnington KE, Kuehn MJ (2016) Outer membrane vesicle production facilitates LPS remodeling and outer membrane maintenance in *Salmonella* during environmental transitions. mBio 7(5):e01532-16
36. Lynch JB, Schwartzman JA, Bennett BD et al (2019) Ambient pH alters the protein content of outer membrane vesicles, driving host development in a beneficial symbiosis. J Bacteriol 201(20):e00319

Chapter 5

Gene Gun-Mediated Transient Gene Expression for Functional Studies in Plant Immunity

Franz Leissing, Anja Reinstädler, Hannah Thieron, and Ralph Panstruga

Abstract

One major threat to plant cultivation are fungal pathogens, which can cause substantial yield losses in agriculture. As an example, cereal powdery mildew fungi such as the barley (*Hordeum vulgare*) pathogen, *Blumeria graminis* f. sp. *hordei* (*Bgh*), are among the ten most relevant fungal plant pathogens in molecular plant pathology and can lead to yield losses of up to 30%. Plant *Mildew resistance Locus O* (*MLO*) genes are required for successful colonization of plants by powdery mildew fungi. Accordingly, loss-of-function *mlo* mutants confer durable resistance against powdery mildew fungi in many plant species. In the case of barley, *mlo*-based resistance has been used for more than 40 years in agriculture without powdery mildew fungi effectively overcoming this kind of immunity. However, the molecular basis of *mlo* resistance and function (s) of the transmembrane Mlo protein(s) are still incompletely understood. The generation of transgenic barley plants to study the plant immune response and the involvement of Mlo therein is time-consuming and challenging. Therefore, transient gene expression via gene gun-mediated particle bombardment became a popular, easy, and efficient tool to investigate different aspects of plant defense responses in barley. Since *Bgh* fails to penetrate leaf epidermal cells of *mlo* mutants, single-cell complementation upon biolistic transformation resulting in (over-)expression of *Mlo* can be used to characterize the Mlo protein functionally in vivo. In this chapter, we describe in detail the gene gun-mediated transient expression of *Mlo* in barley leaf epidermal cells followed by powdery mildew inoculation and the subsequent microscopic evaluation. However, gene gun-mediated transient gene expression may be also used to address other research questions or to transform the epidermal tissues of other plant organs and/or species.

Key words Barley (*Hordeum vulgare*), Biolistic transformation, *Blumeria graminis*, Particle bombardment, β-Glucuronidase (GUS), Mildew resistance Locus O (Mlo), Particle gun, Powdery mildew, Transient expression

1 Introduction

Mildew resistance Locus O (MLO; Mlo in cereals) proteins form a plant-specific protein family whose members are conserved among monocotyledonous and dicotyledonous plant species [1, 2]. The founding member of this protein family was originally described in barley (*Hordeum vulgare*) in the context of resistance to powdery mildew infection [3]. Loss of function of *Mlo* in *H. vulgare*

Thomas A. Kufer and Maria Kaparakis-Liaskos (eds.), *Effector-Triggered Immunity: Methods and Protocols*, Methods in Molecular Biology, vol. 2523, https://doi.org/10.1007/978-1-0716-2449-4_5,

provides durable broad-spectrum immunity against the grass powdery mildew pathogen *Blumeria graminis* f. sp. *hordei* (*Bgh*) [4]. Later it was discovered that powdery mildew resistance is a common feature of plant *mlo* mutants, including the model plant *Arabidopsis thaliana* [5, 6]. Because *mlo*-based resistance in barley has been deployed in European agriculture since 1979 without an apparent loss of its effectiveness [4, 7], demonstrating its unusual longevity, investigating the molecular basis of this conserved type of immunity and understanding the function of plant MLO proteins is of great interest for both scientists and breeders. MLO proteins are heptahelical transmembrane proteins with an extracellular/luminal N-terminus and an intracellular/cytoplasmic C-terminus [8]. The proximal part of the intracellular C-terminus harbors a binding site for the calcium sensor calmodulin (calmodulin-binding domain, CaMBD) that is conserved throughout the protein family [9]. Biochemical binding assays of barley and rice Mlo revealed that calmodulin associates in a calcium-dependent manner with the CaMBD, which is comprised of a stretch of approx. 20 amino acids predicted to form an amphipathic α-helix [9, 10]. Apart from the conserved CaMBD, no well-characterized peptide motifs are consistently present in MLO proteins, and their core biochemical function remains elusive [1]. However, several molecular components have been identified that contribute to *mlo*-mediated resistance. For example, the powdery mildew resistance phenotype of barley *mlo* and *Arabidopsis mlo2* mutants relies on the secretory delivery of toxic metabolites [5, 11, 12]. In particular soluble *N*-ethylmaleimide-sensitive factor attachment receptor (SNARE) proteins (barley Ror2 and SNAP34 as well as Arabidopsis PEN1), implicated in vesicle fusion events at the plasma membrane, have been shown to play a decisive role in *mlo*-based resistance [5, 13, 14]. These proteins, together with their partner SNAREs, focally accumulate at incipient fungal entry sites [15, 16] and are likely involved in the formation of ternary SNARE complexes for immune-related polarized secretion [16, 17]. Despite these insights and although various additional lines of research have been pursued to decipher the molecular pathways involved in *mlo*-mediated resistance in *Arabidopsis* and barley, neither the exact molecular basis of this type of resistance nor the genuine function of MLO proteins has been elucidated so far. One approach to gain information regarding MLO activity is the exploration of structure-function relationships. This requires the generation of site-directed MLO mutant variants and the analysis of their *in planta* functionality, i.e., examining their capacity to complement the *mlo* mutant phenotype and to restore susceptibility to the powdery mildew pathogen. However, despite considerable technical progress [18], the generation of stable transgenic lines remains challenging and is time-consuming in the case of barley. As a convenient alternative, different transient expression systems have been successfully applied for the investigation of genes of interest in monocotyledonous plant species. These comprise (i) virus-mediated (ii) *Agrobacterium tumefaciens*-mediated and (iii) gene gun-mediated transient gene expression in barley [19]. Each of these methods has its advantages and disadvantages (reviewed in [19]), and gene gun-mediated transient gene expression in barley leaves became a popular method especially for the investigation of the interaction between plants and powdery mildew fungi due to several reasons: In 1997, Nelson and Bushnell demonstrated that neither the particle bombardment process itself nor the expression of a reporter gene in barley coleoptile cells interferes with powdery mildew pathogenesis [20]. The combination of particle bombardment-mediated transient gene expression and subsequent inoculation with powdery mildew conidiospores thus turned out to be well-suited to study cereal-powdery mildew interactions at the single-cell level. Germinated *Bgh* spores attempt to penetrate individual cells of the leaf epidermis, which is the primary target tissue of particle bombardment. These fungal penetration attempts may either fail, typically coincident with the formation of a local cell wall reinforcement (papilla) beneath the site of attack [21, 22], or might be successful. In the case of effective host cell entry, the fungal intruder establishes a dedicated feeding structure (haustorium) inside the penetrated epidermal cell. Subsequently, hyphal growth commences epiphytically from sporelings that successfully formed haustoria inside host cells. Fungal pathogenesis is strictly timed and, thus, occurs synchronously for all spores present on the leaf surface [23]. Due to these characteristics, powdery mildew infection success can be easily scored by conventional light microscopy, also in transformed cells (reviewed in [24]).

Shirasu and co-workers were the first to show the complementation of *mlo* resistance via particle bombardment-mediated transient expression of the barley *Mlo* wild-type gene in single epidermal cells of a barley *mlo* mutant, demonstrating that *mlo*-based immunity is a cell-autonomous phenomenon [25]. Subsequently, different aspects regarding *Mlo*'s role in plant (cereal) immunity were analyzed by this experimental setup. This revealed, for example, supersusceptibility of barley wild-type cells upon overexpression of *Mlo* [9]. Additionally, the expression of site-directed mutant variants enabled the identification of various amino acids that are critical for Mlo function [26, 27], including residues within the CaMBD [9] and extracellular cysteines presumed to form two disulfide bridges [28]. Experiments using fluorophore-tagged Mlo showed that the protein focally accumulates at fungal entry sites. Furthermore, the *in planta* interaction dynamics of Mlo and calmodulin was revealed by fluorescence resonance energy transfer (FRET) experiments in barley cells upon particle bombardment [15]. Finally, the procedure can also be used to test the functionality of *Mlo* orthologs [28, 29], to study the role of *Mlo* in nonhost resistance [29], to validate *mlo* candidate mutant plants by successful complementation with wild-type *Mlo* [30, 31], and to investigate components that affect *mlo* resistance [32]. Apart from studying Mlo structure-function relationships, the single-cell transient expression system has been used to explore other aspects of cereal-powdery mildew interactions. These comprise, for example, the analysis of isolate-specific resistance [33], the analysis of *Bgh* effector functions [34], and the examination of gene silencing [35]. Concerning the latter, the now widely exploited phenomenon of host-induced gene silencing (HIGS) was discovered in the context of transient gene silencing experiments in barley [36].

In general, the gene gun-mediated particle bombardment as described in this chapter requires six basic steps (*see* Fig. 1).

1. First, the desired plant material (genotype) has to be grown, in our case barley *mlo* mutant plants, for at least 7 days (note that also other plant species can be biolistically transformed but may require adjustment of growth conditions by species).
2. Evenly grown primary leaves are detached and placed on appropriate agar plates to prevent drying of the plant specimens during the experiment.
3. Next, the microcarriers (gold particles) are washed and coated with the desired plasmid DNA (e.g., *pUBI-HvMlo-nos* as the gene of interest and *pUBI-GUS-nos* as a reporter construct to mark transformed cells). The design of the used plasmids can have a tremendous impact on the expression of genes and, therefore, on the overall outcome of the experiment. For example, other than in dicotyledonous plant species, the cauliflower mosaic virus (CaMV) 35S promoter confers low expression of *Mlo* in barley leaf epidermal cells, whereas the maize *ubiquitin1* promoter mediates strong overexpression of *Mlo* [25].
4. Thereafter, the actual particle bombardment can be performed: the coated microcarriers are shot with high speed onto the barley leaves in the Petri dish by means of a helium-driven biolistic gene gun. In addition to the often performed co-bombardment of a reporter gene and a gene of interest, a bombardment with the reporter gene only should be performed as a control to exclude any effect of the reporter gene in the experiment. We describe here usage of *β-glucuronidase* (*GUS*) as a reporter gene, but other reporters are possible, e.g., the expression of fluorophores such as the green fluorescent protein (GFP) [25, 32].
5. After regeneration of the barley leaves for 1 day (to allow for the decline of any bombardment-induced wound responses and the expression of the gene(s) of interest), the bombarded leaves are inoculated with a suitable powdery mildew (e.g.,

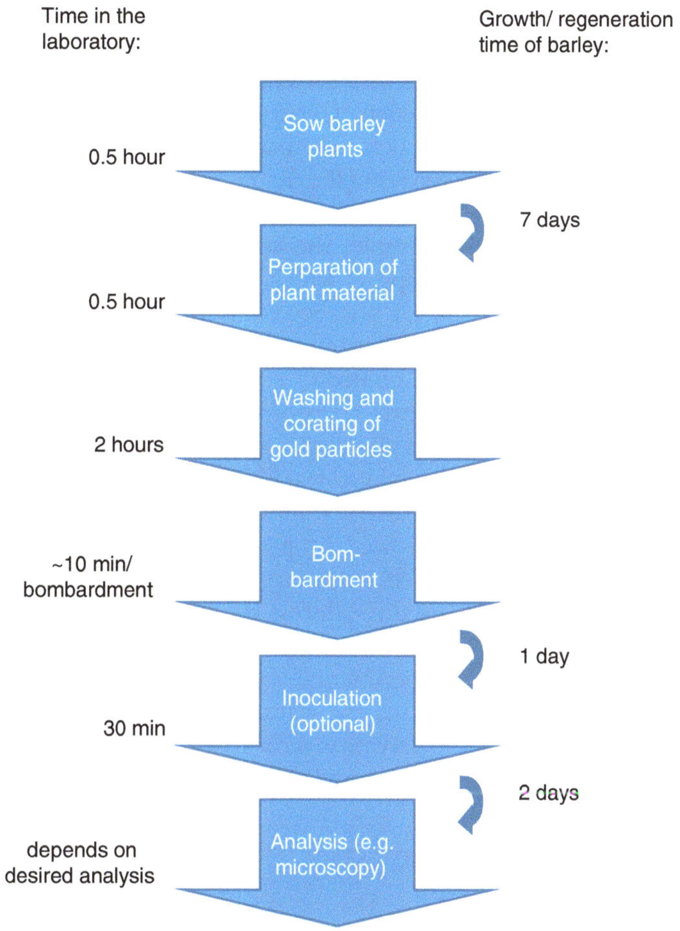

Fig. 1 General workflow of the gene gun-mediated particle bombardment procedure. For further details, see main text

Bgh) isolate. However, this step is dependent on the desired downstream analyses and is therefore optional.

6. The selected evaluation of the experiment depends on the constructs used. Many different applications are possible, e.g., the investigation of protein-protein interactions using Förster resonance energy transfer (FRET) [15] or bimolecular fluorescence complementation (BiFC) [17] and/or the analysis of dynamic changes in protein localization upon powdery mildew infection using fluorescence microscopy [15]. In our case, the transiently transformed and inoculated barley leaves are subjected to GUS staining of transformed cells at 2 days after inoculation (i.e., at 3 days after particle bombardment). Afterward, the penetration rate (ratio of successful and unsuccessful fungal host cell entry events) of transformed cells is determined using standard light microscopy.

In this chapter, we describe the gene gun-mediated particle bombardment using the stationary PDS-1000/He™ Particle Delivery System equipped with a Hepta™ adaptor (Bio-Rad, Feldkirchen, Germany). However, also other devices, e.g., handheld particle bombardment systems such as the Helios™ Gene Gun System (Bio-Rad), may be used. Depending on the research question, each setup has its specific advantages and disadvantages. For example, the stationary system can be operated with a Hepta™ adaptor, which enables an even transformation efficiency and a high number of transformed cells in several leaves upon a single bombardment, but the leaves have to be detached for the process. By contrast, a handheld particle delivery system allows to transiently transform intact leaves that are still attached to the plant, but the target area then is restricted to only a small region of an individual leaf [37]. Although we focus in this chapter on gene gun-mediated transient expression of *Mlo* in barley leaf epidermal cells followed by powdery mildew inoculation to study Mlo's role in plant immunity, gene gun-mediated transient expression may be used to address other research questions or to transform epidermal tissues of other plant organs and/or species. In such instances, the experimental conditions (e.g., the pressure used for the bombardment, the nature and size of the microcarriers, or the distance between macrocarrier and sample) may have to be optimized by the researcher.

2 Materials

We recommend using freshly prepared or aliquoted and non-expired solutions only. Unless indicated otherwise, solutions and materials should be stored and handled as recommended by the manufacturer. In general, solutions should be prepared under sterile conditions using double-distilled water. Waste disposal rules as per laboratory and authority regulations must be followed for all used materials when disposing of waste materials.

2.1 General

1. Biosafety level 1 laboratory for work with BSL 1 genetically modified organisms. We also recommend to autoclave pathogen-infected plant material, even when not transgenic.
2. Standard molecular biology equipment.
3. Medium-speed centrifuges.
4. Autoclave.
5. Reaction tubes: 1.5 mL and 15 mL.
6. Growth chamber/greenhouse for barley cultivation.
7. Platform mixer.
8. Desiccator.

9. Membrane pump (e.g., Laboport TYP N840.3FT.18 (KNF, Freiburg, Germany)).
10. PDS-1000/He™ Particle Delivery System with Hepta™ Adaptor (Bio-Rad).

2.2 Bombardment of Barley Leaves

1. Primary leaves of 7-day-old barley plants (*mlo* mutants): Sow barley seeds 7 days before the bombardment experiment in soil. For each bombardment, at least 8 primary leaves are required (*see* **Note 1**).
2. Plasmid DNA at a concentration of 1 μg/μL (*see* **Note 2**). In this protocol, we use the plasmids *pUBI-Mlo-nos* and *pUBI-GUS-nos* (*see* Fig. 2 and **Note 3**).
3. 2.5-M $CaCl_2$ stock solution: the prepared stock solution should be autoclaved, and can be stored for several months at room temperature.
4. 0.1-M spermidine (N-[3-aminopropyl]-1,4-butanediamine): The stock solution should be sterilized by filtration (0.22 μm), aliquoted, and can be stored at −20 °C for several months. Avoid repeated thawing and freezing cycles.
5. 50% glycerol (v/v) in water, autoclaved.
6. 98% ethanol, extra pure.
7. 70% (v/v) ethanol.
8. 99% 2-propanol, extra pure.
9. 1% agar, 85-μM benzimidazole plates: Prepare a 1% agar solution, autoclave the solution, add benzimidazole to a final concentration of 85 μM when the solution has cooled down to approx. 50 °C, and fill the melted solution under sterile conditions into Petri dishes (Ø 92 mm). Wait until the agar has solidified.

2.2.1 Consumables for the PDS-1000/He™ Particle Delivery System

1. Rupture disks, 900 psi (pounds per square inch) (Bio-Rad).
2. Microcarriers, gold particles, Ø 1.0 μm (Bio-Rad).
3. Macrocarriers (Bio-Rad).
4. Hepta™ stopping screens (Bio-Rad).
5. High-pressure (2400–2600 psi) tank of high purity helium (minimum grade: 4.5).

2.3 Bgh Infection Assay

1. 12–14-day-old susceptible barley plants that are highly infected with a compatible *Bgh* isolate (profusely sporulating fungal colonies present).
2. Inoculation tower or box.

2.4 Evaluation

1. X-Gluc staining solution, 0.1-M Na_2HPO_4/NaH_2PO_4 pH 7.0, 20-mM EDTA (*ethylenediaminetetraacetic acid*),

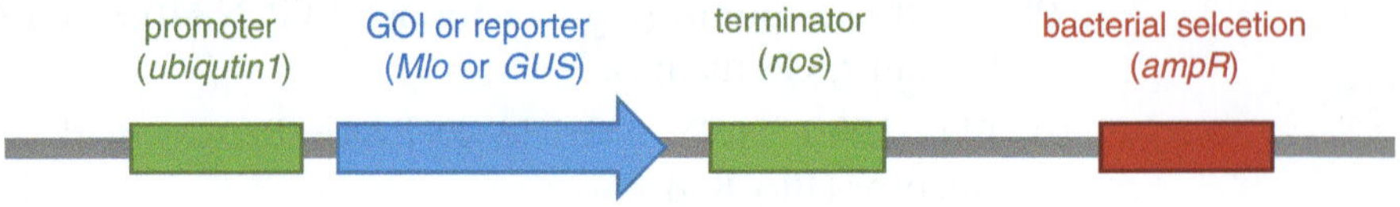

Fig. 2 Schematic overview of plasmid features used in this protocol. *ampR ampicillin resistance*, GOI gene of interest, *Mlo Mildew resistance Locus o, GUS β-glucuronidase*

5-mM $K_3Fe(CN)_6$, 5-mM $K_4Fe(CN)_6$, 1-mg/mL X-Gluc (5-bromo-4-chloro-3-indolyl-ß-D-glucoronide cyclohexylammonium salt), 0.1% (v/v) Triton X-100, 20% (v/v) methanol; prepare 10-mL aliquots and store them at −20 °C. After thawing, the solution can be kept for several days at 4 °C.

2. Destaining solution: 80% (v/v) ethanol.
3. Coomassie staining solution: 0.6% (w/v) Coomassie Brilliant Blue R250 solved in pure ethanol.
4. Light microscope.
5. Glass slides and coverslips.
6. 50% (v/v) glycerol.

3 Methods

3.1 Bombardment of Barley Leaves

Cut off the 7-day-old primary leaves at their base (avoid the lower, inwardly curling part of the leaves), place 8–10 of them with the adaxial side up on benzimidazole-containing agar plates at 2 h before the bombardment, and put the plates back into a light chamber. Proceed with preparing and coating the microcarriers.

3.1.1 Preparation of Microcarriers

1. Place 30 mg of microcarriers (gold particles, Ø 1 μm) in a 1.5-mL reaction tube (suitable for ten bombardments).
2. Add 1 mL of 70% (v/v) ethanol and vortex vigorously for 5 min on a platform mixer (1200 rpm).
3. Allow the microcarriers to settle for 15 min.
4. Pellet the microcarriers by spinning for 5 s at room temperature (max. 6000 g).
5. Discard the supernatant by carefully removing the liquid with a pipette.
6. Add 1 mL of sterile water.
7. Mix for 1 min (600 rpm). Microcarriers must be resuspended.
8. Allow the microcarriers to settle for 1 min.
9. Pellet the microcarriers by spinning for 5 s at room temperature (max. 6000 g).
10. Remove the liquid carefully with a pipette and discard.

11. Repeat **steps 6–10** twice.
12. Add 500-µL 50% glycerol to adjust the microcarrier concentration to 60 mg/mL (*see* **Note 4**).

3.1.2 Coating Microcarriers with DNA

1. Mix the prepared microcarriers (*see* Subheading 3.1.1) for 5 min on a platform mixer (1200 rpm) to disrupt and resuspend agglutinated microcarriers.
2. For one bombardment, remove 50 µL of the prepared microcarriers while mixing and transfer to a 1.5-mL reaction tube.
3. To each 50-µl aliquot to be used for one bombardment, add first 5 µL of plasmid DNA (1 µg/µL) (*see* **Note 5**), then 50 µL of 2.5-M $CaCl_2$, and finally 20 µL of 0.1 M spermidine (*see* **Note 6**). Continue mixing for 3 min (1200 rpm) while constantly mixing.
4. Allow the microcarriers to settle for 1 min.
5. Pellet microcarriers by spinning for 2 s (max. 2000 g).
6. Carefully remove the liquid and discard it.
7. Add 140-µL 70% (v/v) ethanol.
8. Gently resuspend pellet (600 rpm) for 2 s, and spin for 2 s (max. 2000 g) (*see* **Note** 7).
9. Carefully remove the liquid and discard it.
10. Add 140-µL 100% (v/v) ethanol.
11. Gently resuspend pellet (600 rpm) for 2 s, and spin for 2 s (max. 2000 g).
12. Carefully remove the liquid and discard it.
13. Resuspend the coated microcarriers in 60-µL ethanol.
14. Keep on ice until used for bombardment (*see* **Note 8**).

3.1.3 Bombardment

In this section, we describe an optimized protocol for barley using the PDS-1000/He™ Particle Delivery System with a Hepta™ adaptor (Bio-Rad). However, other devices or adaptors may be used, but the exact procedure has to be optimized by the researcher according to the manufacturer's instructions.

1. Turn on the PDS1000/He™ Particle Delivery System and the membrane pump (*see* **Note 9**) and open the helium tank valve.
2. Place each of the seven macrocarriers inside a macrocarrier holder and fix them with a plastic insertion tool.
3. For each macrocarrier, remove a 7-µL aliquot of microcarriers, transferred from a continuously mixed tube, and place it on the center of the respective macrocarrier (*see* **Note 10**).
4. Wait until the ethanol has evaporated completely (approx. 30 s).

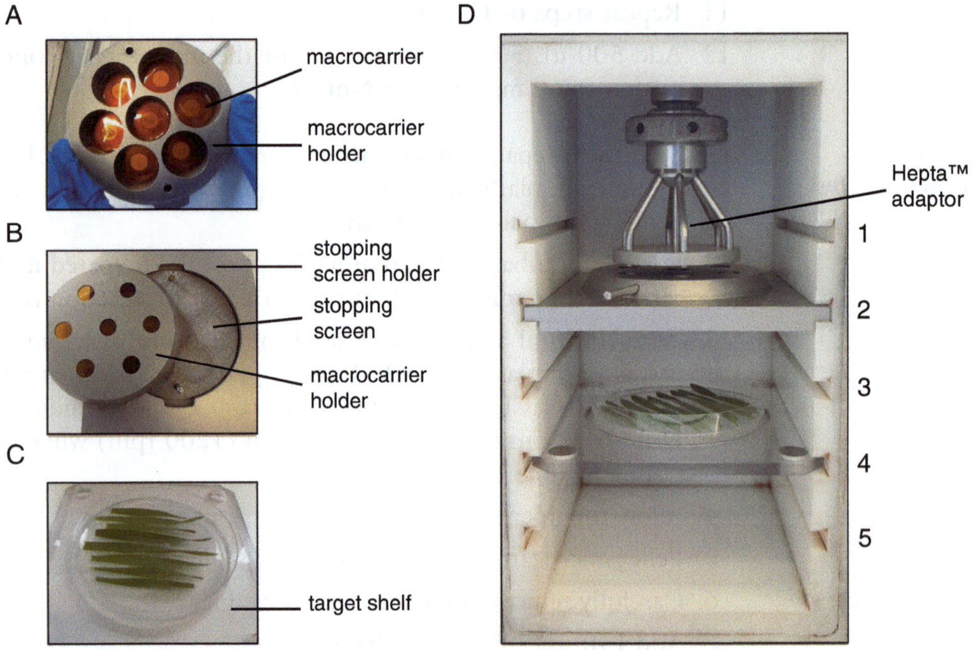

Fig. 3 Overview of the assembled PDS-1000/He™ Particle Delivery System with a Hepta™ adaptor. (**a**) Overview of the macrocarrier holder with inserted macrocarriers and the distributed microcarriers (7 μL each). (**b**) Microcarrier launch assembly: a stopping screen must be placed onto the stopping screen holder. The macrocarrier holder is turned upside down and locked into the stopping screen holder according to the manufacturer's instructions. (**c**) Target shelf with the agar plate containing barley leaves. (**d**) Assembled PDS-1000/He™ Particle Delivery System with a Hepta™ adaptor. The numbers on the right indicate the platform numbering. Note that every outlet of the Hepta™ adaptor is exactly in line with one macrocarrier

5. Briefly wet the rupture disk using 2-propanol (*see* **Note 11**).
6. Load the rupture disk, assemble the microcarrier launch assembly according to the manufacturer's instructions (*see* Fig. 3a–c), and place the device in platform 2 of the chamber (*see* Fig. 3d). Pay attention to the presence of the essential stopping screen, which has to be changed for each new plasmid combination to be bombarded.
7. Align the seven pressure outlets of the Hepta™ adaptor with the centers of the seven holes in the macrocarrier holder exactly above the macrocarriers.
8. Place the target shelf with the Petri dish containing detached leaves at the desired level inside the bombardment chamber (for barley: platform 4; 6 cm, *see* Fig. 3d).
9. Close sample chamber door.
10. Set the vacuum switch to the "VAC" position until 27 inches of mercury is reached and hold this level by quickly pressing the vacuum control to the "HOLD" position.

11. Press and hold the "FIRE" switch to allow the helium pressure to build up inside the gas acceleration tube, which is sealed by the selected rupture disk, until the rupture disk bursts. Release the "FIRE" button (*see* **Note 12**).
12. Release the vacuum in the sample chamber by setting the switch to the "VENT" position.
13. Open the sample chamber door when the vacuum gauge reaches 0 inches of mercury of vacuum.
14. After the bombardment, pick up the leaves (which might have been whirled around in the chamber) and readjust them on the agar surface in the Petri dish.
15. Place the Petri dishes with the detached leaves back into the growth chamber/greenhouse for 1 day (to allow expression of the transformed genes, *see* **Note 13**).
16. After completing all experiments of the day, close the helium tank valve and remove the helium pressure from the system by pressing the "FIRE" button (can be controlled by the helium pressure gauge).

3.2 Infection of Bombarded Barley Leaves

1. Place the Petri dishes containing the bombarded barley leaves in an inoculation tower or a box.
2. Evenly dust *Bgh* conidia from heavily infected barley plants over the Petri dishes in the inoculation tower (*see* **Note 14**). Close the inoculation tower or box and wait for at least 15 min to ensure settling of all conidiospores.
3. Return the Petri dishes with the detached leaves back into the light chamber/greenhouse.
4. Proceed with GUS staining at 2 days post inoculation.

3.3 Evaluation

3.3.1 GUS-Staining of Barley Leaves

1. Thaw X-Gluc aliquots (approx. 10 mL per bombardment).
2. Transfer all leaves of one bombardment in X-Gluc solution in a 15-mL reaction tube.
3. Place the open tubes in a desiccator; apply and release vacuum three times for 10 min each. Ensure that the leaves are well-infiltrated with the X-Gluc solution.
4. Incubate the leaves in X-Gluc solution overnight at 37 °C.
5. Replace the staining solution with the destaining solution and incubate for 3–4 days at room temperature or heat up to 60 °C until leaves are destained (*see* **Note 15**).

3.3.2 Microscopic Evaluation

1. Wash leaves for 10 min in water to remove the destaining solution.
2. Stain the leaves for approx. 30 s in the Coomassie staining solution, rinse with water, and mount individual leaf sections

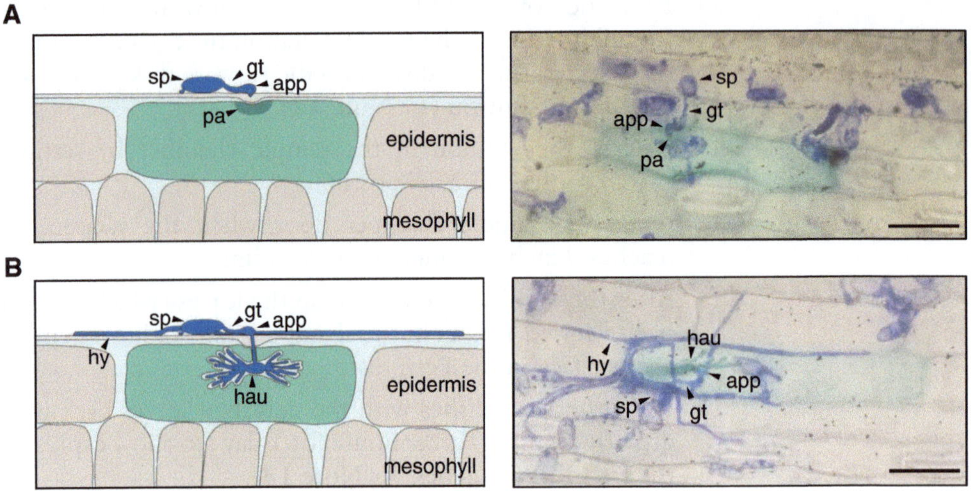

Fig. 4 Schemes (left panels, illustrating transversal leaf sections) and exemplary micrographs (right panels, top view on leaves) of transformed cells with *Bgh* interaction sites. (**a**) The epidermis of a barley leaf (*mlo* genotype) was transformed by gene-gun mediated particle bombardment with the plasmid *pUBI-GUS-nos*. At least one spore (sp) germinated; formed a germ tube (gt), an appressorium (app); and attempted to penetrate the greenish transformed (marker gene-expressing) cell but failed, coincident with the formation of a local cell wall reinforcement (papilla, pa) in the attacked cell. (**b**) The epidermis of a barley leaf (*mlo* genotype) was co-transformed by gene gun-mediated particle bombardment with the plasmids *pUBI-Mlo-nos* and *pUBI-GUS-nos*. At least one germling attempted to penetrate and succeeded in the formation of a haustorium (hau) in the greenish transformed (marker gene-expressing) cell. Additionally, the spore formed hyphae (hy). Fungal infection structures are only present inside or on top of epidermal cells; the underlying mesophyll cells are not affected. Note that neighboring cells of the transformed epidermal cell remain resistant. Scale bar: 50 μm

on a microscopy slide in a droplet of 50% (v/v) glycerol. Add a coverslip and check for fungal structures under a light microscope (*see* **Note 16**).

3. After staining of the specimen with Coomassie, leaf sections will be systematically inspected by microscopy. GUS-positive (greenish) cells will be analyzed for the presence of *Bgh* spores. Attacked cells will be categorized in those where (i) all spores attempted to penetrate the transformed (marker gene-expressing) cell but failed (*see* Fig. 4a) and those where (ii) at least one germling succeeded in the formation of a haustorium (*see* Fig. 4b). A minimum of 100 transformed and *Bgh*-attacked cells should be scored per bombardment (*see* **Note 16**). The fungal entry rate is calculated as the number of transformed cells with at least one haustorium divided by the total number of attacked and transformed cells × 100.

4 Notes

1. We strongly recommend sowing more than eight seeds, because not all seeds may germinate, and also unevenly grown leaves can be discarded. It is easier to place evenly grown barley leaves with the adaxial side up in Petri dishes.
2. It is recommended to use a highly pure plasmid solution (e.g., prepared with a commercially available plasmid isolation kit) since impurities can dramatically decrease the transformation efficiency.
3. For high-level expression of *Mlo* in barley epidermal cells, the maize *ubiquitin1* promoter should be used. The CaMV 35S promoter only drives low-level expression of *Mlo* in barley leaf epidermal cells [25].
4. The prepared microcarriers can be stored at 4 °C for up to 2 weeks.
5. If a co-bombardment with two plasmids is desired, use 2.5 μL of each plasmid solution (1 μL/μg).
6. It is important to add $CaCl_2$ slowly, drop by drop, while mixing vigorously. If added too fast, the microcarriers may aggregate. Aggregates will decrease the transformation efficiency.
7. Residual glycerol may impede the resuspension of microcarriers. Repeat mixing until microcarriers are resuspended completely.
8. Coated microcarriers must be kept on ice since otherwise ethanol will partly evaporate.
9. The first bombardment of the day should be a "dry run" with no microcarriers to ensure that the system is set up properly and the gas pathway is filled with helium (procedure as described but without DNA samples).
10. Ensure constant mixing of the microcarrier suspension to ensure equal amounts of microcarriers on each of the macrocarriers.
11. For optimal transformation efficiency, we recommend using 900-psi rupture disks.
12. As also mentioned in the Bio-Rad instruction manual, the observed helium pressure at the pressure gauge may be higher than the classification of the used rupture disk (e.g., typically ~1100 psi in the case of 900-psi rupture disks).
13. Depending on the constructs used and desired analysis, a shorter regeneration time allowing for the expression of the genes of interest may be possible (we recommend at least 4 h).

14. Ensure a high spore density during the inoculation to enhance the likelihood that a spore will interact with a transformed cell.
15. Destained leaves can be stored in ethanol at room temperature for several months prior to microscopic analysis. For longer storage (more than 6 months), we recommend a mix of ethanol/lactic acid/glycerol (3:1:1 (v/v)) as the destaining solution.
16. We recommend a magnification of 100× to scan the leaf for transformed cells and at least a magnification of 200× for evaluation of the fungal host cell penetration success in GUS-stained (transformed) cells.
17. It is advised to move systematically (not randomly) through the sample to avoid the mis- or duplicate scoring of transformed cells.

References

1. Kusch S, Pesch L, Panstruga R (2016) Comprehensive phylogenetic analysis sheds light on the diversity and origin of the MLO family of integral membrane proteins. Genome Biol Evol 8(3):878–895
2. Acevedo-Garcia J, Kusch S, Panstruga R (2014) Magical mystery tour: MLO proteins in plant immunity and beyond. New Phytol 204(2):273–281
3. Büschges R et al (1997) The barley *Mlo* gene: a novel control element of plant pathogen resistance. Cell 88(5):695–705
4. Jørgensen IH (1992) Discovery, characterization and exploitation of *Mlo* powdery mildew resistance in barley. Euphytica 63:141–152
5. Consonni C et al (2006) Conserved requirement for a plant host cell protein in powdery mildew pathogenesis. Nat Genet 38(6): 716–720
6. Kusch S, Panstruga R (2017) *mlo*-based resistance: an apparently universal “weapon” to defeat powdery mildew disease. Mol Plant-Microbe Interact 30(3):1–47
7. Lyngkjær MF, Newton AC, Atzema JL, Baker SJ (2000) The barley *mlo*-gene: an important powdery mildew resistance source. Agronomie 20(7):745–755
8. Devoto A, Piffanelli P, Nilsson I, Wallin E, Panstruga R, Von Heijne G, Schulze-Lefert P (1999) Topology, subcellular localization, and sequence diversity of the *Mlo* family in plants. J Biol Chem 247(49):34993–35004
9. Kim MC, Panstruga R, Elliott C, Müller J, Devoto A, Yoon HW, Park HC, Cho MJ, Schulze-Lefert P (2002) Calmodulin interacts with MLO protein to regulate defence against mildew in barley. Nature 416(6879):447–451
10. Kim MC et al (2002) Mlo, a modulator of plant defense and cell death, is a novel calmodulin-binding protein. J Biol Chem 277(22): 19304–19314
11. Bednarek P et al (2009) A glucosinolate metabolism pathway in living plant cells mediates broad-spectrum antifungal defense. Science 323(5910):101–106
12. Consonni C, Bednarek P, Humphry M, Francocci F, Ferrari S, Harzen A, Loren V, van Themaat E, Panstruga R (2010) Tryptophan-derived metabolites are required for antifungal defense in the Arabidopsis *mlo2* mutant. Plant Physiol 152(3):1544–1564
13. Collins NC et al (2003) SNARE-protein-mediated disease resistance at the plant cell wall. Nature 425(30):973–977
14. Pajonk S, Kwon C, Clemens N, Panstruga R, Schulze-Lefert P (2008) Activity determinants and functional specialization of Arabidopsis PEN1 syntaxin in innate immunity. J Biol Chem 283(40):26974–26984
15. Bhat RA, Miklis M, Schmelzer E, Schulze-Lefert P, Panstruga R (2005) Recruitment and interaction dynamics of plant penetration resistance components in a plasma membrane microdomain. Proc Natl Acad Sci 102(8): 3135–3140
16. Kwon C et al (2008) Co-option of a default secretory pathway for plant immune responses. Nature 451(7180):835–840
17. Kwaitaal M, Keinath NF, Pajonk S, Biskup C, Panstruga R (2010) Combined bimolecular

fluorescence complementation and förster resonance energy transfer reveals ternary SNARE complex formation in living plant cells. Plant Physiol 152(3):1135

18. Harwood WA (2012) Advances and remaining challenges in the transformation of barley and wheat. J Exp Bot 63(5):1791–1798
19. Lee W-S, Hammond-Kosack KE, Kanyuka K (2015) *In planta* transient expression systems for monocots. Recent Adv Gene Expr Enabling Technol Crop Plants:391–422
20. Nelson AJ, Bushnell WR (1997) Transient expression of anthocyanin genes in barley epidermal cells: potential for use in evaluation of disease. Transgenic Res 244:233–244
21. Hückelhoven R (2014) The effective papilla hypothesis. New Phytol 204(3):438–440
22. Stolzenburg MC, Aist JR, Israel HW (1984) The role of papillae in resistance to powdery mildew conditioned by the ml-o gene in barley. I correlative evidence. Physiol Plant Pathol 25(3):337–346
23. Both M, Csukai M, Stumpf MPH, Spanu PD (2005) Gene expression profiles of *Blumeria graminis* indicate dynamic changes to primary metabolism during development of an obligate biotrophic pathogen. Plant Cell 17(7): 2107–2122
24. Panstruga R (2004) A golden shot: how ballistic single cell transformation boosts the molecular analysis of cereal – mildew interactions. Mol Plant Microbe Interact 5:141–148
25. Shirasu K, Nielsen K, Piffanelli P, Oliver R, Schulze-Lefert P (1999) Cell-autonomous complementation of *mlo* resistance using a biolistic transient expression system. Plant J 17(3): 293–299
26. Acevedo-Garcia J, Spencer D, Thieron H, Reinstädler A, Hammond-Kosack K, Phillips AL, Panstruga R (2017) *mlo*-based powdery mildew resistance in hexaploid bread wheat generated by a non-transgenic TILLING approach. Plant Biotechnol J 15(3):367–378
27. Reinstädler A, Müller J, Czembor JH, Piffanelli P, Panstruga R (2010) Novel induced *mlo* mutant alleles in combination with site-directed mutagenesis reveal functionally important domains in the heptahelical barley Mlo protein. BMC Pant Biol 10:31
28. Elliott C, Müller J, Miklis M, Bhat RA, Schulze-Lefert P, Panstruga R (2005) Conserved extracellular cysteine residues and cytoplasmic loop-loop interplay are required for functionality of the heptahelical MLO protein. Biochem J 385(Pt 1):243–254
29. Elliott C, Zhou F, Spielmeyer W, Panstruga R, Schulze-Lefert P (2002) Functional conservation of wheat and rice *Mlo* orthologs in defense modulation to the powdery mildew fungus. Mol Plant-Microbe Interact 15(10): 1069–1077
30. Ge X, Deng W, Lee ZZ, Lopez-Ruiz FJ, Schweizer P, Ellwood SR (2016) Tempered *mlo* broad-spectrum resistance to barley powdery mildew in an Ethiopian landrace. Sci Rep 6:29558
31. Panstruga R, Molina-Cano JL, Reinstädler A, Müller J (2005) Molecular characterization of *mlo* mutants in North American two- and six-rowed malting barley cultivars. Mol Plant Pathol 6(3):315–320
32. Hückelhoven R, Dechert C, Kogel KH (2003) Overexpression of barley BAX inhibitor 1 induces breakdown of *mlo*-mediated penetration resistance to *Blumeria graminis*. Proc Natl Acad Sci 100(9):5555–5560
33. Seeholzer S et al (2010) Diversity at the *Mla* powdery mildew resistance locus from cultivated barley reveals sites of positive selection. Mol Plant-Microbe Interact 23(4): 497–509
34. Pliego C et al (2013) Host-induced gene silencing in barley powdery mildew reveals a class of ribonuclease-like effectors. Mol Plant-Microbe Interact 26(6):633–642
35. Panstruga R, Kim MC, Cho MJ, Schulze-Lefert P (2003) Testing the efficiency of dsRNAi constructs in vivo: a transient expression assay based on two fluorescent proteins. Mol Biol Rep 30(3):135–140
36. Nowara D, Schweizer P, Gay A, Lacomme C, Shaw J, Ridout C, Douchkov D, Hensel G, Kumlehn J (2010) HIGS: Host-induced gene silencing in the obligate biotrophic fungal pathogen *Blumeria graminis*. Plant Cell 22(9):3130–3141
37. Wahara M, Inoue C, Kohguchi T, Sugai K, Kobayashi K, Nishiguchi M, Yamaoka N, Yaeno T (2017) Improved method for in situ biolistic transformation to analyze barley–powdery mildew interactions. J Gen Plant Pathol 83(3):140–146

Chapter 6

Host-Induced Gene Silencing Using BPMV on Soybean to Study Genes in the Soybean Rust Fungus *Phakopsora pachyrhizi*

Tobias I. Link

Abstract

To obtain direct evidence for the influence of an effector on the virulence or pathogenicity of a pathogen, it is necessary to knock out, knock down, or silence the respective gene. Since genetic transformation is not yet possible for rust fungi, silencing the gene is the only option. Posttranscriptional gene silencing uses RNAi. RNAi in plant pathogens can be accomplished by introducing dsRNA either by direct application of in vitro synthesized dsRNA or through positive-strand or double-strand RNA plant viruses. For studying effectors in *Phakopsora pachyrhizi*, we have implemented a host-induced silencing procedure based on virus-induced gene silencing using the bean pod mottle virus system. Here, procedures and interpretations of results are described and limitations of the system are discussed.

Key words RNAi, HbV, HIGS, VIGS, BPMV, *Phakopsora pachyrhizi*, Soybean

1 Introduction

Probably the most important prerequisite for studying ETI in plants interacting with rust fungi is the identification of effectors.

Effectors in rust fungi can be identified using pipelines that consist of sequencing either the genome or the transcriptome (or both) of the rust fungal species and predicting secreted proteins since effectors – supposedly transferred into the plant – have to be secreted first. After protein prediction, sequences can be checked for signal peptides using software like SignalP [1–3]. This prediction needs to be followed up by predictions of transmembrane domains [4, 5], GPI anchors [6], and ER retention signals to positively identify secreted proteins. In order to include non-classically secreted proteins, these can be predicted using SecretomeP [1].

Among the secreted proteins likely effector candidates are selected using a range of different criteria. The most direct way to

Thomas A. Kufer and Maria Kaparakis-Liaskos (eds.), *Effector-Triggered Immunity: Methods and Protocols*, Methods in Molecular Biology, vol. 2523, https://doi.org/10.1007/978-1-0716-2449-4_6,

identify effectors would be to search for sequence similarity with known effectors. So far, however, the number of bona fide effectors from rust fungi is very limited, so broader criteria need to be applied. These criteria are lineage specificity – meaning either specificity to the species or conservation among rust fungi combined with an absence from other fungi. Also, protein families can be analyzed for expansion or contraction in a given genome. If transcriptomes were sequenced for different developmental stages, effector candidates can also be selected based on differential or stage-specific expression [7]. A whole class of proteins is also considered good effector candidates: the small secreted proteins as well as the small and cysteine-rich secreted proteins. Recently, machine learning has been applied to the prediction of fungal effectors [8, 9]. Ninety-four effectors from 23 fungal plant pathogens (biotrophic and necrotrophic) were used to train the classifier, which was 90% accurate in tests. This classifier can also be applied to select rust fungal effector candidates.

Predicted effectors have to be validated by functional characterization [10]. Proteins can be localized, either in situ by immunofluorescence microscopy [11] or by heterologous expression in the plant as GFP fusion proteins. Alternative screening experiments to identify effector function are also based on heterologous expression in plants and testing for elicitation or suppression of defense responses by the expressed protein. Yet another experimental setup is to express the effector candidate in another plant pathogen, preferentially one that can be easily transformed like phytopathogenic bacteria. A widely used system is *Pseudomonas syringae* pv. *tomato* strain DC3000 (*Pst* DC3000) together with the pEDV6 plasmid that directly delivers candidate effectors into the plant cytoplasm by the type III secretion system [12]. This system can demonstrate suppression of ETI by the effector candidate. *Pseudomonas fluorescens* strain EtHAn [13] together with *Pst* DC3000 and the same plasmid can be used to check for suppression of PTI. For screens that are quicker, the proteins can also be expressed in yeast to check for detrimental effects, or to test for BAX-induced cell death suppression. All these tests are necessary because rust fungi cannot be genetically transformed and have the disadvantage that they can only provide indirect evidence for effector functions. Experimental assays to identify the plant targets of effectors through interaction studies like yeast two-hybrid or pull-down experiments with mass spectrometry provide the only direct information regarding effector function.

Direct evidence for the contribution of effector function to pathogenicity or virulence can only come from removing the effectors from the system. The only current option to achieve this in obligate biotrophs is gene silencing through RNAi.

RNAi is posttranscriptional gene silencing based on double-stranded RNA (dsRNA). dsRNA is recognized by Dicer or Dicer-like proteins [14] that cuts dsRNA into fragments of 21–23 bp in length, the small interfering RNA (siRNA). The siRNA is incorporated into the RISC (RNA-induced silencing complex) [15]. Here one strand of the siRNA is degraded and the other strand remains and mediates binding of the RISC to RNA complementary to the incorporated RNA. This RNA is then degraded by the RISC [16].

RNAi can be accomplished by introducing dsRNA into an organism or a system. One popular way to achieve this is the use of positive-strand or double-strand RNA viruses since these are always linked to dsRNA. The experimental system has been termed virus-induced gene silencing (VIGS) [17].

It has been shown that RNA, either dsRNA or siRNA, can pass between organisms even if they are not related and mediate RNAi in the corresponding other organism. This phenomenon has been termed cross-kingdom RNAi. With this knowledge, methods have been established to introduce dsRNA into plants that can lead to RNAi in pathogens infecting these plants. This method is called host-induced gene silencing (HIGS) [16]. The dsRNA for silencing pathogen genes can either be expressed in the plant from hairpin constructs integrated into the genome or introduced transiently via VIGS. The latter is often either referred to as VIGS or HIGS depending on whether authors want to emphasize that the silencing signal originates from a virus construct or that it comes from the host plant. In my opinion, both terms fall short of unambiguously defining the procedure which should better be called indirect VIGS or better HIGS by VIGS (HbV).

Here I am describing a method for HbV used in soybean to study the Asian soybean rust, *Phakopsora pachyrhizi*. For VIGS in soybean, the bean pod mottle virus (BPMV) system has been established [18, 19]. The two-part genome of the virus was cloned into two plasmids (pBPMV-IA-R1 for RNA1 and pBPMV-IA-R2 for RNA2). Into pBPMV-IA-R2 a *Bam*HI restriction site was introduced allowing insertion of silencing constructs. For silencing soybean genes, it is recommended to use sequences of 300–600 nt in length derived from the 3′ end of the transcript. The plasmids also carry CaMV 35S promoters. This allows biolistic transfer of the plasmids into the plant, where the virus RNA is transcribed and after translation of the virus proteins the virus starts replicating and infecting. To evaluate silencing effects, it is recommended, however, to perform mechanical inoculations using infected material taken from plants that were inoculated with the plasmids. This system has been published in another topic of Methods in Molecular Biology: [20].

We are using the BPMV system, but instead of inserting soybean genes into the viral genome, we use genes of *Phakopsora pachyrhizi*. In order to observe silencing effects, *P. pachyrhizi* is inoculated onto plants 3 weeks after BPMV inoculation when these show systemic infection with the virus. Five and 7 days later, plants are evaluated, both by visual inspection and by RT-qPCR for measuring suppression of *P. pachyrhizi* growth and silencing. We have used the system to silence several genes in *P. pachyrhizi* with mixed success. We are not the only group using the system; Hu et al. [21] have published results for silencing genes expressed early during soybean rust infection.

Since effectors are often redundant in their function, it may be necessary to silence more than one gene at a time. Since the BPMV vector offers limited insertion sites, multiple constructs need to be constructed in a different vector before insertion into pBPMV. I also describe a restriction-free cloning procedure for creating complex constructs.

2 Materials

2.1 Cloning Silencing Constructs

1. Plasmids pBPMV-IA-V1 [19] (best contact Chunquan Zhang, czhang@alcorn.edu) and pUC19 (or any other small plasmid containing *Bam*HI in the cloning site).
2. cDNA from *P. pachyrhizi* (*see* **Note 1**).
3. Restriction enzymes *Bam*HI, *Eco*RI, *Hind*III, and *Dpn*I.
4. Primers: Custom-designed primers for amplifying target gene fragments as inserts with *Bam*HI restriction sites or with part of the different target sequence and pUC sequence included. Primers for screening PCR and/or sequencing from pUC19. F: 5′-CGC CAG GGT TTT CCC AGT CAC-3′, R: 5′-CAC AGG AAA CAG CTA TGA CCA TG-3′. Primers for screening PCR and/or sequencing from pBPMV-IA-V1. pBPMV-F1: 5′-ACA TTC CTG GGA ATT GAT CTT CC-3′, pBPMV-R1: 5′-GAT CGG GGA AAT TCG AGC TAT C-3′.
5. Phusion™ High-Fidelity DNA Polymerase (Thermo Fisher Scientific, Waltham, MA, USA).
6. Competent cells of a standard *E. coli* cloning strain, for example, DH5α.
7. Standard molecular biology lab equipment: PCR machine, DNA electrophoresis equipment, LB medium, plasmid preparation kit, photometer.

2.2 Biolistic Transformation of BPMV into Soybean and Production of Virus Inoculum

1. Growth chamber or greenhouse compartment with good climate control. Set at 16/8 h day/night with 20–25 °C (*see* **Note 2**).
2. Soybean seeds. Any cultivar susceptible to both BPMV and *P. pachyrhizi* can be used.

3. Standard germination soil, all-purpose fertilizer, small plant pots to fit into the gene gun (<8 cm in diameter, <6 cm high), larger plant pots to accommodate full-size soybean plants (i.e., 18 cm in diameter, 15 cm high).
4. Vectors pBPMV-IA-R1M and pBPMV-IA-R2 [19], pBPMV-IA-V1, or pBPMV-IA-V2-based silencing constructs.
5. 1-μm ø gold microcarriers (Bio-Rad Laboratories, Hercules, CA, USA). To prepare the microcarrier suspension for coating with DNA, add 0.5-ml 100% ethanol to 30-mg gold particles, vortex and incubate at room temperature for 10 min, centrifuge briefly, and discard the supernatant. Resuspend the gold in 0.5 ml H_2O, spin down again, and remove the water. Repeat the water wash twice. Again, resuspend in 0.5 ml H_2O and while vortexing remove 25-μl aliquots that can be stored at −20 °C. See also Ref. [22].
6. Spermidine buffer: Freshly prepare for coating microcarriers from 50 μl 50% glycerol, 25 μl 2.5-M $CaCl_2$, and 10 μl 0.1 M spermidine.
7. 50% glycerol, 2.5 M $CaCl_2$, 0.1 M spermidine (0.015 g/ml, frozen in aliquots), isopropanol, 70% ethanol, 100% ethanol.
8. Macrocarrier discs, stopping screens, and 1100-psi rupture discs (Bio-Rad Labs).
9. PDS-1000 biolistic particle delivery system (Bio-Rad Labs), complete with vacuum pump and helium gas bottle. The support for the leaves needs to be custom made. There has to be a slot from one side to the middle to insert the stem of the seedlings. It can be made from plexiglass or any kind of plastic with a smooth surface.
10. Tubes, Drierite, and freeze-dryer.

2.3 Rub Inoculation with BPMV Silencing Constructs, Superinoculation with P. pachyrhizi

1. Greenhouse or growth chamber with good climate control. Set at 16/8 h day/night with 20–25 °C (*see* **Note 2**).
2. Soybean seeds. Use the same cultivar as in Subheading 2.2.
3. KP inoculation buffer: 50 mM potassium phosphate, pH 7.0.
4. Mortar and pestle, carborundum, and cheesecloth.
5. Fresh or dried leaf tissues infected with either wild-type BPMV or any of the constructs (*see* **Note 3**); the material harvested from Subheading 3.2.2.
6. *P. pachyrhizi* inoculation suspension: In an Erlenmeyer flask, mix 200 mg *P. pachyrhizi* urediospores (fresh from −80 °C, do not use spore charges that are too old, i.e., >1 year), 200 ml H_2O, 20 μl Tween 20, and 400 mg milk powder. In the dark, stir vigorously for 30 min using a magnetic stirrer.

2.4 RT-qPCR to Quantify Reductions in Fungal Growth and Gene Silencing

1. Reaction tubes (2 ml screw lid), liquid nitrogen, −80 °C freezer for sample storage; metal balls or ceramic beads and vortex or bead beater for homogenization.
2. Kit for RNA preparation (mini scale), for example, *Quick-*RNA™ Plant Miniprep Kit (Zymo Research Corporation, Irvine, CA); enzyme sets for reverse transcription, for example, RevertAid RT Reverse Transcription Kit (Thermo Scientific, Waltham, MA); Real-time PCR instrument, for example, CFX96™ (Bio-Rad Labs), Real-time PCR chemistry, and plastic consumables.
3. Primers for qPCR [23–26]: Primers for the *P. pachyrhizi* reference gene: CytB1f: 5′-TCA AGA CGC ATC CAA TTC TAG GTC-3′, CytB1r: 5′-GTG TTA CAC CCG TGA TAA TCT GAA T-3′. Primers for the plant reference gene: GmqUKN2f: 5′-GCC TCT GGA TAC CTG CTC AAG-3′, GmqUKN2r: 5′-ACC TCC TCC TCA AAC TCC TCT G-3′. Primer probe set for virus quantification: qBPMV-F1: 5′-CAG GCT ACT GCC AGA TCT TCT-3′, qBPMV-R1: 5′-CAA GCG AAC AAA CGA AAG AA-3′, qBPMV-Probe: FAM-TCC TCA TGC AGA GGA TTC CGC A-TAMRA. Primers for the silenced genes should be chosen in a different part of the transcript sequence than the silencing constructs. The primers should amplify stretches of 100–200 bp, Tm (GC + AT) should be close to 60 °C, and they should be HPLC purified.

3 Methods

3.1 Cloning Silencing Constructs

3.1.1 Cloning Silencing Constructs for a Single Target Gene

1. Choose 200–300 bp regions of target genes for silencing. No data are available for *P. pachyrhizi* so far, but a stretch close to the 3′ end of the transcript is generally recommended.
2. Clone from cDNA (*see* **Note 1**). For single constructs, use PCR amplification with primers that include the *Bam*HI restriction site and clone directly into pBPMV-IA-V1 using the *Bam*HI restriction site.

3.1.2 Cloning Silencing Constructs for Two or Multiple Target Genes (See Note 4)

1. Choose 200–300 bp regions of target genes for silencing. No data are available for *P. pachyrhizi* so far, but a stretch close to the 3′ end of the transcript is generally recommended.
2. Clone from cDNA (*see* **Note 1**). The first insert is amplified using primers including the *Bam*HI restriction site and cloned into pUC19 using the *Bam*HI site. Additional inserts should be inserted using overlap extension cloning [27] or restriction-free cloning. Primers for the second insert are designed according to the RF-Cloning.org website [28] with the left primer binding to the right end of the first insert and the left end of the

second insert and the right primer binding to the right end of the insert and the plasmid at the *Bam*HI site. Primers for the third and further inserts are designed correspondingly with the right primer binding to the left end of the second insert and the left primer again to the plasmid at the *Bam*HI site.

3. Perform PCR using the primers from above on 2.5 μl cDNA in a 50 μl reaction. Thirty-five cycles of amplification are recommended. Purify the PCR product and determine the concentration. (In our experience, gel extraction is not necessary.) Calculate the volumes necessary to obtain a molar insert/plasmid ratio of 20.
4. Perform secondary PCR using this PCR product with the second insert as megaprimer on the pUC19 plasmid already containing the first insert. We normally run two PCR reactions with two different three-step protocols with different annealing temperatures, 60 °C and 70 °C. After a *Dpn*I digest, amplify the plasmid by transforming into *E. coli*. If cloning more than two inserts, perform another secondary PCR using the PCR product with the third insert as megaprimer on the pUC19 plasmid containing the first and second insert, followed by transformation into *E. coli*. With a possible fourth insert, repeat secondary PCR and transformation into *E. coli* for a third time, and so on.
5. To set free the combined insert from the pUC19 plasmid, perform a multiple restriction digest with enzymes *Bam*HI, *Eco*RI, and *Hind*III. The combined insert can then be inserted into pBPMV-IA-V1 using the *Bam*HI restriction site.

3.2 Biolistic Transformation of BPMV into Soybean and Production of Virus Inoculums

The following protocol is adapted from [20].

3.2.1 Growing Plants for Biolistic Transformation

1. Germinate soybean seeds in standard germination soil and in pots small enough to let the whole plant fit below a target tray positioned at the middle level of the gene gun chamber (Fig. 1). Keep the pots at 18–25 °C at a photoperiod of 16 h light and 8 h dark. Fully expanded primary leaves are best for biolistic transformation.

3.2.2 Biolistic Inoculation and Production of Virus Inoculum

1. Mix 2 μg pBPMV-IA-R1M with 2 μg pBPMV-IA-V1 or 2 μg VIGS construct plasmid in a volume of 40 μl maximum.
2. Add 25 μl microcarrier suspension, and vortex briefly; add 85 μl freshly premixed spermidine buffer.
3. Vortex for 1 min; centrifuge briefly to pellet the gold. Discard the supernatant and wash the pellet with 70 μl 70%

Fig. 1 Biolistic transformation. Soybean seedlings with fully expanded primary leaves are inserted into the vacuum chamber of the gene gut setup. The primary leaves are set upon a plexiglass support placed on the third level in the chamber. The metal mesh holding the leaves should neither be too light nor too heavy. (Picture courtesy of Björn Lutsch, previously unpublished)

isopropanol, repeat the wash with 100% isopropanol, and resuspend in 30 μl 100% isopropanol.

4. While vortexing, remove 6 μl aliquots and spread the suspension on a macrocarrier; air-dry for a few minutes.
5. Assemble gene gun shooting set with macrocarriers from **step 4**, stopping screens, and 1100 psi rupture discs (*see* **Note 5**).
6. Place seedlings inside the vacuum chamber and hold the primary leaves flat against a plexiglass support with a metal mesh (Fig. 1). Shoot when the vacuum reaches 25–28 in. Hg. Remove the seedling, mist it with tap water, and keep it for 16 h in a box with 95% humidity.
7. Repot the plants to larger pots. Use standard germination soil. Maintain the plants at the same conditions as mentioned in Subheading 3.2.1; avoid excessive heat (*see* **Note 2**). Fertilize regularly (every week) with diluted all-purpose fertilizer.
8. Three weeks post biolistic inoculation symptoms of infection can be visually observed. Harvest leaves of second trifoliate as well as the upper youngest leaflets.
9. Lyophilize the leaves for 24 h or longer until they are completely dry, crush the dry leaves, and put them into 50 ml tubes containing Drierite at the bottom. Place the tubes at −20 °C for long-term storage.

3.3 Rub Inoculation with BPMV Silencing Constructs, Superinoculation with P. pachyrhizi

3.3.1 Growing Plants for Rub Inoculation

1. Same as described in Subheading 3.2.1 but use larger pots from the beginning. Fully expanded primary leaves are best for rub inoculation.

3.3.2 Rub Inoculation

1. Grind leaf tissue in KP inoculation buffer using a pestle and mortar on ice. The ratio of leaf tissue to KP buffer (w/v) should be 1:10–20 for fresh leaf and 1:20–40 for dry tissue. Dust primary leaves with 600-mesh carborundum.
2. Use a cheesecloth to absorb the sap from the ground tissue and rub inoculate the two primary leaves. Do not rub too hard; rinse the leaves after the inoculation.
3. Maintain the plants as described in Subheading 3.2 (**step 7**). After 3 weeks plants show strong viral infections and they can be inoculated with *P. pachyrhizi*.

3.3.3 Superinoculation with P. pachyrhizi

1. Label second and third trifoliate leaves that are visibly infected with BPMV with paper strips.
2. Spray *P. pachyrhizi* inoculation suspension onto second and third trifoliate leaves using a chromatography dispenser. Spray evenly.
3. Keep plants in the dark at 95% humidity at 18 °C for 12 h. Afterward transfer back to standard growth conditions described above. Try for gradual (2 h) adjustment of the plants to light and reduced humidity.
4. Five days post *P. pachyrhizi* inoculation and again 10 days post *P. pachyrhizi* inoculation, take samples for RNA preparation (*see* **Note 6**).
5. Ten days post *P. pachyrhizi* inoculation, visually assess *P. pachyrhizi* infection. Earlier assessment is not useful because rust symptoms may be not yet visible or indistinguishable from the virus symptoms. Good options for assessments are counting pustules and calculating pustule density per cm^2 or comparing symptoms to a reference picture chart.

3.4 RT-qPCR to Quantify Reductions in Fungal Growth and Gene Silencing

1. Using the top end of a 2-ml tube, cut out three small leaf discs per trifoliate leaf. Immediately freeze them in liquid nitrogen. This should provide close to but below 100 mg tissue.
2. Perform RNA preparation and cDNA synthesis for all samples. We prefer using random hexamer primers for reverse transcription.

3. RT-qPCR should be performed using primers for the following genes: GmUkn2 as a stably expressed plant reference gene, PpCytB as *P. pachyrhizi* reference gene, primers for the silenced genes, and primers for BPMV. This way, it is possible to simultaneously calculate the silencing efficiency, quantify *P. pachyrhizi*, and quantify BPMV.
4. Because of the strong variability, descriptive renderings of the quantitative results should be preferred over calculations of significance. What should be observed are correlations between silencing and reduction of fungal biomass and virus load and both silencing and fungal biomass (*see* **Note 6**).

4 Notes

1. Total RNA can be prepared from pure fungal structures like germinated urediospores or from infected plant material. Especially with effectors that are highly regulated, it is advisable to use the developmental stage during which the effectors are expressed. For reverse transcription, we typically use 1 μg RNA in 20 μl RT reactions.
2. Because the experiment depends on three species and their interactions, the outcomes are highly variable and strongly dependent on environmental conditions. To account for the variability, at least five plants should be used for every construct and all the controls. Also the full HbV experiment should be repeated independently three times. There are two ways to further address the issue: (1) Use growth chambers with full control of all conditions. This should also lead to results that are more uniform. On the other hand, because the optimal conditions could also vary between soybean cultivars and silencing constructs, the optimal conditions could be missed. (2) Use greenhouse chambers. Even with good climate control conditions will slightly vary. This will lead to stronger differences in the results but might also have positive outcomes that could otherwise be missed. It is crucial, however, to avoid excessive heat, since this will both hinder viral and fungal infections and deny any results.
3. Even though Hu et al. [21] describe the use of pBPMV-GFP as a positive control for VIGS, I do not consider this control necessary or useful for HbV, since virus symptoms can be observed directly also without additional bleaching symptoms, because viral infections may differ between different plants and different constructs, and because this is a positive control for silencing in the plant but not for silencing in the fungus. Negative controls should include non-virus inoculated plants, mock virus inoculated plants, plants inoculated with wild-type

BPMV, and, when testing double or multiple constructs, single constructs and constructs minus one of the target genes. Especially comparisons with the empty vector control and the controls to the single constructs are important and may be tricky. This is because compared to non-virus inoculated plants and mock virus inoculated plants, infections with *P. pachyrhizi* can be considerably reduced in plants inoculated with the empty vector. The reason for this phenomenon could either be an induction of plant resistance through the virus infection or less nutrients provided by the plant because of deficiencies caused by the virus.

4. Using pBPMV-IA-V2 is another option for construing constructs with multiple inserts, but in my opinion, the use of restriction-free cloning and transfer of the combined insert is more reliable. An alternative to cloning is DNA synthesis of the whole combined insert, including *Bam*HI restriction sites.
5. The importance of the gap width between rupture disc and macrocarrier in the assembly has been neglected in previous publications. The gap width that gives the highest frequencies of infection is greater than the one that is set with the biggest of the hexagonal tools provided by the manufacturer for setting the gap width. The gap width can best be described as when four threads of the screw thread that is used to move the macrocarrier launch assembly in the supporting plastic tray are visible above the tray. The optimal distance between macrocarrier and stopping screen is the medium distance; there should be one spacer ring above and one spacer ring below the stopping screen support ring.
6. Correlation between virus load and silencing is expected, since the amount of available dsRNA influences the amount of siRNA that ends up in the fungus. Correlation between silencing and fungal growth is also expected, since silencing is supposed to incapacitate the fungus. If there is correlation, this is the best outcome. Missing correlation can be interpreted differently: There can be silencing but no growth reduction. This most likely means that the silenced gene is not essential and silencing does not incapacitate the fungus. There also can be growth reduction without silencing. This could be due to reactions of the fungus against the silencing. There have been frequent observations that silenced genes show increased transcript levels, which could be due to overexpression of the genes after the fungus registers the deficiency. This is also the reason why I recommend two time points for sampling to catch the silencing. An additional sampling at 3 dpi could also be useful; time points still earlier are not recommended because there is too little fungal biomass to allow reliable RT-qPCR quantification of fungal genes.

References

1. Bendtsen JD, Jensen LJ, Blom N et al (2004) Feature-based prediction of non-classical and leaderless protein secretion. Protein Eng Des Sel 17:349–356
2. Bendtsen JD, Nielsen H, von Heijne G et al (2004) Improved prediction of signal peptides: SignalP 3.0. J Mol Biol 340:783–795
3. Armenteros JJA, Tsirigos KD, Sønderby CK et al (2019) SignalP 5.0 improves signal peptide predictions using deep neural networks. Nat Biotechnol 37:420–423
4. Käll L, Krogh A, Sonnhammer EL (2004) A combined transmembrane topology and signal peptide prediction method. J Mol Biol 338: 1027–1036
5. Sonnhammer EL, von Heijne G, Krogh A (1998) A hidden Markov model for predicting transmembrane helices in protein sequences. In: Proceedings of the sixth international conference on intelligent systems for molecular biology
6. Pierleoni A, Martelli PL, Casadio R (2008) PredGPI: a GPI-anchor predictor. BMC Bioinform 9:392
7. Saunders DG, Win J, Cano LM et al (2012) Using hierarchical clustering of secreted protein families to classify and rank candidate effectors of rust fungi. PLoS One 7:e29847
8. Sperschneider J, Gardiner DM, Dodds PN et al (2016) EffectorP: predicting fungal effector proteins from secretomes using machine learning. New Phytol 210:743–761
9. Sperschneider J, Dodds PN, Gardiner DM et al (2018) Improved prediction of fungal effector proteins from secretomes with EffectorP 2.0. Mol Plant Pathol 19:2094–2110
10. Qi M, Grayczyk JP, Seitz JM et al (2018) Suppression or activation of immune responses by predicted secreted proteins of the soybean rust pathogen *Phakopsora pachyrhizi*. Mol Plant-Microbe Interact 31:163–174
11. Kemen E, Mendgen K, Voegele RT (2011) Immunolocalization of pathogen effectors. In: McDowell JM (ed) Plant immunity: methods and protocols. Methods in molecular biology, vol 712. Humana Press/Springer Science + Business Media, New York, pp 211–225
12. Sohn KH, Lei R, Nemri A et al (2007) The downy mildew effector proteins ATR1 and ATR13 promote disease susceptibility in *Arabidopsis thaliana*. Plant Cell 19:4077–4090
13. Thomas WJ, Thireault CA, Kimbrel JA et al (2009) Recombineering and stable integration of the *Pseudomonas syringae* pv. *syringae* 61 hrp/hrc cluster into the genome of the soil bacterium *Pseudomonas fluorescens* Pf0-1. Plant J 60:919–928
14. Bernstein E, Caudy AA, Hammond SM et al (2001) Role for a bidentate ribonuclease in the initiation step of RNA interference. Nature 409:363–366
15. Hammond SM, Bernstein E, Beach D et al (2000) An RNA-directed nuclease mediates post-transcriptional gene silencing in *Drosophila* cells. Nature 404:293–296
16. Koch A, Kogel KH (2014) New wind in the sails: improving the agronomic value of crop plants through RNAi-mediated gene silencing. Plant Biotechnol J 12:821–831
17. Purkayastha A, Dasgupta I (2009) Virus-induced gene silencing: a versatile tool for discovery of gene functions in plants. Plant Physiol Biochem 47:967–976
18. Zhang C, Yang C, Whitham SA et al (2009) Development and use of an efficient DNA-based viral gene silencing vector for soybean. Mol Plant-Microbe Interact 22:123–131
19. Zhang C, Bradshaw JD, Whitham SA et al (2010) The development of an efficient multipurpose bean pod mottle virus viral vector set for foreign gene expression and RNA silencing. Plant Physiol 153:52–65
20. Zhang C, Whitham SA, Hill JH (2013) Virus-induced gene silencing in soybean and common bean. In: Becker A (ed) Virus-induced gene silencing. Methods in molecular biology (methods and protocols), vol 975. Humana Press, Totowa, pp 149–156
21. Hu D, Chen Z-Y, Zhang C et al (2020) Reduction of *Phakopsora pachyrhizi* infection on soybean through host- and spray-induced gene silencing. Mol Plant Pathol 21:794–807
22. BioRad (1995) Biolistic PDC-1000/He particle delivery system. www.bio-rad.com
23. Hirschburger D, Müller M, Voegele R et al (2015) Reference genes in the pathosystem *Phakopsora pachyrhizi*/soybean suitable for normalization in transcript profiling. Int J Mol Sci 16:23057–23075
24. Hu R, Fan C, Li H et al (2009) Evaluation of putative reference genes for gene expression normalization in soybean by quantitative real-time RT-PCR. BMC Mol Biol 10:93
25. Schmitz HK, Medeiros CA, Craig IR et al (2014) Sensitivity of *Phakopsora pachyrhizi* towards quinone-outside-inhibitors and demethylation-inhibitors, and corresponding resistance mechanisms. Pest Manag Sci 70: 378–388

26. Müller M (2014) Etablierung eines Wirts-induzierten RNAi-Systems für die Kontrolle des Asiatischen Sojabohnenrostes *Phakopsora pachyrhizi*. Dissertation, University of Hohenheim
27. Bryksin AV, Matsumura I (2010) Overlap extension PCR cloning: a simple and reliable way to create recombinant plasmids. BioTechniques 48:463–465
28. Bond SR, Naus CC (2012) RF-Cloning.org: an online tool for the design of restriction-free cloning projects. Nucleic Acids Res 40: W209–W213

Chapter 7

Virus-Mediated Protein Overexpression (VOX) in Monocots to Identify and Functionally Characterize Fungal Effectors

Kostya Kanyuka

Abstract

One of the important armories that pathogens utilize to successfully colonize the plants is small secreted effector proteins, which could perform a variety of functions from suppression of plant innate immunity to manipulation of plant physiology in favor of the disease. Plants, on the other hand, evolved disease resistance genes that recognize some of the effectors or avirulence (Avr) proteins. Both, identification of the Avr proteins and understanding of the mechanisms of action of other effectors, are important areas of research in the molecular plant-pathogen interactions field as this knowledge is critical for the development of new effective pathogen control measures. To enable functional analysis of the effectors, it is desirable to be able to overexpress them readily in the host plants. Here we describe detailed experimental protocols for transient effector overexpression in wheat and other monocots using binary *Barley stripe mosaic virus* (BSMV)- and *Foxtail mosaic virus* (FoMV)-derived vectors. This functional genomics tool, better known as VOX (Virus-mediated protein OvereXpression), is rapid and relatively simple and inexpensive.

Key words VOX, BSMV, FoMV, Effectors, Cereals, Wheat, Functional genomics

1 Introduction

Plant pathogenic microorganisms, including bacteria, fungi, and oomycetes, secrete a number of proteins, known as "effectors," which are thought to facilitate disease by suppressing host immunity and/or altering host metabolism and physiology [1]. Similarly, plant pests, such as insects and nematodes, also utilize effectors to manipulate host cell physiology to counteract plant defenses and access the nutrients [2, 3]. Plants, on the other hand, evolved disease resistance genes whose products could recognize some of the pathogen effectors, called avirulence (Avr) proteins, and trigger defense responses which either prevent or substantially reduce disease burden [4].

Recent advances in next-generation sequencing technologies and marked reduction in speed and the associated costs permit rapid and accurate whole genome sequencing of pathogen and

Thomas A. Kufer and Maria Kaparakis-Liaskos (eds.), *Effector-Triggered Immunity: Methods and Protocols*, Methods in Molecular Biology, vol. 2523, https://doi.org/10.1007/978-1-0716-2449-4_7,

pest genomes [5–9]. In parallel, powerful bioinformatics and computational tools have been developing for the construction of chromosome-length genome assemblies, gene space annotations, and machine learning-aided prediction of pathogen effector genes [10, 11].

However, the validation of the effector gene candidates, including Avr genes, and their functional characterization remains a huge challenge and is especially difficult for the agriculturally important, monocot crops-infecting obligate biotrophic pathogen species, e.g., rust and powdery mildew fungi, which cannot be subcultured in vitro, away from the plant. Although the molecular tools are being developed, many of these pathogens cannot be or are very difficult to manipulate genetically, for example, by knocking out or overexpressing the corresponding candidate effector genes. On the other hand, although knockout of Avr or necrotrophic effectors determining host genotype specificity in the culturable and transformable pathogens (e.g., wheat pathogenic fungi *Zymoseptoria tritici* and *Parastagonospora nodorum*) is very effective as a gene validation tool, knockout of other effectors (i.e., those that are not recognized by immune receptors or targeting host necrotrophic effector sensitivity genes) often results in no phenotypic change due to the functional redundancy among these genes [12, 13]. For these reasons, overexpression of effectors in their native form or as fusions with short peptide tags (e.g., cMyc, HA, or 6xHis) or longer, fluorescent protein tags (e.g., GFP) in the fungal species of interest followed by functional analyses *in planta* post fungal inoculation, is often used as a complementary strategy. Nevertheless, achieving a high level of protein expression in these experiments, sufficient for downstream analysis, could be challenging. Therefore, as an alternative, effectors could be produced in heterologous expression systems such as *Escherichia coli* and/or *Pichia pastoris* and then delivered to the leaves of host plants, for example, using infiltration with a needleless syringe, for functional analyses. This methodology is widely and successfully used, for instance, for the investigation of necrotrophic effectors in *P. nodorum* [14]. However, it may be difficult or even impossible to express every effector of interests using this approach. Moreover, some heterologous produced effectors may remain inactive either due to misfolding, lack of appropriate posttranslational modification, or hyperglycosylation. Thus, the expression of biologically active effectors directly in the host plants is desirable.

One of the most common and highly efficient approaches for the effector function analysis used in dicot systems is the *Agrobacterium tumefaciens*-mediated transient overexpression in plant leaves known as "agroinfiltration," often utilizing a model plant *Nicotiana benthamiana* [15]. This method is useful for the validation of candidate Avr proteins, as co-expression of *Avr* with the corresponding plant resistance gene induces macroscopically visible cell death in the agroinfiltrated leaf patches. This approach, therefore, could be used for the validation of Avr effectors from the monocot-infecting pathogens, but importantly it requires the isolation of the corresponding resistance genes, which is a serious undertaking in cereal crops whose genomes are very large and complex. Unfortunately, agroinfiltration either does not work or is extremely inefficient in monocots, as these species are not natural hosts for *A. tumefaciens* [16].

Plant viruses, due to simplicity of their genomes, ease of genome manipulation and inoculation, and ability to rapidly spread systemically throughout plant tissues, have attracted considerable interest as vectors for *in planta* gene expression. The first successful attempt to manipulate the plant viral genome for expression of foreign sequences was made almost four decades ago [17], and *Potato virus X*-based vector was the first vector used for the expression of fungal effector proteins in plants [18]. Since those early seminal studies, many plant DNA and RNA viruses have been converted to vectors for overexpression of heterologous proteins in various plant species, but primarily in dicots.

Here we describe the detailed protocols for expression of pathogen effector proteins in wheat (*Triticum aestivum*), which utilize binary expression vectors based on monopartite *Foxtail mosaic virus* (FoMV) and tripartite *Barley stripe mosaic virus* (BSMV) we developed or redesigned, respectively [19–21]. Each of these two VOX vectors has their own advantages but also disadvantages that need to be considered.

BSMV is particularly useful for expression of small, up to approximately 150 amino acids long proteins in wheat and barley, and at both seedlings and adult plant stage. We have also been successful in expressing fungal effectors and/or other heterologous proteins in barley (*Hordeum vulgare*), triticale (× *Triticosecale*), and black-grass (*Alopecurus myosuroides*) using BSMV, and there is evidence that this virus could infect other monocots including durum wheat (*T. durum*), oat (*Avena sativa*), rye (*Secale cereale*), *Brachypodium distachyon*, *Aegilops tauschii*, and *T. monococcum* [19, 22–24]. Larger inserts are unstable in BSMV, and in some host genotypes, BSMV could induce mild to severe symptoms, thus potentially preventing accurate phenotyping of effects induced by the heterologous protein it is designed to express. FoMV, on the other hand, could accommodate and relatively stably express larger proteins, of up to 600 amino acids long, and induces no or only very mild symptoms in wheat, barley, maize (*Zea mays*), proso millet (*Panicum miliaceum*), foxtail millet (*Setaria italica* and *S. viridis*), *B. distachyon*, sorghum (*Sorghum bicolor* and *S. sudanense*), black-grass, and rice (*Oryza sativa*) [23, 25–29], and thus is well suited for expression of effectors of practically any size from different pathogen species [21, 30–32]. It is important to note, however, that genotypes within the given monocot species could differ in their susceptibility to BSMV or FoMV, and some genotypes may be completely recalcitrant to virus infection, for example, due to the presence of natural virus resistance genes [33–35].

2 Materials

2.1 Plants and Plant Growth Materials

1. Seeds of *Nicotiana benthamiana*.
2. Seeds of monocot species (e.g., *T. aestivum* cv. Cadenza).
3. Plant propagators with vented lids.
4. Small pots (e.g., 5 cm × 5 cm × 8 cm) for raising *N. benthamiana* plants.
5. Larger pots (e.g., 9 cm × 9 cm × 10 cm) for raising wheat or other monocot crops.
6. Low-nutrient compost for raising *N. benthamiana* plants.
7. Nutrient-rich soil mix suitable for raising wheat or other monocot crops.
8. Pot labels and plant tags.
9. Temperature-, light-, and humidity-controlled plant growth rooms/cabinets set to run at the following conditions: 16 h light (~130 μmol m^{-2} s^{-1} for *N. benthamiana* and ~180–220 μmol m^{-2} s^{-1} for wheat and other monocots), day/night temperature of 20 °C/23 °C or 26.7 °C/21.1 °C for experiments involving BSMV and FoMV, respectively, and relative humidity 60–65%.

2.2 General Materials and Apparatus

1. Thermocycler.
2. Microfuge.
3. Benchtop centrifuge for spinning universal 28-mL glass bottles or 50-mL conical centrifuge tubes.
4. Incubators for Petri dishes set to 37 °C and 28 °C.
5. Shaking incubators for universal bottles set to 37 °C and 28 °C.
6. Nano μL Spectrophotometer.
7. Equipment for DNA electrophoresis.
8. Fridge/cold room (4–8 °C) and freezers (−20 °C and −80 °C).
9. Pipets and sterile pipet tips and filter tips.
10. 0.2-mL PCR tubes.
11. Sterile 1.5-mL microtubes.
12. Sterile inoculation loops and spreaders.
13. Laboratory Wrapping Film.

14. Sterile 28-mL glass bottles with rubber-lined aluminum screw caps.
15. Sterile round-bottom 14-mL polypropylene tubes.
16. Sterile Petri dishes.
17. Sterile wooden toothpicks.
18. Plastic, disposable semi-micro (1 mL) spectrophotometer cuvettes.
19. Agarose.
20. DNA electrophoresis buffers, e.g., TAE (Tris-acetate-EDTA pH 8.3; 40-mM Tris base, 20-mM acetic acid, and 1-mM EDTA) and TBE (Tris-borate-EDTA pH 8.3; 90-mM Tris base, 90-mM boric acid, and 2-mM EDTA).
21. 1 kb DNA ladder.
22. SOC liquid medium (0.5% yeast extract, 2% tryptone, 10-mM NaCl, 2.5-mM KCl, 20-mM $MgSO_4$, and 20-mM glucose).
23. Low salt LB broth and LB agar.
24. High salt LB broth and LB agar.
25. Stock solutions of antibiotics kanamycin (50 mg/mL in water), gentamicin (25 mg/mL in water), and tetracycline (10 mg/mL in ethanol).
26. Sterile 50% glycerol in water.
27. Nuclease-free, molecular biology grade water.
28. Low error rate DNA Polymerase, e.g., Phusion DNA Polymerase (Fisher Scientific UK Ltd).
29. 5-mM dNTPs mix.
30. Standard DNA Polymerase master mix.
31. DNA Gel Extraction Kit.
32. Plasmid Miniprep Kit.

2.3 Materials for the Development of BSMV VOX Constructs

1. Binary vectors pCaBS-α and pCaBS-β [36] directing expression of the BSMV genomic RNAs α and β, respectively, and the vector pCa-γb2A-LIC directing expression of the modified RNA γ that is used for insertion of foreign sequences targeted for *in planta* expression (Fig. 1) [20] (*see* **Note 1**). pCaBS-α and pCaBS-β and pCa-γb2A-LIC are available upon request from Kostya Kanyuka, subject to material transfer agreement (MTA).
2. Restriction endonuclease *Apa*I.
3. Primers for PCR amplification of effector gene coding sequences, which should have the appropriate 5′-extensions required for the ligation-independent cloning (LIC) into pCa-γb2A-LIC (see below).

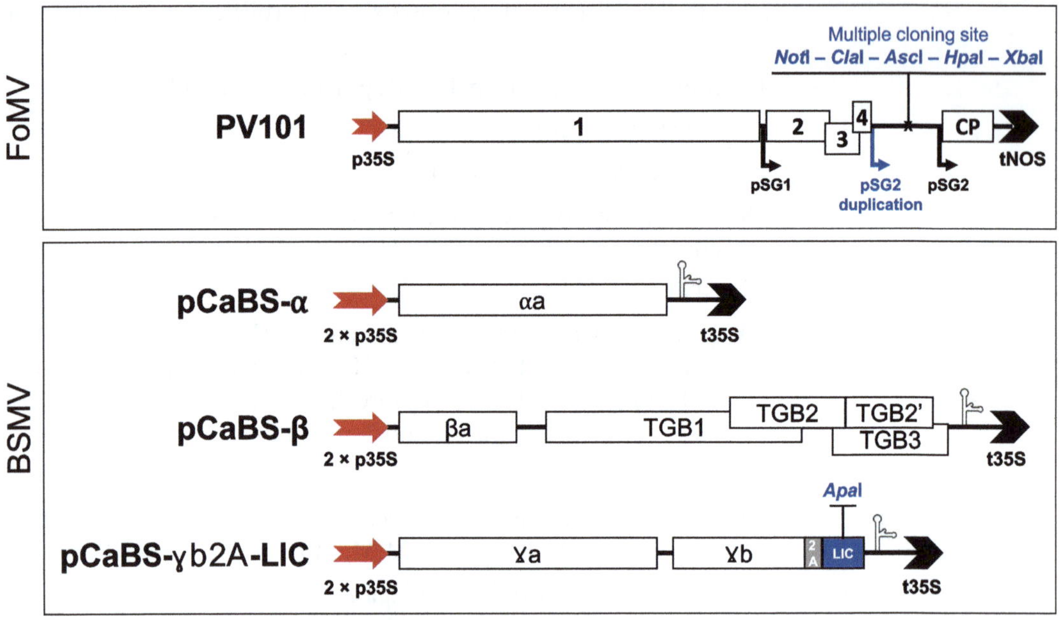

Fig. 1 Schematic representation of the FoMV and BSMV VOX vectors. Coding sequences of the effector genes of interest could be inserted into the FoMV vector PV101 downstream of the duplicated subgenomic promoter 2 (pSG2) of the viral coat protein (CP) cistron using a single or a combination of unique restriction enzyme sites present in the multiple cloning site. Coding sequences of the effector genes could be inserted into the BSMV vector pCa-γb2A-LIC using a ligation-independent cloning (LIC) method following restriction digestion with *Apa*I, a recognition site for which is located in the LIC site (a blue rectangle). Effector sequences are inserted immediately downstream of the sequence coding for the FMDV 2A peptide (2A). The effector is therefore initially translated as the C-terminal fusion with the BSMV γb-2A polypeptide, with the latter being co-translationally cleaved off the effector due to the self-cleavage ability of 2A peptide at its C-terminus

4. T4 DNA Polymerase.
5. dATP and dTTP.
6. Transformation competent *E. coli* strain JM109.
7. Primers for colony-PCR and verification of constructs by Sanger sequencing: 2235.F (5′- GATCAACTGCCAATCGT GAGTA-3′) and 2615.R (5′-CCAATTCAGGCATCGTTTTC -3′).
8. LB-Miller agar plates supplemented with kanamycin (50 μg/ mL).

2.4 Materials for the Development of FoMV VOX Constructs

1. Vector PV101 (Fig. 1) [21] available from Kostya Kanyuka upon request, subject to MTA.
2. Binary vector pBIN61-p19 for *in planta* expression of the *Tomato bushy stunt virus* silencing suppressor protein p19 [21] available from Kostya Kanyuka upon request.
3. Restriction endonucleases *Not*I (or *Cla*I) and *Xba*I.

4. Primers for PCR amplification of effector gene coding sequences, which should have the appropriate 5′-extensions (see below).
5. T4 DNA Ligase.
6. DH 5-alpha electrocompetent *E. coli* (subcloning efficiency).
7. LB-Miller agar plates supplemented with kanamycin (50 μg/mL).
8. Primers for colony-PCR and verification of constructs by Sanger sequencing: cons5073F (5′-TCCTCACACAGCCATATC TAGC -3′) and cons5558R (5′-TAGCTGCTTGAACAAAGGCC-3′).

2.5 Materials for Transformation of VOX Vectors into *Agrobacterium tumefaciens*

1. Electroporation competent *A. tumefaciens* strain GV3101 (pMP90) [37] or GV3101 pSa-Rep [38] cells for transformation of BSMV and FoMV vectors, respectively.
2. Electroporator.
3. Electroporation cuvettes, 1-mm gap.
4. LB-Lennox agar plates supplemented with kanamycin (50 μg/mL), gentamicin (25 μg/mL), and tetracycline (5 μg/mL).

2.6 Materials for *Agrobacterium*-Mediated Inoculation of *N. benthamiana* Plants

1. Young, approximately 25–30-day-old *N. benthamiana* plants.
2. *A. tumefaciens* strains carrying pCaBS-α, pCaBS-β, and pCa-γb2A-LIC derivatives for BSMV VOX.
3. *A. tumefaciens* strains carrying pBIN61-p19 and PV101 derivatives for FoMV VOX.
5. LB-Lennox liquid medium supplemented with kanamycin (50 μg/mL) and gentamicin (25 μg/mL) or kanamycin (50 μg/mL), gentamicin (25 μg/mL), and tetracycline (5 μg/mL) for growing *Agrobacterium* strains carrying BSMV vectors and pBIN61-p19 or PV101 derivatives, respectively.
4. Infiltration medium: 10-mM MES (2-(N-morpholino)ethanesulfonic acid) buffer pH 5.6, 10-mM $MgCl_2$, 150-μM acetosyringone (3′,5′-dimethoxy-4′-hydroxyacetophenone) (*see* **Note 2**).
5. Sterile, disposable 1-mL plastic syringes.

2.7 Virus (Rub) Inoculation of Monocot Plant Leaves

1. Young, two-leaf stage wheat plants (*see* **Note 3**).
2. Sterile pestles and mortars.
3. Miracloth (VWR International Ltd, Lutterworth, UK).
4. 10-mM potassium phosphate buffer, pH 7.
5. Celite 545 acid washed (Fisher Scientific UK Ltd).
6. Water mist spray bottle.
7. Large plastic boxes with leads or large plastic bags.

3 Methods

3.1 BSMV VOX Constructs Development

1. Identify the coding sequence (CDS) of the selected pathogen effector gene for the ligation-independent cloning (LIC) into the pCa-γb2A-LIC vector [39] (*see* **Note 4**).
2. Amplify the entire CDS of the selected effector gene by PCR using Phusion DNA Polymerase or another low error rate DNA Polymerase and sequence-specific primers carrying the following 5′-extensions required for LIC into pCa-γb2A-LIC: 5'-CCAACCCAGGACCGTTG*ATGATG*-3′ (for the forward primer; the italicized *ATG* is the translation start codon of the effector gene; *see* **Note 5**) and 5′-AACCACCACCACCGTTA-3′ (for the reverse primer) from an appropriate template (*see* **Notes 6** and 7).
3. Purify the resulting PCR product following agarose gel electrophoresis using a Gel Extraction Kit.
4. Digest the pCa-γb2A-LIC with *Apa*I (using 5 U of the enzyme per 1 μg of the plasmid DNA) for 2 h or longer at 25 °C. Verify the digestion is complete by analyzing an aliquot of the reaction using agarose gel electrophoresis. Then incubate the reaction mix at 65 °C for 20 min to inactivate the enzyme.
5. Mix 0.5 μg of pCa-γb2A-LIC that was digested with *Apa*I to completion with 3 U of T4 DNA Polymerase and 5-mM dTTP in the reaction buffer in a 0.2-mL PCR tube in a total volume of 50 μL. Mix by pipetting and incubate in a thermocycler for 30 min at 22 °C. Then heat-inactivate the enzyme by incubating the tube at 75 °C for 15 min in a thermocycler.
6. In a separate tube, incubate 100–250 ng of the gel purified PCR product of the effector CDS obtained in **step 3** (above) with 0.6 U of T4 DNA Polymerase and 5-mM dATP in the reaction buffer (supplied with the enzyme) in a total volume of 10 μL. Mix by pipetting and incubate for 30 min at 22 °C. Then heat-inactivate the enzyme by incubating the tube at 75 °C for 15 min in a thermocycler.
7. Combine the T4 DNA Polymerase-treated PCR product (10 μl) from **step 6** and the *Apa*I-linearized and T4 DNA Polymerase-treated pCa-γb2A-LIC vector (2 μl) from **step 5** (*see* **Note 8**). Incubate the mixture at 65 °C for 2 min followed by 10–15-min incubation at room temperature to allow the sticky ends of the PCR product to anneal with the complementary sticky ends of the vector.
8. Use 2–3 μL of the annealed products for transformation of chemically competent *E. coli* JM109 (50 μl) following the manufacturer's instructions. Spread the entire transformation mixture onto one or two LB-Miller agar plates supplemented

with kanamycin (50 μg/mL) and incubate at 37 °C overnight (*see* **Note 9**).

9. Most of the colonies that will appear on the selection plates the next day will carry the desired pCa-γb2A-LIC-derived VOX construct. However, to verify this, some six to eight randomly picked bacterial colonies need to be analyzed by colony PCR using GoTaq Green Master Mix (Promega) or similar standard DNA Polymerase master mix and primers 2235.F and 2615.R flanking the LIC site in the vector (*see* **Note 10**).
10. Select two or more independent transformants verified by colony PCR and grow them in sterile 28-mL glass universal bottles containing 5-mL LB-Miller broth supplemented with kanamycin (50 μg/mL) overnight at 37 °C in a shaking incubator.
11. Use overnight cultures for extracting the plasmid DNA using a Miniprep Kit.
12. Estimate the concentration of plasmid DNA using a spectrophotometer.
13. Verify the effector sequences cloned into pCa-γb2A-LIC by Sanger sequencing using primers 2235.F and 2615.R.

3.2 FoMV VOX Constructs Development

1. Identify the coding sequence (CDS) of the pathogen effector gene for cloning into the PV101 vector using a standard restriction digest and ligation method.
2. Amplify the entire CDS of the selected gene by PCR using Phusion DNA Polymerase (Fisher Scientific UK Ltd) or another low error rate DNA Polymerase and sequence-specific forward and reverse primers carrying the short 5′-extensions incorporating *Not*I (or *Cla*I) and *Xba*I cloning site, respectively, and two to four random bases upstream of restriction sites to improve efficiency of restriction digestion from the appropriate template (*see* **Notes 11** and **12**).
3. Digest both the PCR fragment and PV101, with *Not*I (or *Cla*I) and *Xba*I (using 5 U of each enzyme per 1-μg plasmid DNA or 100–200-ng PCR product) for 2 h or longer at 37 °C in the appropriate restriction enzyme reaction buffer that permits 100% activity of each of the two restriction enzymes (*see* **Note 13**).
4. Purify the restriction enzyme digested PCR product and the linearized vector following an agarose gel electrophoresis using a Gel Extraction Kit.
5. Ligate the restriction enzyme digested and gel purified PV101 (15–30 ng) and the PCR product (75–150 ng) using T4 DNA Ligase following the manufacturer's instructions.

9. Use 2–3 μL of the ligation reaction mixture for transformation of 50 μl of competent (subcloning efficiency) *E. coli* cells, e.g., NEB 5-alpha (New England Biolabs), following the manufacturer instructions. Spread 50–100-μL aliquots of the transformed *E. coli* cell suspension onto LB-Miller agar plates supplemented with kanamycin (50 μg/mL) and incubate at 37 °C overnight.
10. To identify the recombinants, analyze 8–12 bacterial colonies by colony PCR using a standard DNA Polymerase master mix and primers cons5073F and cons5558R flanking a multiple cloning site in the vector PV101 (*see* **Note 14**).
11. Select two or more independent transformants verified by colony PCR and grow them in sterile 28-mL glass universal bottles containing 5-mL LB-Miller broth supplemented with kanamycin (50 μg/mL) overnight at 37 °C in a shaking incubator.
14. Use overnight cultures for extracting the plasmid DNA using a Plasmid Miniprep Kit.
15. Estimate concentration of plasmid DNA using a spectrophotometer.
16. Verify the effector sequences cloned into the PV101 vector by Sanger sequencing using primers cons5073F and cons5558R.

3.3 Preparing Agrobacterium tumefaciens Strains for Agroinfiltration

1. Use 10 ng of plasmid DNA for transformation of 50-μL electrocompetent *A. tumefaciens* cells using an electroporator and 0.1-cm gap electroporation cuvettes pre-chilled on ice following the manufacturer's instructions. Immediately post transformation transfer bacterial cells to a sterile 14-mL tube and allow bacteria to grow for 1.5 h at 28 °C in a shaking incubator.
2. Spread the transformation mixture onto LB-Lennox agar plates supplemented with the antibiotics kanamycin (50 μg/mL) plus gentamicin (25 μg/mL) for *A. tumefaciens* strain GV3101 (pMP90) transformed with the binary BSMV plasmids or pBIN61-p19, or kanamycin (50 μg/mL), gentamicin (25 μg/mL), and tetracycline (5 μg/mL) for *A. tumefaciens* strain GV3101 pSa-Rep transformed with PV101-derived VOX constructs and incubate at 28 °C for 3–4 days.
3. Pick two to three bacterial colonies using a sterile loop from the freshly cultured plates and inoculate each into universal bottles containing 5-mL LB-Lennox broth supplemented with the appropriate antibiotics. Allow agrobacteria to grow overnight at 28 °C in a shaking incubator (*see* **Note 15**).
4. Take a small aliquot of the overnight bacterial culture to prepare glycerol stocks (final concentration of glycerol c. 25%) for long-term storage at −80 °C.

5. Collect the rest of the bacterial cells by spinning the universal bottles at 2500 *g* for 15–20 min at 4–12 °C in a benchtop centrifuge.
6. Carefully remove the supernatant and resuspend the bacterial cells in 5 mL of the infiltration buffer supplemented with 150-μM acetosyringone (*see* **Note 16**).
7. Estimate the bacterial density using spectrophotometer at O.D.$_{600\ nm}$ and 1-mL disposable cuvettes. Dilute the bacterial suspensions containing pCaBS-α, pCaBS-β, and a pCa-γb2A-LIC derivatives to the final OD_{600} of 1.5, and those containing PV101 derivatives or pBIN61-p19 to the final OD_{600} of 0.6 and 0.3, respectively.
8. For BSMV VOX, combine equal volumes (1:1:1 ratio) of the three *A. tumefaciens* suspensions containing pCaBS-α, pCaBS-β, and a pCa-γbLIC derivative in a sterile universal bottle and mix gently by turning the tube upside down a few times. For FoMV VOX, combine equal volumes (1:1 ratio) of the two bacterial cell suspensions containing a PV101 derivative and pBIN61-p19.
9. Incubate the bacterial suspensions at room temperature in the dark or low light for 3 h or longer before agroinfiltration (*see* **Note 17**). No shaking is required.

3.4 Agroinfiltration of *N. benthamiana* Leaves

1. Healthy, 3–5-week-old *N. benthamiana* plants are used for agroinfiltration. Generously water plants approximately 30–60 min before agroinfiltration to stimulate stomata to open.
2. Use a 1-mL needleless syringe loaded with the *A. tumefaciens* cells suspension to infiltrate the fully expanded leaves (two to four per plant) by placing the syringe tip firmly against the small incision made by gentle scratching with a pipette tip on the abaxial side of a leaf. Inject the bacterial cells slowly but firmly aiming to infiltrate as large area of the leaf as possible (*see* **Notes 18** and **19**).
3. After agroinfiltration return the plants to a growth chamber maintained at 23–25 °C and 16-h photoperiod (light intensity of ~130 μmol m^{-2} s^{-1}).
4. Collect agroinfiltrated leaves at 5–7 days post agroinfiltration by cutting them off the plant with sharp scissors. Then, quickly weigh the snapped leaves (remembering to record their fresh weight) before wrapping them up in an aluminum foil and dropping inside a Dewar filled up with liquid nitrogen. The sampled, virus containing *N. benthamiana* leaves can be stored in a −80 °C freezer before used for preparing virus inoculum for inoculation of monocot plants (*see* **Note 20**).

3.5 Wheat Inoculation

1. Pre-germinate seeds of wheat (or other monocot species) on a filter paper or heat sterilized sand soaked with water for a few days. To maximize growth uniformity among test plants, pot only the germinated seeds with one seed per pot. Grow plants until they produce two to three leaves prior to virus inoculation (*see* **Note 3**).
2. Prepare virus inoculum by grinding agroinfiltrated *N. benthamiana* leaves in 10-mM potassium phosphate buffer, pH 7.0, using pre-chilled pestle and mortar. For each 1 g of leaf tissue, use 1.5 mL of buffer.
3. Squeeze the viruliferous sap through two layers of Miracloth, add 1–2% (w/v) of Celite 545 AW to serve as an abrasive, and mix well before rub-inoculation.
4. Dip your gloved fingers in the sap, then firmly squeeze a leaf between your thumb and an index finger close to the base of the leaf, and apply the inoculum by gently scrubbing the leaf in an upward stroke. You may hear a slight squeaking sound as the wax is being scraped off the leaf. Repeat this step two to three times, starting each stroke from the base of the leaf until the leaf is fully wetted in the inoculum (*see* **Note 21**).
5. Inoculate at least ten individual plants per expression construct (*see* **Note 22**) in each experiment and aim to repeat the experiment at least one additional time. Allow the inoculated plants to stand at room temperature for 5–10 min to adsorb the virus and then gently wash the excess inoculum and residual celite off the inoculated leaves using tap water in a spray bottle (*see* **Note 23**).
6. Place the inoculated plants inside a large plastic box or cover them with a large plastic bag shaping the bag in the form of a dome above the plants. Keep the plants overnight in darkness or low light to allow them to recover from injury. Next day unbox/unbag the plants and grow under the environmental conditions specified in Subheading 2.1 (*see* **Note 24**).
7. First virus symptoms (i.e., pale green to yellow streaks dotted along the leaf blade) should become visible 7–10 days post inoculation (dpi) in the upper uninoculated leaves of fully susceptible plant genotypes (*see* **Notes 25** and **26**).
8. Expression of the full-length effector in plant tissues can be verified, albeit indirectly, using RT-PCR with primers 2235.F and 2615.R (for BSMV VOX constructs) or cons5073F and cons5558R (for FoMV VOX constructs).
9. If available, specific antibodies could be used in a western blot for direct detection and quantification of the effector protein expression. Effectors expressed as either C- or N-terminal fusions with short peptide tags or fluorescent reporter proteins

could be detected *in planta* using commercially available anti-tag antibodies (*see* **Note 27**).

4 Notes

1. BSMV contains a tripartite single-strand, positive-sense RNA genome comprising RNA α, RNA β, and RNA γ. Each of the genomic RNAs encodes essential proteins, and therefore all three RNA genomes are required for infection.
2. 100-mM stock solution could be prepared by dissolving 0.196-g acetosyringone in 10-mL dimethyl sulfoxide. We recommend storing the stock solution in small aliquots in a −20 °C freezer and adding it to the infiltration buffer immediately before use.
3. Young plants are more susceptible to the virus infection than older or adult plants. If inoculation of adult plants is desirable, we recommend inoculating two youngest fully expanded leaves. But expect some plants to escape the infection. The BSMV VOX vectors described here are infectious on a wide range of wheat genotypes, as well as on a range of other monocot species, but are not infectious on maize. The FoMV VOX vector PV101, on the other hand, could infect a number of maize genotypes, and several other monocot species including wheat. However, some wheat genotypes, e.g., cv. Bobwhite, appear to be resistant to PV101 [21]. This needs to be taken into consideration when designing BSMV/FoMV VOX experiments.
4. The main advantage of LIC is that it requires no specific restriction sites to be present for cloning of DNA inserts. This method utilizes the 3′→5′ exonuclease activity of T4 DNA Polymerase in the absence of one or more dNTPs to generate recessed 5′-ends complementary between the vector and the insert. Adding specifically designed 5′-extensions to the PCR primers allows to obtain overhangs which can be efficiently annealed with the complementary overhangs in the vector and joined together after transformation into *E. coli* by the bacterial DNA recombination and repair machinery.
5. The pCa-γb2A-LIC vector permits expression of heterologous proteins as C-terminal fusions to the BSMV γb protein via the *Foot-and-mouth disease virus* (FMDV) self-cleaving 2A peptide bridge [20]. Therefore, it is essential that the CDS of the effector gene is in the same reading frame with the open reading frame coding for BSMV γb-FMDV 2A protein fusion, and the 5′-extension to the forward primer is specifically designed to achieve this.

6. CDS of the effector gene could be amplified by PCR from the cDNA clone (if available), first-strand cDNA derived from the total pathogen RNA or the pathogen-infected plant tissue, or pathogen DNA (for intronless effector genes). For the extracellular located effectors, we recommend using an entire CDS, or replacing the native secretion signal peptide (SP) encoding sequence with that of the well-characterized extracellular plant proteins such as PATHOGENESIS-RELATED-1 (PR-1). For the intracellular effectors, DNA sequences corresponding to the mature peptides (i.e., those lacking the SP sequences) need to be selected for cloning into the expression vectors. In this latter case, replace the sequence corresponding to SP with an ATG start codon for translation.
7. Effectors could also be expressed using BSMV and FoMV VOX vectors as either C- or N-terminal fusions with short peptide tags, e.g., cMyc, HA, or FLAG. As the FoMV VOX vector PV101 could accommodate large inserts, of at least up to 1.8 kb, this vector system could be used to express effectors as either C- or N-terminal fusions with fluorescent reporter proteins, e.g., GFP/YFP, mRFP, and mCherry. The tagged effector sequences could be produced using gene synthesis with the help of commercial service providers or in house, using a variety of methods, e.g., Gibson assembly [40].
8. The remaining *Apa*I-digested and T4 DNA Polymerase-treated BSMV VOX vector pCa-γb2A-LIC can be stored at −20 °C for future use.
9. To reduce the volume, transfer the post-transformation *E. coli* culture (1 mL) into a sterile 1.5-mL tube and spin for 1 min at 4000 rpm in a microfuge. Discard the supernatant and resuspend the pelleted cells in 50–100-μL SOC media.
10. Primers 2235.F and 2615.R are located 290-bp upstream and 97-bp downstream of the LIC site in the BSMV VOX vector pCa-γb2A-LIC.
11. A multiple cloning site (MCS) in the FoMV VOX vector PV101 contains sites for the following restriction enzymes starting from the 5′-end of MCS in the following order: *Not*I, *Cla*I, *Asc*I, *Hpa*I, and *Xba*I (Fig. 1). We prefer directionally cloning the effector sequences between *Not*I and *Xba*I, or between *Cla*I and *Xba*I. However, if any of these restriction enzyme sites are present in the effector sequences (or the corresponding tags), a different combination or even a single digestion site present in MCS could be selected for cloning into PV101.
12. As an easier alternative to PCR, the desired effector coding gene sequence with or without N- or C-terminal tags and the appropriate flanking restriction enzyme digestion sites could be

produced using gene synthesis with the help of commercial service providers or in house, using a variety of methods, e.g., Gibson assembly [40].

13. As an example, the following enzymes from New England Biolabs Ltd. are fully active (100%) in the CutSmart buffer supplied with the enzymes: *Not*I-HF, *Cla*I, and *Xba*I.
14. Primers cons5073F and cons5558R are located 286-bp upstream and 259-bp downstream of the MCS site in the FoMV VOX vector PV101.
15. Generally, 5 mL of *A. tumefaciens* culture is sufficient to infiltrate all fully expanded leaves of seven to ten young (3–5 weeks old) *N. benthamiana* plants. The culture volume can be scaled up based on the number of plants to be used for agroinfiltration.
16. Acetosyringone stimulates expression of the *A. tumefaciens* virulence genes, which improves transformation efficiency of plant cells.
17. A minimum of 3-h incubation of bacterial cell suspensions in the presence of acetosyringone ensures sufficient activation of the *A. tumefaciens* virulence genes.
18. With practice, an entire small leaf could be infiltrated using a single injection, whereas for larger leaves, several injections initiated in different parts of the leaf may be required.
19. Care should be taken to avoid cross-contamination when performing experiments using multiple expression constructs. Change gloves and syringes before agroinfiltration with a new expression construct.
20. It usually takes longer than 5–7 days for BSMV and FoMV to spread from the agroinfiltrated leaves to the young upper uninoculated leaves in *N. benthamiana* where the virus symptoms could be observed. We therefore recommend keeping the plants for another few days after harvesting the agroinfiltrated leaves to allow for the appearance of virus symptoms, which serves as a useful indicator of the successful agroinfiltration and accumulation of the virus capable of spreading systemically throughout the plant tissues. However, it is important to note that FoMV induces much less prominent symptoms than BSMV, which could be difficult to spot (Fig. 2).
21. The actual rubbing technique is important. Be gentle, but not too gentle when rubbing. This is a critical step and it takes practicing to apply the correct pressure to the leaf as both, too gentle or too hard rubbing, will result in no virus infection at all or low frequency of infection. Rubbing too gently may not allow the virus to get inside the plant cells in the quantity sufficient for initiating infection. On the other hand, rubbing

Fig. 2 Mosaic symptoms induced by BSMV and FoMV in the upper uninoculated ("systemic") leaves of agroinfiltrated *Nicotiana benthamiana* plants. (Photos are taken at 12 days post agroinfiltration)

the leaf surface too hard could severely damage or even kill the leaf, again resulting in no or substantially decreased efficiency of virus infection.

22. It is advisable to include control constructs in every overexpression experiment. For BSMV VOX, these could be an empty virus vector and/or BSMV::mcs4d containing a 275-bp noncoding sequence corresponding to the MCS of the vector pBluescript II KS [41]. For FoMV VOX, we suggest including PV101::GFP, which expresses GFP. This construct could also be useful when assessing previously untested monocot genotypes for their sensitivity to FoMV.
23. When multiple expression constructs are being tested, care should be taken to carefully remove all used material; disinfect the bench, mortars, and pestles; and change the gloves each time before proceeding with a new construct inoculation.
24. Correctly rub-inoculated leaves of test plants should display some (minimal) signs of injury the next day after inoculation.
25. BSMV induces striking symptoms by approximately 10–14 days post inoculation, which could be described as moderate to severe chlorotic and sometimes even necrotic stripes/mosaic, depending on the host genotype and age of the plant at time of inoculation. However, later in the infection, the symptoms become much milder. This infection profile is thought to be associated with the presence and expression of a strong silencing suppressor of RNAi (the γb protein) and the ability of BSMV to invade plant meristems and be transmitted vertically (through seed) [42]. For these reasons, BSMV VOX vectors seem to be better suited for the identification or validation of Avr proteins which could be expected to induce severe cell death in the inoculated leaves and or trailing cell death in the newly emerging leaves specifically in genotypes carrying the

Fig. 3 Plant virus-mediated overexpression (VOX) of pathogen effector proteins in barley and wheat. (**a**) BSMV-mediated expression of the avirulence protein Nip1 from *Rhynchosporium commune* induces trailing cell death on local and systemic leaves of barley cultivar Atlas 46 containing the corresponding disease resistance gene *Rrs1*. A single plant at 14 dpi with BSMV::Nip1 is shown on the left, and leaves detached from two other infected plants are shown. (**b**) FoMV-mediated expression of the necrotrophic effector ToxA from *Parastagonospora nodorum* induces necrotic streaks on local leaves of wheat cultivar Halberd containing the corresponding sensitivity gene *Tsn1*, but not in cultivar Riband which is insensitive to ToxA. The same leaves were photographed at 7 dpi and 18 dpi with PV101::ToxA. Note that the virus-induced mosaic symptoms in cultivar Riband, which is fully susceptible to FoMV, become more prominent with time

corresponding disease resistance genes (Fig. 3a). This type of symptom was previously observed in barley possessing the disease resistance gene *Rrs1*, when the corresponding Avr protein Nip1 of the fungus *Rhynchosporium commune* was expressed from BSMV [19]. More recently, we also found that BSMV VOX vectors expressing Avr proteins of the stem rust fungus *Puccinia graminis* f. sp. *tritici* were unable to infect wheat and/or triticale genotypes possessing the corresponding disease resistance genes [24, 43].

26. FoMV infection profile is different to that of BSMV, and this virus induces no or very mild symptoms in many different monocot species. In wheat, the FoMV-induced symptoms gradually become more prominent with time on the leaves previously showing only slight signs of infection (Fig. 3b). This indicates that FoMV possibly encodes a weak silencing suppressor of RNAi, and hence the PV101 vector-mediated effector protein expression could be fairly low at early time points post inoculation. This makes FoMV VOX well suited for the functional analyses of a wide variety of effectors beside those that induce host genotype-dependent cell death (Fig. 3b).

27. Short peptide or longer fluorescent protein tags fused to the N- or C-terminus of the effector could be used for a variety of applications including an affinity purification of the effector for downstream analyses, e.g., for testing its suspected enzymatic activity. Tags could also enable identification of plant proteins targeted by the effectors, for example, by using co-immunoprecipitation followed by gas chromatography–mass spectrometry. In addition, fluorescent protein tags could also be used for analyzing subcellular localization of the effectors.

Acknowledgments

This work was supported by the Institute Strategic Program Grant "Designing Future Wheat" (BB/P016855/1) from the Biotechnology and Biological Sciences Research Council (BBSRC) of the UK, and by the Rothamsted-Syngenta Alliance (RoSy).

References

1. Asai S, Shirasu K (2015) Plant cells under siege: plant immune system versus pathogen effectors. Curr Opin Plant Biol 28:1–8
2. Kaloshian I, Walling LL (2016) Hemipteran and dipteran pests: effectors and plant host immune regulators. J Integr Plant Biol 58: 350–361
3. Vieira P, Gleason C (2019) Plant-parasitic nematode effectors - insights into their diversity and new tools for their identification. Curr Opin Plant Biol 50:37–43
4. Hammond-Kosack KE, Jones JD (1997) Plant disease resistance genes. Annu Rev Plant Physiol Plant Mol Biol 48:575–607
5. Aylward J, Steenkamp ET, Dreyer LL, Roets F, Wingfield BD, Wingfield MJ (2017) A plant pathology perspective of fungal genome sequencing. IMA Fungus 8:1–15
6. Stajich JE (2017) Fungal genomes and insights into the evolution of the kingdom. Microbiol Spectr 5(4). https://doi.org/10.1128/microbiolspec.FUNK-0055-2016
7. Slatko BE, Gardner AF, Ausubel FM (2018) Overview of next-generation sequencing technologies. Curr Protoc Mol Biol 122:e59
8. Li F, Zhao X, Li M, He K, Huang C, Zhou Y, Li Z, Walters JR (2019) Insect genomes: progress and challenges. Insect Mol Biol 28: 739–758
9. Xu J, Wang N (2019) Where are we going with genomics in plant pathogenic bacteria? Genomics 111:729–736
10. Sperschneider J, Dodds PN, Gardiner DM, Singh KB, Taylor JM (2018) Improved prediction of fungal effector proteins from secretomes with EffectorP 2.0. Mol Plant Pathol 19:2094–2110
11. Sperschneider J, Dodds PN, Singh KB, Taylor JM (2018) ApoplastP: prediction of effectors and plant proteins in the apoplast using machine learning. New Phytol 217: 1764–1778
12. Lo Presti L, Lanver D, Schweizer G, Tanaka S, Liang L, Tollot M, Zuccaro A, Reissmann S, Kahmann R (2015) Fungal effectors and plant susceptibility. Annu Rev Plant Biol 66: 513–545
13. Selin C, de Kievit TR, Belmonte MF, Fernando WG (2016) Elucidating the role of effectors in plant-fungal interactions: progress and challenges. Front Microbiol 7:600
14. McDonald MC, Solomon PS (2018) Just the surface: advances in the discovery and characterization of necrotrophic wheat effectors. Curr Opin Microbiol 46:14–18
15. Bally J, Jung H, Mortimer C, Naim F, Philips JG, Hellens R, Bombarely A, Goodin MM, Waterhouse PM (2018) The rise and rise of *Nicotiana benthamiana*: a plant for all reasons. Annu Rev Phytopathol 56:405–426

16. De Cleene M, De Ley J (1976) The host range of crown gall. Bot Rev 42:389–466
17. Brisson N, Paszkowski J, Penswick JR, Gronenborn B, Potrykus I, Hohn T (1984) Expression of a bacterial gene in plants by using a viral vector. Nature 310:511–514
18. Hammond-Kosack KE, Staskawicz BJ, Jones JD, Baulcombe DC (1995) Functional expression of a fungal avirulence gene from a modified *Potato virus X* genome. Mol Plant-Microbe Interact 8:181–185
19. Lee W-S, Hammond-Kosack KE, Kanyuka K (2012) *Barley stripe mosaic virus*-mediated tools for investigating gene function in cereal plants and their pathogens: virus-induced gene silencing, host-mediated gene silencing, and virus-mediated overexpression of heterologous protein. Plant Physiol 160:582–590
20. Franco-Orozco B, Berepiki A, Ruiz O, Gamble L, Griffe LL, Wang S, Birch PRJ, Kanyuka K, Avrova A (2017) A new proteinaceous pathogen-associated molecular pattern (PAMP) identified in Ascomycete fungi induces cell death in Solanaceae. New Phytol 214:1657–1672
21. Bouton C, King RC, Chen H, Azhakanandam K, Bieri S, Hammond-Kosack KE, Kanyuka K (2018) *Foxtail mosaic virus*: a viral vector for protein expression in cereals. Plant Physiol 177:1352–1367
22. Groszyk J, Kowalczyk M, Yanushevska Y, Stochmal A, Rakoczy-Tro-janowska M, Orczyk W (2017) Identification and VIGS-based characterization of *Bx1* ortholog in rye (*Secale cereale* L.). PLoS One 12:1–21
23. Mellado-Sánchez M, McDiarmid F, Cardoso V, Kanyuka K, MacGregor DR (2020) Virus-mediated transient expression techniques enable gene function studies in black-grass. Plant Physiol 183:455–459
24. Upadhyaya NM, Mago R, Panwar V, Hewitt T, Luo M, Chen J, Sperschneider J, Nguyen-Phuc H, Wang A, Ortiz D, Hac L, Bhatt D, Li F, Zhang J, Ayliffe M, Figueroa M, Kanyuka K, Ellis JG, Dodds PN (2021) Genomics accelerated isolation of a new stem rust avirulence gene – wheat resistance gene pair. Nat Plants 7:1220–1228
25. Liu N, Xie K, Jia Q, Zhao J, Chen T, Li H, Wei X, Diao X, Hong Y, Liu Y (2016) *Foxtail mosaic virus*-induced gene silencing in monocot plants. Plant Physiol 171:1801–1807
26. Mei Y, Beernink BM, Ellison EE, Konečná E, Neelakandan AK, Voytas DF, Whitham SA (2019) Protein expression and gene editing in monocots using *Foxtail mosaic virus* vectors. Plant Direct 3:e00181
27. Huang YW, Chang CY, Hsu YH (2020) Virus-induced gene silencing in Poaceae using a *Foxtail mosaic virus* vector. Methods Mol Biol 2172:15–25
28. Yuan C, Li H, Qin C, Zhang X, Chen Q, Zhang P, Xu X, He M, Zhang X, Tör M, Xue D, Wang H, Jackson S, He Y, Liu Y, Shi N, Hong Y (2020) *Foxtail mosaic virus*-induced flowering assays in monocot crops. J Exp Bot 71:3012–3023
29. Torti S, Schlesier R, Thümmler A, Bartels D, Römer P, Koch B, Werner S, Panwar V, Kanyuka K, Wirén NV, Jones JDG, Hause G, Giritch A, Gleba Y (2021) Transient reprogramming of crop plants for agronomic performance. Nat Plants 7:159–171
30. Montenegro Alonso AP, Ali S, Song X, Linning R, Bakkeren G (2020) UhAVR1, an HR-triggering avirulence effector of *Ustilago hordei*, is secreted via the ER-Golgi pathway, localizes to the cytosol of barley cells during in planta-expression, and contributes to virulence early in infection. J Fungi (Basel) 6:178
31. Darino M, Chia KS, Marques J, Aleksza D, Soto-Jiménez LM, Saado I, Uhse S, Borg M, Betz R, Bindics J, Zienkiewicz K, Feussner I, Petit-Houdenot Y, Djamei A (2021) *Ustilago maydis* effector Jsi1 interacts with Topless corepressor, hijacking plant jasmonate/ethylene signaling. New Phytol 229:3393–3407
32. Navarrete F, Grujic N, Stirnberg A, Saado I, Aleksza D, Gallei M, Adi H, Alcântara A, Khan M, Bindics J, Trujillo M, Djamei A (2021) The Pleiades are a cluster of fungal effectors that inhibit host defenses. PLoS Pathog 17:e1009641
33. Edwards MC, Bragg J, Jackson AO (2006) Natural resistance mechanisms to viruses in barley. In: Loegenstein G, Carr JP (eds) Natural resistance mechanisms of plants to viruses. Springer, Berlin, pp 465–501
34. Ji Q, Yang B, Lee M, Chen Y, Lübberstedt T (2010) Mapping of quantitative trait loci/locus conferring resistance to *Foxtail mosaic virus* in maize using the intermated B73 × Mo17 population. Plant Breed 129:721–723
35. Cui Y, Lee MY, Huo N, Bragg J, Yan L, Yuan C, Li C, Holditch SJ, Xie J, Luo MC, Li D, Yu J, Martin J, Schackwitz W, Gu YQ, Vogel JP, Jackson AO, Liu Z, Garvin DF (2012) Fine mapping of the *Bsr1 Barley stripe mosaic virus* resistance gene in the model grass *Brachypodium distachyon*. PLoS One 7:e38333
36. Yuan C, Li C, Yan L, Jackson AO, Liu Z, Han C, Yu J, Li D (2011) A high throughput *Barley stripe mosaic virus* vector for virus induced gene silencing in monocots and dicots. PLoS One 6:e26468

37. Koncz C, Schell J (1986) The promoter of T_L-DNA gene 5 controls the tissue-specific expression of chimaeric genes carried by a novel type of Agrobacterium binary vector. Mol Gen Genet 204:383–396
38. Hellens RP, Edwards EA, Leyland NR, Bean S, Mullineaux PM (2000) pGreen: a versatile and flexible binary Ti vector for Agrobacterium-mediated plant transformation. Plant Mol Biol 42:819–832
39. Aslanidis C, de Jong PJ (1990) Ligation-independent cloning of PCR products (LIC-PCR). Nucleic Acids Res 18:6069–6074
40. Gibson DG, Young L, Chuang RY, Venter JC, Hutchison CA, Smith HO (2009) Enzymatic assembly of DNA molecules up to several hundred kilobases. Nat Methods 6:343–345
41. Hen-Avivi S, Savin O, Racovita RC, Lee WS, Adamski NM, Malitsky S, Almekias-Siegl E, Levy M, Vautrin S, Bergès H, Friedlander G, Kartvelishvily E, Ben-Zvi G, Alkan N, Uauy C, Kanyuka K, Jetter R, Distelfeld A, Aharoni A (2016) A metabolic gene cluster in the wheat *W1* and the barley *Cer-cqu* loci determines β-diketone biosynthesis and glaucousness. Plant Cell 28:1440–1460
42. Bradamante G, Mittelsten Scheid O, Incarbone M (2021) Under siege: virus control in plant meristems and progeny. Plant Cell 33: 2523–2537
43. Chen J, Upadhyaya NM, Ortiz D, Sperschneider J, Li F, Bouton C, Breen S, Dong C, Xu B, Zhang X, Mago R, Newell K, Xia X, Bernoux M, Taylor JM, Steffenson B, Jin Y, Zhang P, Kanyuka K, Figueroa M, Ellis JG, Park RF, Dodds PN (2017) Loss of *AvrSr50* by somatic exchange in stem rust leads to virulence for *Sr50* resistance in wheat. Science 358:1607–1610

Chapter 8

Time-Resolved Fluorescence Microscopy Screens on Host Protein Subversion During Bacterial Cell Invasion

Lisa Sanchez, Yuen-Yan Chang, Nora Mellouk, and Jost Enninga

Abstract

Intracellular bacterial pathogens have evolved a plethora of strategies to invade eukaryotic cells. By manipulating host signaling pathways, in particular vesicular trafficking, these microbes subvert host functions to promote their internalization and to establish an intracellular niche. During these events, host endomembrane compartments are dynamically reorganized. *Shigella flexneri*, the causative agent of bacillary dysentery, recruits components of the host recycling pathway and the exocyst of non-phagocytic enterocytes in the vicinity of its entry site to facilitate its access to the host cytosol. These factors are either dynamically tethered to in situ formed macropinosomes or to the bacteria-containing vacuole itself. The underlying interactions cannot readily be monitored as individual bacterial infection events take place without synchronicity using cellular infection models. Therefore, time-resolved screens by fluorescence microscopy represent a powerful tool for the study of host subversion. Such screens can be performed with libraries of fluorescently tagged host factors. Using the cytosolic pathogenic agent *Shigella flexneri* as a model, we provide detailed protocols for such medium-to-high throughput multidimensional imaging screening of the dynamic host-pathogen cross talk. Our workflow is designed to be easily adapted for the study of different host factor libraries and different pathogen models.

Key words Intracellular bacterial pathogens, Membrane trafficking, High-content screening, *Shigella flexneri*, Bacterial infection, BAR domain-containing proteins, Rab GTPases

1 Introduction

Through a myriad of molecular mechanisms, bacterial pathogens invade non-phagocytic cells by injecting effector proteins using specialized secretory systems. This strategy thus enables them to survive and thrive within their host by profiting from and reprogramming host pathways. Despite bacterial invasion occurring by a variety of mechanisms, similar themes are found with regard to the subverted host components and machineries. First, following host cell contact, bacteria swiftly trigger their internalization within a

Yuen-Yan Chang and Nora Mellouk contributed equally with all other contributors.

Thomas A. Kufer and Maria Kaparakis-Liaskos (eds.), *Effector-Triggered Immunity: Methods and Protocols*, Methods in Molecular Biology, vol. 2523, https://doi.org/10.1007/978-1-0716-2449-4_8,

vacuole. Then, the invading pathogens reach their preferred intracellular niche by either (i) triggering vacuole rupture and egress into the cytosol (*Shigella flexneri*, *Listeria monocytogenes*) or (ii) by reprogramming the bacterial vacuole into a specialized bacterial compartment (*Salmonella typhimurium*, *Yersinia pseudotuberculosis*) [1]. Numerous host and pathogen factors have been identified to be involved during this cross talk. Nevertheless, the dynamic implications of the host molecular pathways controlling the subversion of host trafficking pathways remain often unclear. Identifying at which specific step of the infection host factors are manipulated by bacterial pathogens poses a challenge. This is due to heterogenicity in the behavior of the individual bacteria and the targeted host cells, and the nonsynchronous nature of the bacterial invasion process. In cellular infection models for instance, *Salmonella* entry varies greatly as individual bacteria exert near-surface swimming and "test" several cells before eventually invading a target cell [2]. In the case of *Listeria* entry, homogenous invasion behavior cannot be readily assessed experimentally as bacterial adhesion and its subsequent internalization reportedly depend on the individual host cell heterogeneity [3]. Moreover, the rapidity of *Listeria* cell-to-cell spread varies hugely [4]. Other factors such as bacteria cooperation (e.g., *Salmonella*) and the specific bacterial load of a given experimental condition also cause variability in the timing of infection events and in the process of internalization, thus hampering a straightforward interpretation of the obtained data [2, 5, 6]. Adding to this complexity, some bacteria possess multiple intracellular lifestyles. Subsequent to its entry, *Salmonella* can either (i) break free from its vacuole and replicate at high rates inside the host cytosol, or the bacteria (ii) mature the *Salmonella*-containing vacuole, or they (iii) remain dormant within their vacuoles [7–9]. The complexity of synchronizing bacterial infection therefore highlights the necessity of developing assays with dynamic readouts at the single cell level, and adapting them for screening for the identification of novel roles of the investigated factors.

Enabling the visualization of the highly dynamic events occurring during bacterial invasion, time-resolved imaging is an appropriate technique. Coupled with fluorescence, time-lapse microscopy allows for targeting specific "candidates" via fluorescence labeling including both cellular and bacterial compartments. Genetically encoded fluorescent proteins, such as the green fluorescent protein (GFP), are widely used enabling the visualization of numerous factors with minimal functional interference. This can be applied for the tagging of specific infection markers targeting bacterial compartments. Consequently, fluorescence time-lapse microscopy is a robust tool to decipher the molecular pathways exploited by pathogens. Current automated microscope setups allow for time-lapse acquisitions in the range of seconds (sometimes even within milliseconds) and minutes, which suits very well the internalization dynamics of different bacterial pathogens. Equipping these systems with motorized stages piloted through a dedicated software and robust systems for maintaining the focus during automated acquisitions enables the long-term monitoring of multiple experimental conditions in parallel. These advances have been further boosted in recent years through microscope innovations, especially through the development of novel detectors that include electron-multiplying charge-coupled device (EMCCD) or scientific complementary metal-oxide-semiconductor (sCMOS) cameras, as well as Yokogawa Confocal Spinning Disk setups. More recently, lattice light-sheet microscopes have provided the means for an ever-increasing gain in temporal resolution during image acquisition. Finally, miniaturization through patterning or microfluidics coupled with dynamic imaging is a powerful development to provide the means to increase the throughput of fluorescence microscopy.

The internalization of *Shigella* into non-phagocytic colonic epithelial cells is a rapid process triggered by the injection of effector proteins through a type 3 secretion system (T3SS) [10, 11]. Within a few seconds to minutes after coming into contact with the host cell, *Shigella* triggers plasma membrane and actin cytoskeleton rearrangements [1, 11]. This leads to (i) its uptake within a tight vacuole, termed the bacteria-containing vacuole (BCV), and simultaneously (ii) the formation of large vesicles, called infection-associated macropinosomes (IAMs) [1, 12]. Afterward, *Shigella* forms sometimes a dynamic actin cage around the BCV [11, 13]. Subsequently, within 7–10 min of entering, *Shigella* ruptures its vacuole and escapes into the host cytosol [14]. After establishing its niche, the bacterium forms an actin comet tail around 10–40 min post-entry and spreads to neighboring cells [14]. Investigating these invasion steps using cellular models, a large variability in the timing of the successive events is observed with cell-to-cell spreading of bacteria in some cells and early internalization events in other target cells occurring simultaneously. Therefore, approaches with high temporal resolution are key to studying the precise sequence of *Shigella* internalization.

The use of genetically encoded fluorescent protein libraries has allowed these recent years to clarify the contribution of host trafficking factors in *Shigella* invasion. In a microscopy-based functional assay, several Rab GTPases, key regulators of vesicular trafficking, were identified as potential host factors involved in *Shigella*-containing vacuole rupture [15]. Among them the Rab GTPase Rab11A, involved in endosomal recycling, was shown to be recruited to the *Shigella*-induced macropinosomes ([12, 15] and Fig. 1). Indeed, *Shigella* was found to sequester Rab11A to macropinosomes but not to the BCV through the modification of phosphatidylinositol phosphates by the bacterial effector IpgD, an inositol phosphatase [12, 15, 16]. Moreover, time-lapse

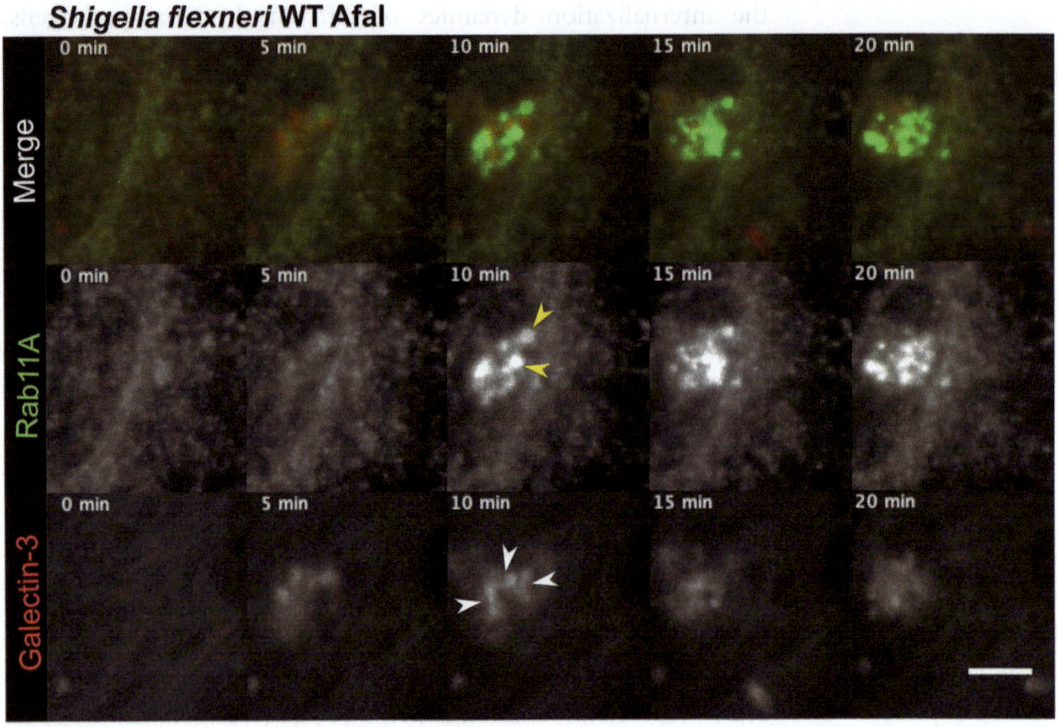

Fig. 1 Re-localization of Rab11A during *Shigella* M90T AfaI infection of HeLa cells as a useful tool to determine the dynamics of the bacterial invasion process. Recruitment of Rab11A to the *Shigella*-IAMs (yellow arrowheads) can be distinctively observed during BCV rupture (white arrowheads show the galectin-3-mOrange signal around the bacteria) and remains present there during the BCV disassembly process 20 min post-infection. Time-point 0 indicates the addition of the bacteria to the cells. Scale bar is 5 μm

microscopy analysis of host factors identified by a *Shigella*-IAMs proteomic assay showed that the vesicular tethering exocyst complex drives the clustering of IAMs around the BCV [17]. These events were found to be crucial for efficient BCV lysis and *Shigella* cytosolic escape. In addition to these trafficking factors, *Shigella* was found to subvert other endocytic factors. Indeed, shortly after internalization, by reprogramming actin signaling pathways, some *Shigella*-BCVs are surrounded by a thick dynamic structure distinct from any known cellular actin structure termed the actin cocoon [13, 15]. Although the function of this bacteria-driven actin structure remains to be elucidated, it is presumed to act as a gatekeeper of IAM-BCV contacts, and thus delays vacuolar rupture [13]. These novel strategies of manipulation of host pathways highlight the importance of using dynamic microscopy setups in conjunction with host factor localization subversion screens to further elucidate the molecular mechanism of *Shigella* invasion.

Fluorescence-based screens enable the observation of a protein re-localization from its physiological environment. These advantages have been essential at decoding the bacterial invasion mechanism through screening assays. Small inhibitor molecules [18] and bacterial effectors [12] have been crucial tools used for effectively decrypting molecular pathways subverted by bacteria though screens. However, these indirect assays often impact multiple signaling pathways which may pose a challenge in identifying specific molecular pathways. More powerful and direct screens such as loss-of-function-based assays relying on RNA interference or CRISPR-Cas9 knockout by specifically targeting host proteins may provide a clearer molecular interpretation [19]. Moreover, these types of screens can be adapted for genome-wide coverage of host factors [20]. These loss-of-function screens require however a robust read-out for interpretation of the results and often have been adapted for fixed experimental conditions. By contrast, dynamic screens have the advantage of allowing the direct observation of the subversion of cellular factors through the use of genetically encoded fluorescent probes or dyes. Coupling these assays with automated microscopy has allowed the development of dynamic screens for high throughput [21, 22]. Medium-sized screens covering a chosen subset of fluorescently labeled host factors represent a compromise for live microscope as their generation and handling of these subgenomic expression vector libraries of fluorescently tagged host factors can be done relatively easily. Here, we detail an optimized protocol for a medium-sized multichannel time-lapse screen using *Shigella* as a model, which is readily adaptable for numerous host-pathogen interactions or chemical insults.

2 Materials, Solutions, Cell Lines, and Bacterial Strains

2.1 Bacterial Strains

1. *Shigella flexneri* wild-type strain M90T [23] expressing the adhesin AfaI [24].
2. The noninvasive strains *Shigella flexneri* mutant M90T Δ*mxiD* AfaI [25] and the *Shigella flexneri* derivative lacking the virulence plasmid BS176 AfaI [26] as controls.
3. Chemically competent *E. coli* DH5 alpha [27].

2.2 Bacterial Culture

1. Trypticase casein soy broth medium (TCSB): 17-g/L casein (pancreatic digest), 3-g/L soya peptone (papaic digest), 5-g/L NaCl, 2.5-g/L K_2HPO_4, 2.5-g/L dextrose.
2. TCSB agar (to the TCSB composition: 15-g/L agar; pH 7.3 ± 0.2) supplemented with 0.01% w/v Congo Red and 50-μg/mL ampicillin.
3. LB medium: 10-g/L Bacto tryptone, 5-g/L Bacto yeast extract, 10-g/L NaCl, pH: 7.0 ± 0.2.
4. LB agar: add to the LB medium composition 15-g/L agar.

2.3 Cell Culture

1. Dulbecco's Modified Eagle Medium (DMEM).
2. Fetal Bovine Serum (FBS).
3. HeLa ATCC CCL-2.
4. 96-well black cell culture microplate, PS walls (polystyrene), borosilicate bottom, F-bottom (chimney well), μCLEAR, advanced TC (Greiner Bio-One).

2.4 Plasmid Library Maintenance, Purification, and Transfection

1. Plasmid stocks in sterile TE buffer (10-mM Tris, 1 mM EDTA, pH 8.0) or DNase-free water (plasmid library).
2. Glycerol 50% (v/v).
3. Plasmid Transfection-grade Mini kit for ultrapure plasmid DNA.
4. X-tremeGENE 9 DNA transfection reagent (Roche).
5. Opti-MEM Reduced Serum Medium (Gibco).
6. Water bath or heating block.

2.5 Infection and Microscopy

1. EM buffer: 120-mM NaCl, 7-mM KCl, 1.8-mM $CaCl_2$, 0.8-mM $MgCl_2$, 5-mM glucose, 25-mM HEPES, pH 7.3.
2. Nikon Ti-E inverted epifluorescence microscope enclosed in a heating chamber with an integrated Perfect Focus System (TI-ND6-PFS Perfect Focus Unit) and equipped with the Ti-S-E motorized stage. The NIS-Elements AR (Advanced Research) microscope imaging software version 4.60 piloted the microscope, and a Nikon Plan Fluor 40× air objective (NA 0.75, WD 0.66) or equivalent material.
3. Filter cubes: FITC (Ex, 465–495 nm; Em, 515–555 nm; Mirror, 505 to ∞ nm) and TRITC (Ex, 528–553 nm; Em, 590–650 nm; Mirror, 565 to ∞ nm).

2.6 Data Processing and Analysis

1. Fiji (Fiji Is Just ImageJ) 2.1.0/1.53c (download: https://imagej.net/Fiji/Downloads) and Icy (download: http://icy.bioimageanalysis.org/download).
2. Microsoft Excel.

3 Methods

In comparison to single-well live cell imaging, time-lapse screening requires important adjustments along with a well-designed experimental workflow. To address a specific question, key experimental parameters must be evaluated prior to proceeding to a screen. The weight of each parameter depends on the specific question to be investigated. Generally, the critical parameters to be assessed include the spatial resolution, the temporal resolution, and the number of events necessary for proper interpretation of the results. Medium-sized screens are achievable by multiplexing, and the size of the screen depends on the parameters analyzed (the compartments of interest, the swiftness of the event). Pre-screen experiments are therefore important to determine the optimal experimental parameters. The following protocol was designed for *Shigella* infections and optimized for our microscope and objectives. As the subversion of Rab GTPases, such as Rab11A by *Shigella* is already well characterized, using this host factor as setup control comes very handy for experimental fine-tuning (Fig. 1). During these experiments co-transfection with a vacuolar rupture marker galectin-3 fused to mOrange is useful as it allows correlations between the progress of bacterial invasion and the recruitment of the Rab11A protein [28]. We suggest to use this control to go through the entire protocol (Subheadings 3.1, 3.2, 3.3, 3.4, and 3.5) first as it provides a frame for the following screen. The described protocol can also be adapted to perform screens with other bacterial pathogens (*see* **Note 1**).

3.1 Plasmid Library Preparation and Handling

Our protocol has been successfully performed on a published library of expression vectors for fluorescently labeled proteins encoding the BAR domains involved in trafficking and membrane curvature [29]. This library contains 66 proteins, and it was maintained as follows:

1. Thaw 80 μL of chemically competent DH5 alpha *E. coli* on ice for 30 min from frozen −80 °C stocks for each plasmid of the library.
2. Add 0.5–3 μL (around 100 ng) of each plasmid DNA and incubate the bacteria on ice for 30 min.
3. Perform a 30-s heat shock at 42 °C in a water bath (or a heating block).
4. Recover the bacteria in 950 μL of LB medium for 1 h at 37 °C shaking at 220 rpm (*see* **Note 2**).
5. Plate 100 μL of the culture onto an LB agar plate (containing 30-μg/ml kanamycin) and incubate at 37 °C overnight.
6. Prepare an overnight culture from one of the obtained colonies (the plate can be kept for a month at 4 °C if prepared freshly). A bacterial stock can be prepared with 500 μL of an overnight culture and 500 μL of 50% glycerol (v/v), and it can be stored at −80 °C as a transformed bacterial stock.
7. The following day, take 4 mL of each bacterial overnight culture (from one of the obtained colonies) and purify the plasmid DNA of the 66 plasmids (as described in the manufacturer's instructions of the plasmid DNA purification kit). Elute DNA in 50 μL of DNase-free sterile water (*see* **Note 3**).

8. Measure DNA quantity using a nanodrop spectrophotometer.
9. The plasmid DNA can be kept at −20 °C for long-term storage. A working stock at a concentration of 200 ng/L can be stored at 4 °C and is stable for at least 1 h.

3.2 Cell Culture and Transfection

The localization of each transfected protein "candidate" of the plasmid library as well as the level of expression of the protein must be verified and optimized if necessary before performing a full screen. For this, one can follow the transfection steps below and move directly to the imaging described in Subheading 3.5 to inspect the localization of the transfected proteins (*see* **Note 4**).

1. Three days prior to the experiment, seed HeLa cells into a black 96-well plate (*see* **Notes 5** and **6**). HeLa cells are seeded at a density of 8000 cells per well in DMEM supplemented with 10% FBS and incubated overnight at 37 °C, 5% CO_2 (*see* **Note 7**). The volume of cells seeded is 100 μL to ensure an adequate amount of nutrients to the cells until the experimental day.
2. The next day, transfect the cells using X-tremeGENE 9 DNA reagent at a 3:1 transfection reagent to DNA ratio (following the manufacturer's instructions). Keeping all DNA working stocks at the same concentration (200 μg/μL) allows for down-scaling the transfection mix.
3. Incubate at room temperature and add 5 μL of transfection mix to each well followed by a 24- to 48-h incubation at 37 °C with 5% CO_2. The transfection can be optimized following the recommendations of the manufacturer (*see* **Note 8**).

3.3 Bacterial Inoculum Preparation

Shigella flexneri is a human pathogen. All experimental steps with *Shigella* require a dedicated laboratory environment following the specific guidelines in place for handling pathogens. In most countries, work with *Shigella* is restricted to at least BSL-2 lab space.

1. On the day of the transfection, streak the M90T AfaI strain on a TCSB agar plate supplemented with Congo Red and ampicillin (50 μg/mL) and incubate overnight at 37 °C. The next day, for *Shigella* three to five red colonies (indicative of virulent *Shigella* [30]) are inoculated in TCSB broth supplemented with ampicillin (50 μg/mL) and incubated overnight at 37 °C in a 220-rpm shaking incubator on a 45° inclined tray.
2. Prepare a 1:100 dilution of bacterial overnight culture in ampicillin-supplemented TCSB media (50 μg/mL). The subculture is incubated for 2 h at 37 °C shaking at 220 rpm.
3. During the growth of the subculture, EM buffer is warmed at 37 °C.

4. After 2 h of shaking, measure the optical density 600 (OD_{600}) of the bacterial subculture. An OD_{600} of $\approx$0.4–0.7 should be reached. 1 mL of subculture is spun down at 6000 g in a tabletop centrifuge for 1 min and the bacterial pellet is washed once with 1 mL of EM buffer (*see* **Note 9**).
5. Prepare 1 mL of bacterial inoculum at a multiplicity of infection (MOI) of 20 in EM buffer for an estimated number of 64,000 host cells in each well using the following equation: concentration of bacteria = 20(MOI) × [number of cells per flask]/[volume of infection per flask]. The optical density of the bacterial culture can be estimated with $1(OD_{600}) = 8 \times 10^8$ bacteria/ml to dilute it accordingly.
6. The bacterial inoculum is kept at 37 °C until the bacterial infection (*see* **Note 10**).

3.4 Bacterial Infection

The bacterial infection steps in the following section are optimized for a *Shigella* screen. Nonetheless, these steps can be adapted to other pathogens. Particular attention should be drawn to the infection pre-screen. The bacterial load (MOI and volume of bacterial inoculum) should be carefully pre-screened. The following steps can be used as generic tests; however, we recommend testing bacterial infections at different MOIs to ensure an adequate number of infection events for the pre-determined optimal number of seeded cells:

1. Wash the HeLa cells three times with 37 °C EM buffer and replace the medium with 50 μL of warm EM buffer (*see* **Note 11**).
2. Place the microplate onto a 37 °C pre-warmed Nikon Ti-E microscope stage (*see* **Note 12**).
3. Set the exposure parameters using the NIS-Elements AR software for an N-dimensional acquisition (ND acquisition window) using the Time-Lapse T and Multi-point XY acquisition dimensions (*see* Subheading 3.5 on microscope acquisition for more details). At this point, a test run is performed to ensure sufficient time to acquire all positions (*see* **Note 13**).
4. Manually choose different positions for the time-lapse screening setup (*see* **Note 14**).
5. Add 40 μL of the bacterial inoculum to each well (keep the plate immobile to maintain the XY for each position), cover the plate to prevent evaporation of the medium (to keep the focus set on the positions), and launch the run (*see* **Notes 15** and **16**). The volume of bacteria added to the well may be adjusted to optimize the start time of the infections (*see* **Notes 17** and **18**).

3.5 Image Acquisition

In the following workflow, we detail a possible setup for the image acquisition (Fig. 2). We stress that the acquisition parameters mentioned below are to be carefully predetermined according to the microscope that will be used for the screen. Concerning our screen, in addition to the BAR domain protein library which is eGFP-tagged, we also use galectin-3 [28], coupled to an mOrange fluorophore as used in our setup experiments with Rab11A (*see* Fig. 3). This marker allows for the monitoring of the precise onset of the BCV rupture and the fate of the membrane remnants during the time course of infection. Using two fluorescent reporters, attention needs to be paid to the use of proper filter cubes to avoid cross talk of signals between the channels (*see* **Note 19**).

1. To screen for the 66 BAR domain proteins, divide the library for transfection into a layout of seven plates, each containing ten BAR domain proteins (Fig. 2).
2. Perform cell preparation, transfection, and infection on each of these seven plates, one plate after the other following the descriptions of Subheadings 3.2, 3.3, and 3.4. Two adjustments of the described protocol need to be followed. A. the transfection mix needs to contain the plasmid of the BAR domain protein of interest plus the galactin-3 reporter plasmid in equal amount. B. the transfection mixes should be prepared in sterile plastic 96-well plates in the same way as the cell layout (*see* Fig. 2) using a multichannel pipette to allow simple administration of the mixes to the cells.
3. Close the plates containing the transfection mixes with sealing film before vortexing.
4. Add the transfection mix to the cells using a multichannel pipette and incubate for 48 h before infection (*see* Subheading 3.4).
5. Each of the seven plates is then imaged one after the other using FITC (eGFP-tagged library) and TRITC (galectin-3-mOrange) filter cubes (*see* Subheading 2.5, **item 2**).
6. Set the exposure time and the binning of the camera (*see* **Note 20**) to 200–300 ms and 1 × 1, respectively. It is important to test several exposure times prior to the screen as it will affect the number of positions acquired. Differences in protein expression among the screened factors need to be taken into account and should be considered when pooling the specific subsets for the successive microscope runs (*see* **Note 21**).
7. Set the focus on the cells with the PFS (Perfect Focus System) using a 40× air objective (although higher magnification long working distance air objectives may also be used). It is important to set the focus prior to adding the bacteria; otherwise, the focus may drift from the cells to the bacteria. Moreover, setting

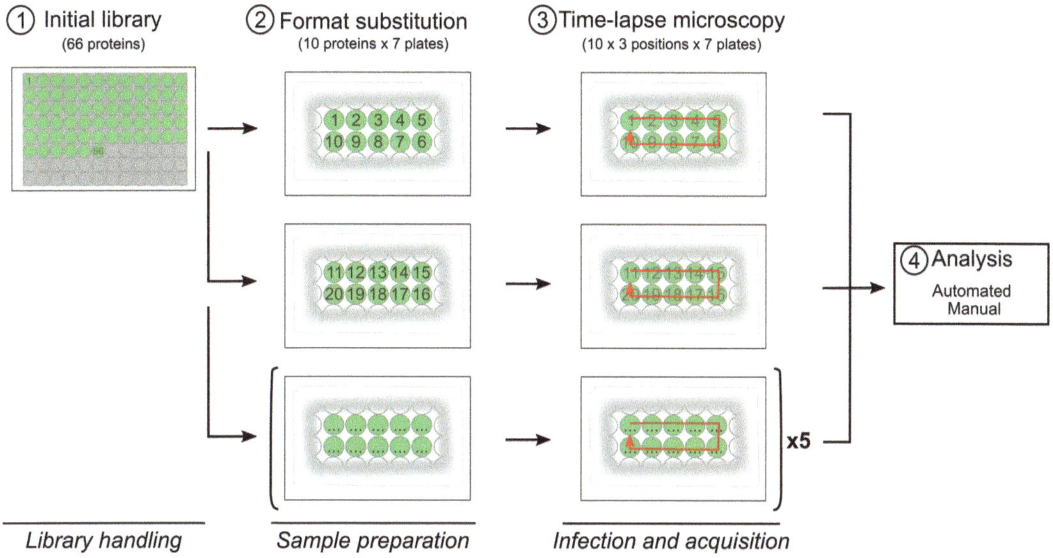

Fig. 2 Workflow example of a medium-sized time-resolved screen. The initial plasmid library (66 proteins) is prepared and stored (1). Afterward the samples that are used for imaging with the library are subdivided (the specific subdivision being determined by pre-screen experiments) (2). Following this, the cells are infected and the protein "candidates" are screened by time-lapse microscopy following a path from well to well in an optimal way (also respecting the addition of bacteria to the wells for synchronicity) (3). The obtained time-lapse series are analyzed either manually or using analytical image analysis tools (4)

the focus prior to choosing the positions allows the PFS position to be recorded when selecting the positions in the NIS-Elements software.

8. Set the time-lapse to 2 min for a 2-h-long acquisition. Depending on the event-of-interest, this parameter can be optimized, but a compromise has to be made between the number of events to observe and the number of positions to acquire.
9. XY positions are selected based on brightness and phenotype using the FITC channel (the signal-of-interest). Per plate, three positions per well are selected manually. Each position should contain multiple cells co-transfected with the protein of interest (*see* **Note 22**). The order of acquisition of each well is chosen in a way to allow that the motorized stage can travel along the fastest path to minimize delays between image acquisitions.

3.6 Data Analysis of Bacterial Infection Foci and Bacteria-Triggered Compartments

Through this example, we present tools to analyze time-lapse imaging data for a medium-size protein screen. Nonetheless, it should be noted that a negative control for protein recruitment (such as eGFP only) and a positive control (such as the described Rab11A, Fig. 1) are crucial for accurately identifying positive and negative hits. Non-infecting bacterial mutants, in the case of a *Shigella*: *ΔmxiD* or BS176 strains (or other), represent excellent negative controls for infection to properly analyze such a large data

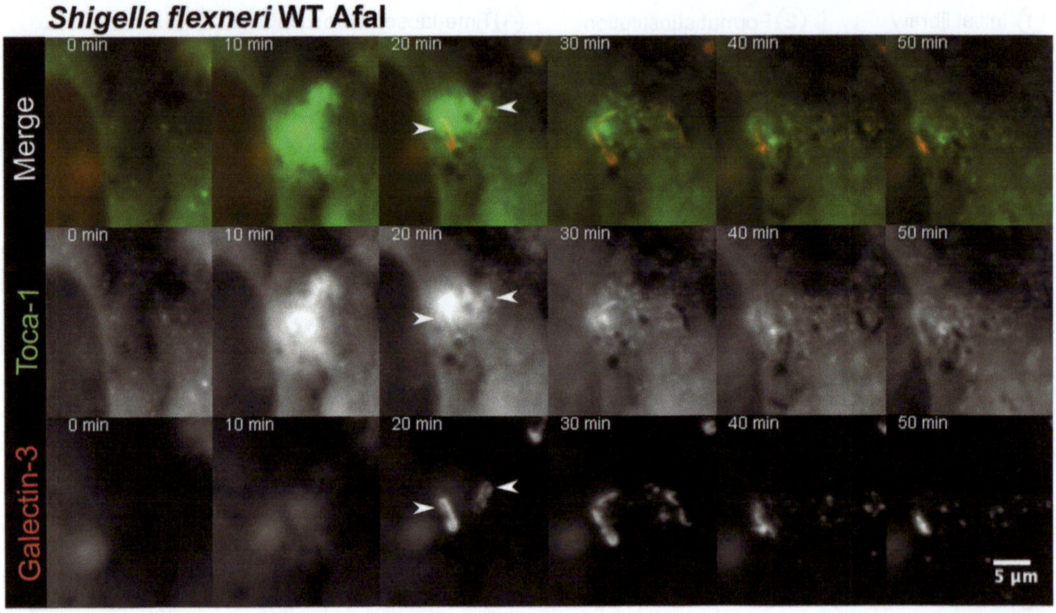

Fig. 3 BAR-domain protein candidate Toca-1 (F-BAR family) enrichment to a *Shigella* M90T AfaI infection site in real time (2-min time-lapse). Two bacteria are shown entering HeLa cells (indicated by the arrowheads). Both BCVs are shown ruptured at 20 min post-infection by the galectin-3-mOrange reporter. During bacterial entry, Toca-1-eGFP is recruited to the entry foci near the bacteria; the signal decreases in later time-points. The BCV membrane remnants can be seen being peeled-off after vacuolar rupture starting from the 30 min time-point as the galectin signal becomes loose not resembling the outline of a bacterium enclosed in a broken vacuole. Time-point 0 indicates the addition of the bacteria to the cells

library. The fluorescence images contain very rich information that can be analyzed either manually, or with the help of image analysis software, or combining both. As heterogenous phenotypes need to be anticipated in relation to the impact of bacterial invasion on the investigated genes present in the library, it is not possible to present one dedicated protocol for their analysis. Nevertheless, as some BAR domain proteins get recruited to macropinosomes in the vicinity of the entering bacteria, these phenotypes can be quantified as described in detail in [12].

1. The data obtained from the Nikon NIS-Elements software are generated in ND2 format (*see* **Note 23**). It can be processed using the open-source software Fiji and Icy.
2. The appearance of ghostlike galectin-3 reporter signals indicating the time-point of vacuolar damage is manually scored (Figs. 1 and 3).
3. The galectin-3 signals can also be handy to track bacterial entry with this reporter. This is achieved enhancing the contrast of the signal in either Fiji or Icy, where a wavelike perturbation of

the signal at the plasma membrane indicates the time-point of bacterial entry.

4. Protein recruitment of BAR domain proteins to the sites of bacterial entry is then scored with the help of the Icy Spot Detector tool [31], which automatically detects bright spots in fluorescence images (*see* **Note 24**).
5. For this purpose, a specific region of interest (ROI) around the previously determined galectin-3 signal is analyzed. This is useful to detect the recruitment of proteins enriched at cellular structures at the invasion site or (at a higher magnification) to bacteria-triggered compartments such as macropinosomes. Size exclusion is used to avoid nonspecific cross talk. See [12] for more details.
6. In case galectin-3 remnants remain associated with bacteria after vacuolar rupture, individual bacterial movements can be tracked using the ImageJ plugin TrackMate [32] (*see also* **Note 25**).

Regardless, there are no direct automated methods to investigate the specific recruitment of proteins to a bacterial compartment. Hence, manual inspection remains necessary to investigate more complex phenotypes of the host factor subversion during the dynamic entry process. This is only feasible when using libraries containing up to 100 different conditions. In case of larger libraries, we recommend developing dedicated detection algorithms.

4 Notes

1. Although this screen has been optimized for *Shigella* infection, this workflow has successfully worked with *Salmonella* and other bacterial pathogens and can be combined with more complex dynamic setups (e.g., molecular inhibitors). If other pathogens are used, we recommend carefully testing the infection (infectivity, number of events, etc.) using our infection setup. For *Salmonella* infections, we suggest being very stringent with the temperature as everything must be at 37 °C prior to the infection (*see* **Note 15**) and adjusting the bacterial load (MOI and inoculum volume) to optimize the infection.
2. In case transformations of plasmids do not work well, we suggest to use commercially available competent Top10 bacteria, which in our hands show a better transformation efficiency.
3. The eluted plasmids must be endotoxin-free to avoid unspecific immunogenic reactions upon transfection. Therefore, the DNA extraction kit must thus have an ethanol or isopropanol step. This step is critical for proper interpretation of the obtained data.

4. The overexpression of proteins can sometimes lead to mis-localization of the protein from its host cellular compartment. This can also lead to artifacts. Thus, verifying the localization of the protein of interest is necessary. The comparison to previously reported observations in online protein databases such as the Human Protein Atlas and UniProt is a quick way to confirm its proper localization. The localization of the protein can also be inspected by immunofluorescence. Additionally, for proteins for which the expression levels strongly impact cellular mechanisms (such as GTPases), verifying gene expression is crucial. Western blot using an anti-GFP antibody or RT-qPCR is an excellent option to compare the overexpression level of the protein to the endogenous protein.
5. Dark polystyrene plates with a clear bottom are best suited for fluorescence imaging as this prevents the light from contaminating the other wells.
6. We have used plates with either optical-grade polystyrene, cycloolefin, or a borosilicate bottom (175-micrometer thickness). Better performance and a higher resolution are achieved with glass-bottom dishes with a thickness adjusted for high-resolution objectives. Nevertheless, for the fluorophores used in our study, we have also obtained excellent results with plates with the plastic polymer bottoms which are much cheaper. We suggest to test the limits of the different plates before carrying out the screen. Furthermore, a water or oil objective can provide excellent resolution; however, due to the immersion media, fewer conditions can be analyzed per acquisition run. In this case a compromise must be made between the "size" of the screen and the spatial resolution desired.
7. In this screen, we have used HeLa cells. A particular attention should be drawn to optimizing the density of cells; we recommend testing a range of 5000–10,000 HeLa cells per well for the 96-well plate format. Our protocol can be adapted for any adherent cells (optimization of cell seeding, DNA transfection, infection, and acquisition parameters may be required). At a given density, we roughly obtained ten fluorescent cells per position with our microscope setup. Therefore, if using different cells and for a specific microscope setup, we recommend adjusting cell seeding at satisfying density if the cell growth rate differs from HeLa cells.
8. The transfection incubation time should be optimized based on changes in cellular morphology as well as on the desired fluorescence level of the protein of interest. The DNA transfection should be tested in advance to determine the optimal transfection conditions. Also, the transfection mix preparation can be downscaled to a fifth of the manufacturer's

recommended 3:1 ratio reagent/DNA volumes—this may be helpful for a 96-well format. For cells that are difficult to transfect, we recommend using XtremeGENE HP (Promega).

9. *Shigella* colonies and bacterial cultures require a minimum of 16 h for adequate growth. This is critical for the subculture to reach an OD_{600} of 0.4–0.7 in 2 h. The agar plate can be preserved at 4 °C after colony growth and is usable for infection experiments for up to 7 days.
10. If kept at 37 °C, the bacteria are infectious for up to 1 h after preparation. However, we recommend using it within 15 min of preparing them to ensure reproducibility of the results.
11. For infection performed in EM buffer, we suggest to perform time-courses up to 3 h long to remain cells in healthy conditions. In addition, washing and keeping the cells prior to the infection with pre-warmed buffer also ensures their well-being. The addition of 37 °C liquid paraffin (we recommend 12 μL/well) after bacterial infection helps further to avoid evaporation.
12. We recommend the acquisition order of the wells to follow the bacterial inoculum to each well. This ensures that the acquisition occurs with minimal delay in the wells. It is especially important if the events of interest occur early during the infection.
13. Simultaneous multicolor microscopy can be used to increase the screening throughput. This can be achieved using two cameras and a dichroic beam-splitter. This setup exposes the sample only once while capturing simultaneously two fluorescence signals, thanks to the un-mixing of the signals by the beam splitter.
14. Transfected cells with a healthy morphology are selected. Such cells are well spread on the support with a discernable nucleus and cytoplasm. Additionally, the chosen cells should express the protein of interest at low to medium levels. This is crucial as recruitment leads to a gain of signal. If the fluorescence level is too high before infection, saturation may occur. Moreover, a high expression of the candidate protein might lead to changes in morphology and nonphysiological behavior of the protein.
15. Bacteria should be rapidly added to each well prior to launching the acquisition, especially if the invasion process starts rapidly. In case of rapid events, multichannel pipettes help to increase synchronicity.
16. Importantly, all buffers, as well as the cells, must be pre-warmed and kept at 37 °C. The *Shigella* T3SS is expressed at 37 °C, and if the bacteria are not properly kept at this temperature, they may lose their invasion capacity. In our

hands, this is even more important for other bacterial pathogens like *Salmonella*.

17. The volume of bacterial inoculum affects the time delay between the inoculum and the initial contact between the bacteria and the host cell. Testing different volumes during the pre-screening experiments is useful. For *Salmonella* infections, we recommend lowering the volume to 30 μL for a quicker invasion as otherwise infections start much later during the acquisition. It is also important to note that one must in all cases keep everything at 37 °C and launch rapidly the bacterial infection.
18. Plates can be "recycled" by using the unused wells if kept sterile or the wells can be decontaminated between the runs.
19. To reduce bleed-through, the exposure time should be carefully set. If bleed-through is unavoidable, we recommend in this case lowering the exposure time to observe a minimal amount of cross talk.
20. The differential expression levels of fluorescent protein candidates can hamper the proper acquisition of some of the tested proteins. Therefore, we suggest to subdivide the library of expression vectors in accordance with expression level of transfected protein, with a pooling of similarly expressed proteins in each sub-experiment.
21. It is critical for a time-lapse acquisition to adjust the exposure times and the power of the light source in a way to avoid extensive photobleaching throughout the acquisition. For a general protein recruitment screen during *Shigella* (or *Salmonella*) entry into host cells, we recommend 1- to 2-min time-lapses for an exploratory primary screen. This setup allows a good observation of the *Shigella* invasion events.
22. The number of positions acquired should be a compromise when screening a medium-sized library. For this reason, the size of the microscope field of view and the speed of the motorized stage are important.
23. We recommend saving the acquisition files in .nd2 format as this format also records the acquisition metadata.
24. The varying physiological localization of host factor recruitment to entering bacteria can make analysis difficult to interpret. Consequently, automated time-resolved imaging analysis is often challenging, and we advise to start the analysis of medium-sized screens manually as it is likely to be more efficient in comparison to the time it takes to implement robust automated analytical pipelines for time-lapse image series.
25. Upon initial screening using the library of fluorescently tagged host factors in conjunction with the galectin-3 reporter, we

suggest to perform subsequent time-lapse series using the identified "hit" proteins together with either fluorescently labeled bacteria or co-transfecting fluorescently labeled actin. These additional experiments provide a better context between the bacterial invasion dynamics and the implication of the identified host proteins.

Acknowledgments

The authors of this manuscript wish to thank the entire DIHP unit and particularly the following lab members Laurent AUDRY, Laura BARRIO, and Magdalena GIL and Sonja KUHN for their fruitful discussion, constructive criticism, technical support, and mentorship during this project. We are particularly grateful to Emmanuel Boucrot for sharing the BAR-domain library for our work (Chan Wah Hak, 2018). The DIHP unit is supported by an ERC grant (ERC-CoG-Endosubvert), and the ANR (StopBugEntry, HBPSensing), and it is part of the two LabExes IBEID and Milieu Interieur.

References

1. Cossart P, Sansonetti PJ (2004) Bacterial invasion: the paradigms of enteroinvasive pathogens. Science. https://doi.org/10.1126/science.1090124
2. Voznica J, Gardella C, Belotserkovsky I, Dufour A, Enninga J, Stévenin V (2017) Identification of parameters of host cell vulnerability during salmonella infection by quantitative image analysis and modeling. Infect Immun. https://doi.org/10.1128/IAI.00644-17
3. Rengarajan M, Theriot JA (2020) Rapidly dynamic host cell heterogeneity in bacterial adhesion governs susceptibility to infection by *Listeria monocytogenes*. Mol Biol Cell. https://doi.org/10.1091/mbc.E19-08-0454
4. Ortega FE, Koslover EF, Theriot JA (2019) *Listeria monocytogenes* cell-to-cell spread in epithelia is heterogeneous and dominated by rare pioneer bacteria. elife. https://doi.org/10.7554/eLife.40032
5. Lorkowski M, Felipe-López A, Danzer CA, Hansmeier N, Hensel M (2014) Salmonella enterica invasion of polarized epithelial cells is a highly cooperative effort. Infect Immun. https://doi.org/10.1128/IAI.00023-14
6. Fredlund J, Santos JC, Stévenin V, Weiner A, Latour-Lambert P, Rechav K, Mallet A, Krijnse-Locker J, Elbaum M, Enninga J (2018) The entry of Salmonella in a distinct tight compartment revealed at high temporal and ultrastructural resolution. Cell Microbiol. https://doi.org/10.1111/cmi.12816
7. Malik-Kale P, Winfree S, Steele-Mortimer O (2012) The bimodal lifestyle of intracellular Salmonella in epithelial cells: replication in the cytosol obscures defects in vacuolar replication. PLoS One. https://doi.org/10.1371/journal.pone.0038732
8. Stévenin V, Chang YY, Le Toquin Y, Duchateau M, Gianetto QG, Luk CH, Salles A, Sohst V, Matondo M, Reiling N, Enninga J (2019) Dynamic growth and shrinkage of the Salmonella-containing vacuole determines the intracellular pathogen niche. Cell Rep. https://doi.org/10.1016/j.celrep.2019.11.049
9. Luk CH, Valenzuela C, Gil M, Swistak L, Bomme P, Chang YY, Mallet A, Enninga J (2021) Salmonella enters a dormant state within human epithelial cells for persistent infection. PLoS Pathog. https://doi.org/10.1371/journal.ppat.1009550
10. Schroeder GN, Hilbi H (2008) Molecular pathogenesis of Shigella spp.: controlling host cell signaling, invasion, and death by type III secretion. Clin Microbiol Rev. https://doi.org/10.1128/CMR.00032-07
11. Ehsani S, Santos JC, Rodrigues CD, Henriques R, Audry L, Zimmer C, Sansonetti P, Tran Van Nhieu G, Enninga J

(2012) Hierarchies of host factor dynamics at the entry site of Shigella flexneri during host cell invasion. Infect Immun. https://doi.org/10.1128/IAI.06391-11

12. Weiner A, Mellouk N, Lopez-Montero N, Chang YY, Souque C, Schmitt C, Enninga J (2016) Macropinosomes are key players in early Shigella invasion and vacuolar escape in epithelial cells. PLoS Pathog. https://doi.org/10.1371/journal.ppat.1005602
13. Kühn S, Bergqvist J, Gil M, Valenzuela C, Barrio L, Lebreton S, Zurzolo C, Enninga J (2020) Actin assembly around the Shigella-containing vacuole promotes successful infection. Cell Rep. https://doi.org/10.1016/j.celrep.2020.107638
14. Ray K, Bobard A, Danckaert A, Paz-Haftel I, Clair C, Ehsani S, Tang C, Sansonetti P, Tran GV, Enninga J (2010) Tracking the dynamic interplay between bacterial and host factors during pathogen-induced vacuole rupture in real time. Cell Microbiol. https://doi.org/10.1111/j.1462-5822.2010.01428.x
15. Mellouk N, Weiner A, Aulner N, Schmitt C, Elbaum M, Shorte SL, Danckaert A, Enninga J (2014) Shigella subverts the host recycling compartment to rupture its vacuole. Cell Host Microbe. https://doi.org/10.1016/j.chom.2014.09.005
16. Brumell JH, Scidmore MA (2007) Manipulation of rab GTPase function by intracellular bacterial pathogens. Microbiol Mol Biol Rev. https://doi.org/10.1128/MMBR.00023-07
17. Chang YY, Stévenin V, Duchateau M, Giai Gianetto Q, Hourdel V, Rodrigues CD, Matondo M, Reiling N, Enninga J (2020) Shigella hijacks the exocyst to cluster macropinosomes for efficient vacuolar escape. PLoS Pathog. https://doi.org/10.1371/journal.ppat.1008822
18. Perlman ZE, Slack MD, Feng Y, Mitchison TJ, Wu LF, Altschuler SJ (2004) Multidimensional drug profiling by automated microscopy. Science. https://doi.org/10.1126/science.1100709
19. Conrad C, Gerlich DW (2010) Automated microscopy for high-content RNAi screening. J Cell Biol. https://doi.org/10.1083/jcb.200910105
20. Mohr S, Bakal C, Perrimon N (2010) Genomic screening with RNAi: results and challenges. Annu Rev Biochem. https://doi.org/10.1146/annurev-biochem-060408-092949
21. Pepperkok R, Ellenberg J (2006) High-throughput fluorescence microscopy for systems biology. Nat Rev Mol Cell Biol. https://doi.org/10.1038/nrm1979
22. Sönnichsen B, Koski LB, Walsh A, Marschall P, Neumann B, Brehm M, Alleaume AM, Artelt J, Bettencourt P, Cassin E, Hewitson M, Holz C, Khan M, Lazik S, Martin C, Nitzsche B, Ruer M, Stamford J, Winzi M, Heinkel R, Röder M, Finell J, Häntsch H, Jones SJ, Jones M, Piano F, Gunsalus KC, Oegema K, Gönczy P, Coulson A, Hyman AA, Echeverri CJ (2005) Full-genome RNAi profiling of early embryogenesis in Caenorhabditis elegans. Nature. https://doi.org/10.1038/nature03353
23. Sansonetti PJ, Kopecko DJ, Formal SB (1982) Involvement of a plasmid in the invasive ability of Shigella flexneri. Infect Immun. https://doi.org/10.1128/iai.35.3.852-860.1982
24. Nowicki B, Hart A, Coyne KE, Lublin DM, Nowicki S (1993) Short consensus repeat-3 domain of recombinant decay-accelerating factor is recognized by Escherichia coli recombinant Dr adhesin in a model of a cell-cell interaction. J Exp Med. https://doi.org/10.1084/jem.178.6.2115
25. Allaoui A, Sansonetti PJ, Parsot C (1993) MxiD, an outer membrane protein necessary for the secretion of the Shigella flexneri lpa invasins. Mol Microbiol. https://doi.org/10.1111/j.1365-2958.1993.tb01097.x
26. Sansonetti PJ, Ryter A, Clerc P, Maurelli AT, Mounier J (1986) Multiplication of Shigella flexneri within HeLa cells: lysis of the phagocytic vacuole and plasmid-mediated contact hemolysis. Infect Immun. https://doi.org/10.1128/iai.51.2.461-469.1986
27. Chan Wah Hak L, Khan S, Di Meglio I, Law AL, Lucken-Ardjomande Häsler S, Quintaneiro LM, Ferreira APA, Krause M, McMahon HT, Boucrot E (2018) FBP17 and CIP4 recruit SHIP2 and lamellipodin to prime the plasma membrane for fast endophilin-mediated endocytosis. Nat Cell Biol. https://doi.org/10.1038/s41556-018-0146-8
28. Paz I, Sachse M, Dupont N, Mounier J, Cederfur C, Enninga J, Leffler H, Poirier F, Prevost MC, Lafont F, Sansonetti P (2010) Galectin-3, a marker for vacuole lysis by invasive pathogens. Cell Microbiol. https://doi.org/10.1111/j.1462-5822.2009.01415.x
29. Green MR, Sambrook J (2018) The Hanahan method for preparation and transformation of competent *Escherichia coli*: high-efficiency transformation. Cold Spring Harb Protoc. https://doi.org/10.1101/pdb.prot101188
30. Qadri F, Hossain SA, Ciznár I, Haider K, Ljungh A, Wadstrom T, Sack DA (1988) Congo red binding and salt aggregation as indicators of virulence in Shigella species. J

Clin Microbiol. https://doi.org/10.1128/jcm.26.7.1343-1348.1988

31. Olivo-Marin J-C (2002) Extraction of spots in biological images using multiscale products. Pattern Recogn 35:1989–1996

32. Tinevez JY, Perry N, Schindelin J, Hoopes GM, Reynolds GD, Laplantine E, Bednarek SY, Shorte SL, Eliceiri KW (2017) TrackMate: an open and extensible platform for single-particle tracking. Methods. https://doi.org/10.1016/j.ymeth.2016.09.016

Chapter 9

Assaying RIPK2 Activation by Complex Formation

Heidrun Steinle, Kornelia Ellwanger, and Thomas A. Kufer

Abstract

The receptor-interacting serine/threonine-protein kinase-2 (RIPK2, RIP2) is a key player in downstream signaling of nuclear oligomerization domain (NOD)-like receptor (NLR)-mediated innate immune response against bacterial infections. RIPK2 is recruited following activation of the pattern recognition receptors (PRRs) NOD1 and NOD2 by sensing bacterial peptidoglycans leading to activation of NF-κB and MAPK pathways and the production of pro-inflammatory cytokines. Upon NOD1/2 activation, RIPK2 forms complexes in the cytoplasm of human cells, also called RIPosomes. These can be induced by *Shigella flexneri* or by the inhibition of RIPK2 by small compounds, such as GSK583 and gefitinib.

In this chapter, we describe fluorescent light microscopic and Western blot approaches to analyze the cytoplasmic aggregation of RIPK2 upon infection with the invasive, Gram-negative bacterial pathogen *Shigella flexneri*, or by the treatment with RIPK2 inhibitors. This method is based on HeLa cells stably expressing eGFP-tagged RIPK2 and describes a protocol to induce and visualize RIPosome formation. The described method is useful to study the deposition of RIPK2 in speck-like structures, also in living cells, using live cell imaging and can be adopted for the study of other inhibitory proteins or to further analyze the process of RIPosome structure assembly.

Key words RIP2, RIPK2, RIPosomes formation, *Shigella flexneri*, Bacterial infection, Western blot, Immunostaining, Fluorescence microscopy, Live cell imaging

1 Introduction

The adaptor protein receptor-interacting serine/threonine-protein kinase-2 (RIPK2 kinase, also known as RIP2, RICK, CARDIAK, and CARD3) is a multidomain, dual-specificity kinase that is expressed ubiquitously and is recruited and activated by NOD-like receptors (NLRs) nucleotide-binding oligomerization domain-containing protein 1/2 (NOD1/NOD2). NOD1 and NOD2 are well-characterized pattern recognition receptors (PRRs) involved in sensing bacterial infection [1, 2]. The kinase domain of RIPK2 undergoes autophosphorylation and subsequent ubiquitination, enabling downstream signaling through activation of nuclear factor kappa-light-chain-enhancer of activated B cells

Thomas A. Kufer and Maria Kaparakis-Liaskos (eds.), *Effector-Triggered Immunity: Methods and Protocols*,
Methods in Molecular Biology, vol. 2523, https://doi.org/10.1007/978-1-0716-2449-4_9,

(NF-κB) and mitogen-activated protein (MAP) kinases, resulting in the transcription of multiple inflammatory cytokine genes [3, 4]. Moreover, RIPK2 tyrosine kinase activity has also been proposed to play a direct role in activating NOD-dependent autophagy and might be linked to the induction of adaptive immunity [5].

In vitro, RIPK2 can form long filaments mediated by its caspase recruitment domain (CARD). Upon stimulation of NOD1/2, CARD-CARD interactions of NOD1/2 and RIPK2 mediate the formation of helical RIPK2 filaments, termed RIPosomes [6, 7]. In living cells, RIPK2 complexes are induced upon bacterial infection with *Shigella flexneri*, a Gram-negative bacterial pathogen that activates the NOD1 pathway. Once *S. flexneri* activates NOD1, RIPK2 complexes are formed after NF-kB activation downstream of NOD1/2 activation [8]. Autophosphorylation of RIPK2 at multiple sites controls activation and oligomer formation [9]. Thereby, autophosphorylation of RIPK2 at tyrosine 474 is essential for NOD signaling [10] and for RIPK2 complex formation [8]. Furthermore, ubiquitination by the E3 ubiquitin ligase X-linked inhibitor of apoptosis (XIAP) is essential to allow downstream activation of inflammatory signaling and plays a crucial role in the recruitment and activation of RIPK2 [11]. RIPK2 complex formation is also induced by inhibition of XIAP, suggesting that they are not a prerequisite for NF-KB activation and subsequent pro-inflammatory responses [8]. RIPK2 complexes might thus represent an inactive form of RIPK2. The disruption of the XIAP-RIPK2 interaction by blocking RIPK2 ubiquitination [12] promotes the formation of RIPK2 complexes and can be induced by RIPK2 inhibitors, such as GSK583 and the tyrosine kinase inhibitor gefitinib [10, 13, 14]. It was shown that the treatment with GSK583 leads to spontaneous formation of RIPosomes [12]; however, the function of formation of the oligomeric structure is yet not known. Upon bacterial infection, RIPK2 recruits Erlin1/2 proteins and is released from 14-3-3 binding [15], suggesting that RIPK2 complexes might have additional functions in acting as signaling hubs for other pathways associated with bacterial infection, such as cell-intrinsic stress responses and autophagy [16].

RIPK2 inhibitors that pharmacologically target the NOD/RIPK2 signaling pathway are investigated for the therapeutic treatment of autoinflammatory diseases which lack specific and effective therapies [10]. Dysregulation of this signaling pathway leads to defects in bacterial detection and in some cases to autoimmune disease and severe inflammatory diseases. For example, for the inflammatory bowel disease Crohn's disease (CD) which is an increasingly frequent disorder in the western world, one of the major genetic susceptibility factors is loss-of-function single nucleotide polymorphisms (SNPs) in NOD2 that results in impaired epithelial mucosal barrier function. Hence, NOD signaling pathways play an important role in health and disease [17]. However, there is still an incomplete understanding of RIPK2 activation downstream of NOD1/2, albeit this is important for the future development of new therapies for clinical conditions such as Crohn's disease.

Here, we present a method based on using RIPK2 complex formation as a readout for RIPK2 activation upon infection with *Shigella*. This is based on the use of HeLa cells stably expressing eGFP-tagged RIPK2. The detection of labeled RIPK2 by standard experimental methods is applied to aid further studying the process and function of RIPosome structure assembly: fluorescent light microscopy and live cell imaging enables the detection of eGFP-RIPK2 in its soluble and aggregated form at different time points of infection/inhibition in fixed and living cells, and with Western blot analysis, a molecular weight upshift that is associated with RIPK2 activation can be used as easy accessible readout.

2 Materials

Prepare all solutions using ultrapure water (prepared by purifying deionized water, to attain a sensitivity of 18 MΩ-cm at 25 °C) and analytical grade reagents. For cell culture, use sterile-filtered (0.2 μm) or autoclaved material. Prepare and use all reagents at room temperature, and store at 4 °C (unless indicated otherwise).

2.1 Cell Culture

1. HeLa Flp-In T-REx stable inducible eGFP-RIPK2 cells: generated according to Flp-In™ T-REx™ Core Kit K650001 (Thermo Fisher Scientific). HeLa eGFP-RIPK2 cells are also available upon request from the authors on a collaborative basis [8]. pcDNA5/FRT/TO-EGFP-RIP2 is available by Addgene (#131201).
2. Dulbecco's Modified Eagle's Medium (DMEM).
3. Dulbecco's phosphate-buffered saline (DPBS).
4. 1 mg/mL doxycycline in H_2O, sterile filtered, stored at −20 °C. Dilute to final concentration of 1 μg/mL in culture medium for cell culture experiments.
5. Cultivation medium: DMEM supplemented with 10 % heat-inactivated (30 min, 56 °C) fetal bovine serum, 1 % penicillin-streptomycin (p/s) (10,000 U/mL), 10 μg/mL blasticidin, and 500 μg/mL hygromycin B Gold.
6. 10× trypsin/EDTA solution (0.5 % trypsin) diluted in DPBS without magnesium and calcium to 1×.
7. 15 mL tubes, T75 cell culture flasks, 24-well cell culture plates.

8. Glass coverslips (ISO 8255-1, 12 mm in diameter, thickness: 0.13–0.16 mm), autoclaved.
9. CELLview™ dish or four-compartment dish with glass-bottom (thickness of 175 μm ± 15 μm) or equivalent chamber slides with glass or polymer bottom that meet the optical requirements for microscopic imaging.
10. 1 mM GSK583 in DMSO, sterile filtered. Store at −20 °C.
11. 1 mM gefitinib in DMSO, sterile filtered. Store at −20 °C.
12. Dimethyl sulfoxide (DMSO), sterile filtered.

2.2 Bacterial Strains and Reagents for Culture

1. *Shigella flexneri* M90T *afaE+* (spectinomycin^R), containing a pIL22-afaE (spectinomycin^R) plasmid, encoding the afimbrial adhesion factor afaE from uropathogenic *Escherichia coli* [18] or *S. flexneri* M90T Sm available from ATCC.
2. Casein soy (CASO) bouillon: 17 g/L casein peptone (pancreas hydrolysate), 3 g/L soy peptone (papain hydrolysate), 2.5 g/L dipotassium bicarbonate, 5 g/L sodium chloride (NaCl), 2.5 g/L glucose monohydrate, pH 7.3, autoclaved.
3. 1 % Congo Red solution in autoclaved water.
4. Congo Red agar plates: 1.5 % Agar-Agar in CASO bouillon, supplemented with 200 μg/mL spectinomycin and 0.01 % Congo Red (diluted from 1 % Congo Red).
5. 50 mg/mL spectinomycin in H_2O, sterile filtered.
6. 25 mg/mL gentamycin in H_2O, sterile filtered.
7. Round-bottom polypropylene culture tubes with vented snap cap.
8. 1.5 mL tubes.
9. OD_{600} photometer.

2.3 Polyacrylamide Gel Electrophoresis (PAGE)

1. Separation gel buffer: 1.5 M Tris–HCl, pH 8.8. Add about 100 mL water to a 1 L graduated cylinder or a glass beaker. Weigh 181.7 g Tris–HCl and transfer to the cylinder. Add water to a volume of 900 mL. Mix and adjust pH with 1 M HCl. Add water up to 1 L.
2. Stacking gel buffer: 0.5 M Tris–HCl, pH 6.8. Weigh 60.6 g Tris–HCl and prepare a 1 L solution as in previous step.
3. Acrylamide/bisacrylamide solution (37.5:1): 30 % acrylamide solution 0.8 % bisacrylamide.
4. 10 % ammonium persulfate (APS) in water. Aliquots can be stored at −20 °C.
5. N, N, N, N′-Tetramethylethylenediamine (TEMED) 99 % p. a., store at 4 °C.
6. Prestained molecular weight protein standards.

7. 10 × PAGE running buffer: 0.025 M Tris–HCl, pH 8.3, 0.192 M glycine, 0.1 % SDS. For 1× running buffer, dilute 10× PAGE running buffer 1:10 in water. Do not autoclave.
8. SDS lysis buffer (5×): 1 M Tris–HCl (pH 6.8), 10 % SDS, 25 % β-mercaptoethanol, 0.05 % bromophenol blue (BPB), 25 % glycerol. Leave one aliquot at 4 °C for current use and store the remaining aliquots at −20 °C. SDS precipitates at 4 °C. Therefore, lysis buffer needs to be pre-warmed to room temperature prior to use. Preparation of 2× lysis buffer: dilute 5× buffer 1:2.5 in water.

2.4 Western Blot (WB)

1. Nitrocellulose membrane (0.2 μm).
2. Phosphate-buffered saline (PBS): 137 mM NaCl, 2.7 mM KCl, 10 mM Na_2HPO_4, 2 mM KH_2PO_4 at pH 7.4.
3. Western blot transfer buffer: 0.025 M Tris–HCl, 0.192 M glycine, 20 % methanol.
4. Protein electrophoresis system, power supply.
5. Whatman filter paper.
6. Automated high sensitivity camera system.

2.5 Immunostaining and Chemiluminescent Detection of Proteins

1. Ponceau staining solution: 0.2 % Ponceau S, 3 % acetic acid in water. Ponceau staining solution is stored at room temperature and can be reused.
2. Washing buffer: 0.05 % Tween 20 in PBS (PBST).
3. Blocking solution: 5 % skimmed milk powder in PBS (*see* **Note 1**).
4. Antibody diluent solution: 0.5 % Roche blocking solution in PBS (*see* **Note 2**). Dilute 2.5 mL 10× Roche blocking (aliquot and store at −20 °C) in 47.5 mL PBS and store at 4 °C.
5. Primary antibodies: anti-RIPK2 antibody ((D10B11), #4142, Cell Signaling Technology), anti-β-actin (sc-47,778, Santa Cruz).
6. Secondary antibodies for ECL detection: anti-mouse/rabbit IgG HRP-conjugated antibody.
7. Homemade ECL substrate (hECL):

   Solution A: 50 mg luminol in 200 mL 0.1 M Tris–HCl, pH 8.6.

   Solution B: 11 mg para-hydroxycoumaric acid in 10 mL DMSO.

   Mix 1 mL Solution A and 100 μL solution B and add 1 μL 30 % H_2O_2 per blotting membrane and incubate for 2 min prior to detection (*see* **Note 3**).

2.6 Fluorescence Microscopy and Live Cell Imaging

1. Phenol red free DMEM or FluoroBrite DMEM (Thermo Fisher Scientific). As the pH of DMEM relies on a CO_2-bicarbonate-based buffer system, exogenous CO_2 regulation is required. For CO_2-independent pH regulation, 25 mM HEPES can be added to the culture medium for short time incubations.
2. PBS: 137 mM NaCl, 2.7 mM KCl, 10 mM Na_2HPO_4, 2 mM KH_2PO_4, pH 7.4.
3. 4 % paraformaldehyde (PFA) in PBS: heat to maximal 60 °C and stir until PFA is dissolved, aliquot, and store at −20 °C.
4. Mowiol-Hoechst33258: prepare Mowiol 4–88 according to manufacturers' instructions and supplement with 10 μg/mL Hoechst 33258 (Merck). Aliquot and store at −20 °C protected from light.
5. Microscope slides.
6. 3 cm glass-bottom petri dish with four subdivisions.
7. Inverted wide-field epifluorescence microscope with an incubation chamber, a light source, and filter set that is appropriate for the used fluorescent proteins and a high-aperture oil or water immersion 63× or 100× objective. The system used here consists of a Leica DMi8 microscope with an external light source; HC PL apo63×/1.40 oil objective; a filter set for DAPI, FITC, Y3, and Y5; an ORCA-Flash 4.0 digital camera (Hamamatsu) for image acquisition; and for live cell imaging, an incubation chamber, a digital temperature control, and a CO_2 control via a heated stage insert (*see* **Note 4**).

3 Methods

The method presented here describes how RIPK2 complex (RIPosome) formation can be induced in stable inducible HeLa eGFP-RIPK2 cells by the infection with *S. flexneri* or by inhibition of RIPK2 by inhibitor treatment with GSK583 or gefitinib, and how RIPosomes can be visualized by immunostaining of transferred RIPK2 on Western blotting (WB) membranes or by fluorescence microscopy imaging and live cell imaging (LCI) (Fig. 1).

3.1 Cell Culture

1. HeLa cells are maintained at 37 °C and 5 % CO_2 in T75 flasks until they reach 80–90 % confluency.
2. To prepare single cell suspension for seeding, cells are washed once with PBS, following addition of 3 mL trypsin/EDTA and incubation for 5 min at 37 °C and 5 % CO_2 to detach the cells.
3. Cells are resuspended in 7 mL DMEM culture medium, transferred to a 15 mL tube, and centrifuged for 3 min at 290 × g.

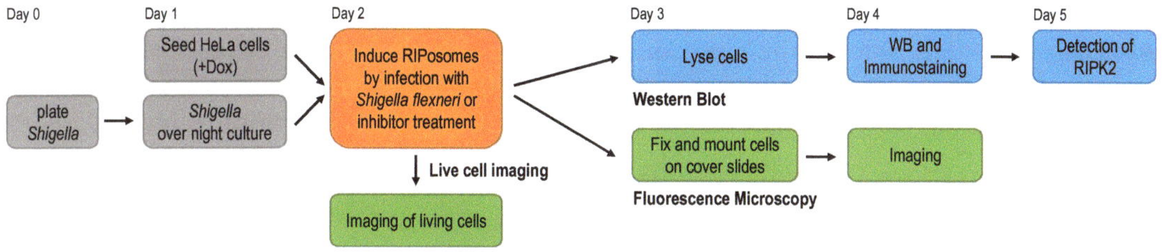

Fig. 1 Schematic flowchart of experimental methods for visualization of RIPosome formation. HeLa eGFP-RIPK2 cells are activated with either *Shigella flexneri* or the RIPK2 inhibitors, GSK583 and gefitinib, to assemble into multimeric structures (RIPosomes). These structures can be visualized in living cells via live cell imaging (LCI) or in fixated cells with fluorescence microscopy. RIPK2 in its complexed form causes a molecular weight upshift compared to soluble RIPK2 which can be seen in immunologically stained WB membranes of cell lysates of HeLa eGFP-RIPK2

4. After centrifugation, the supernatant is discarded and the cell pellet is resuspended in 10 mL culture medium, following cell counting, e.g., with a Neubauer hemocytometer.
5. Prepare dishes for cell seeding according to readout (*see* **Note 5**) and induce RIPK2 gene expression with 1 μg/mL doxycycline in culture medium (*see* **Note 6**):
   - *Western blot* (*see* Subheading 3.4): plate 1.5×10^5 cells/well by adjusting the concentration of the cell suspension to 3×10^5 cells/mL in culture medium with added doxycycline and transfer 500 μL/well in 24-well plates.
   - *Fluorescence microscopy imaging* (*see* Subheading 3.5): place one single sterile glass coverslip into each well of a 24-well plate and seed 1.2×10^5 cells/well in 500 μL doxycycline containing culture medium. Cell confluency should not exceed a maximum of 80 %. After plating, the coverslips are pressed to the bottom of the dishes with sterile pipette tips to remove eventual air bubbles underneath.
   - *Live cell imaging* (*see* Subheading 3.6): plate cells in appropriate live cell imaging dishes to provide optimal optical characteristics. Using 3 cm glass-bottom petri dish with four subdivisions, 1×10^5 cells per subdivision are seeded in 500 μL medium containing doxycycline to reach a cell density of maximal 80 % cell confluency at the time of imaging.
6. Incubate cells overnight at 37 °C in 5 % CO_2.
7. The next day, the culture medium is removed and the cells are washed twice with pre-warmed DMEM without supplements.
8. 250 μL/well pre-warmed DMEM without supplements is added per well and incubated for at least 30 min at 37 °C and 5 % CO_2 prior to infection (*see* **Note 7**).

3.2 Induction of RIPosomes by Bacterial Infection

1. One day before seeding the cells and preparing the bacteria for the experimental settings, the invasive *S. flexneri* M90T *afaE+* strain or *S. flexneri* M90T Sm is streaked out on Congo Red agar plates, containing 200 μg/mL spectinomycin.
2. The bacteria are incubated overnight at 37 °C (*see* **Note 8**).
3. The next day, one red single colony per strain is picked and inoculated in 5 mL CASO broth, containing 200 μg/mL spectinomycin, following by incubation at 37 °C at 180 rpm overnight.
4. Congo Red agar plates are sealed with parafilm and can be stored for reuse at 4 °C for a maximum of about 2 weeks.
5. In the morning, the *Shigella* overnight cultures are diluted in 5 mL CASO broth at different ratios (1:15, 1:25, and 1:50) and incubated for 1.5 h at 37 °C and 180 rpm until bacteria reach their exponential growth phase.
6. For infection, a multiplicity of infection (MOI) of 10 is used. OD_{600} of the bacterial dilutions is determined. The culture with an OD_{600} between 0.3 and 0.6 is selected.

   Calculation of MOI: $OD_{600} = 0.3$ equals about 3×10^8 bacteria/ml (*see* **Note 9**).
7. To adjust the MOI of 10, 1 mL of the bacterial culture is transferred to a 1.5 mL tube, centrifuged for 5 min at 560 × g at room temperature, the cell pellet is resuspended and diluted in DMEM without supplements to obtain a suitable bacterial stock concentration (*see* **Note 5**).
8. Remove the culture medium from the cells and add 250 μL of the bacterial suspension from **step** 7 to obtain a MOI of 10 (*see* **Note 10**).
9. For sedimentation of bacteria, plates are set aside for 15 min at room temperature. The infection is started by transferring the plate to 37 °C in a 5 % CO_2 atmosphere (*see* **Note 11**).
10. 30 min post infection (p.i.), *Shigella* containing medium is changed to 500 μL DMEM without supplements containing 100 μg/mL gentamycin to kill extracellular bacteria and prevent reinfection of cells.
11. Depending on the scientific question and the experimental setting, the cells can be lysed for WB or fixed at different time points for microscopy. As a starting point, 0.5 h, 1 h, 2 h, 4 h, and 6 h can be used (*see* **Note 12**). To measure *Shigella*-induced RIPosome formation, sampling is typically done at two time points: 2–3 h and 4–6 h p.i. (*see* **Note 13**).

3.3 Induction of RIPosomes by Inhibitor Treatment

The treatment with the RIPK2 inhibitors GSK583 and gefitinib can be performed in parallel to the infection with *Shigella* as a comparative analysis of RIPosome formation. This can also be used as a positive control.

1. After removing the culture medium and washing the cells with pre-warmed DMEM without supplements (*see* Subheading 3.1), 450 μL of fresh DMEM without supplements is added.
2. From a stock solution of 1 mM, the inhibitors are diluted in DMEM without supplements to obtain a final concentration of 1 μM. Prepare 100 μM dilution (1:10 from 1 mM) in 50 μL DMEM per well.
3. 50 μL of the inhibitor solution is added per well resulting in a total volume of 500 μL per well and cells are incubated at 37 °C and 5 % CO_2 for 4 h (*see* **Note 14**).

3.4 Electromobility Change of RIPK2 via Western Blot Analysis

1. *Shigella* infected (from Subheading 3.2), RIPK2 inhibitor treated (from Subheading 3.3), and untreated HeLa eGFP-RIPK2 cells (control) are washed once with PBS and lysed with 50 μL/ well 2× SDS lysis buffer directly after each time point.
2. Cell lysis is performed by scraping the cells with a pipet tip in SDS lysis buffer from the plate surface and multiple up-and-down pipetting. Transfer the samples in 1.5 mL safe-lock tubes and heat them at 95 °C for 5 min. Centrifuge shortly to bring down the condensate and insoluble debris. Samples can be stored here at −20 °C for several weeks if needed.
3. Acrylamide separating gel preparation: Mix 2.5 mL of separation gel buffer, 3.33 mL of acrylamide mixture, and 4.1 mL water in a 15 mL conical tube. Add 60 μL of 10 % APS and 13 μL of TEMED and cast gel. Allow space for stacking gel and gently overlay with isopropanol (*see* **Note 15**). Wait for about 10–15 min until complete polymerization.
4. Prepare the stacking gel by mixing 1.25 mL of stacking gel buffer, 0.7 mL of acrylamide mixture, and 3 mL water in a 15 mL conical tube. Add 25 μL of 10 % APS and 20 μL of TEMED. Remove isopropanol from separation gel. Overlay separation gel with the solution and insert a 15-well gel comb immediately without introducing air bubbles.
5. Assemble the gel and running chamber and add 1× running buffer according to the gel system manufacturer's instructions.
6. Load 10 μL of each sample and protein standard in one lane on the gel.
7. Run electrophoresis at 80–100 V (open current) until the samples have reached the separation gel and then continue at 120–140 V until the dye front (from the BPB dye in the samples) has reached the bottom of the gel.

8. Cut a nitrocellulose membrane and 8× Whatman filter to the size of the gel and immerse in 1× transfer buffer.
9. Following electrophoresis, open the gel plates and transfer the gel carefully to a container with 1× transfer buffer onto 4× cut and equilibrated filters.
10. Gently lift the gel-membrane sandwich from the glass plate and lay a nitrocellulose membrane, soaked in transfer buffer on top of the gel. Next, place 4× equilibrated filters on top of the nitrocellulose membrane.
11. Place the nitrocellulose-gel-filter paper sandwich between the electrodes, lock the cassette (*see* **Note 16**), and run semidry transfer according to the device instructions.
12. Immediately after blotting, transfer the membrane into a container with 10–20 mL Ponceau staining solution and stain for 3–5 min at room temperature, shaking.
13. Rinse 3× with water for 1–3 min each, until bands are visible in red and the background is cleared (white). Detect total protein load with transillumination detector.
14. Next, wash 3× with PBST for 5 min each to completely remove Ponceau staining and continue with immunostaining.
15. For immunostaining, first, block the membrane with blocking solution (*see* **Note 1**) for about 1 h.
16. Add primary anti-RIPK2 antibody (1:1000 dilution, diluted in antibody diluent solution) and incubate overnight at 4 °C or for 2 h at room temperature. Antibody can be stored at 4 °C and reused multiple times (add sodium azide) (*see* **Note 2**).
17. After washing 3× with PBST, 5 min each time, anti-IgG HRP conjugate (1:3000 dilution, diluted in antibody diluent solution) is added and incubated for 1–2 h at room temperature. Then, the membrane is washed 3× with PBST for 5 min.
18. Place membrane between plastic foil cover and add sufficient amount of (e.g., 1 mL hECL) ECL substrate solution evenly distributed onto the membrane. Incubate for 2 min and immediately continue with the detection of bands with chemiluminescence detection system (*see* **Note 3**).
19. Wash 3× with PBST for 5 min and proceed with the next primary antibody to be probed. Add anti-β-actin (1:1000 dilution, diluted in diluent solution) and incubate over night at 4 °C or for 2 h at room temperature. Repeat **steps 17–19** for protein detection.
20. For storage, rinse two to three times with deionized water and freeze membrane if to be reused within the next weeks, or let membrane dry for long-term storage at RT. Blotting membranes can be reprobed with specific antibodies at a later time

point (for this, rehydrate blotting membrane again with blocking solution and add antibodies).

The detection of total amount of protein with Ponceau prior to the immune staining with antibodies helps to analyze protein input and compare treated and untreated cells. The multimeric assembly of RIPK2 after infection with *Shigella* causes an upshift after separation of proteins in the gel, shown on the immunologically stained WB membrane (Fig. 2a).

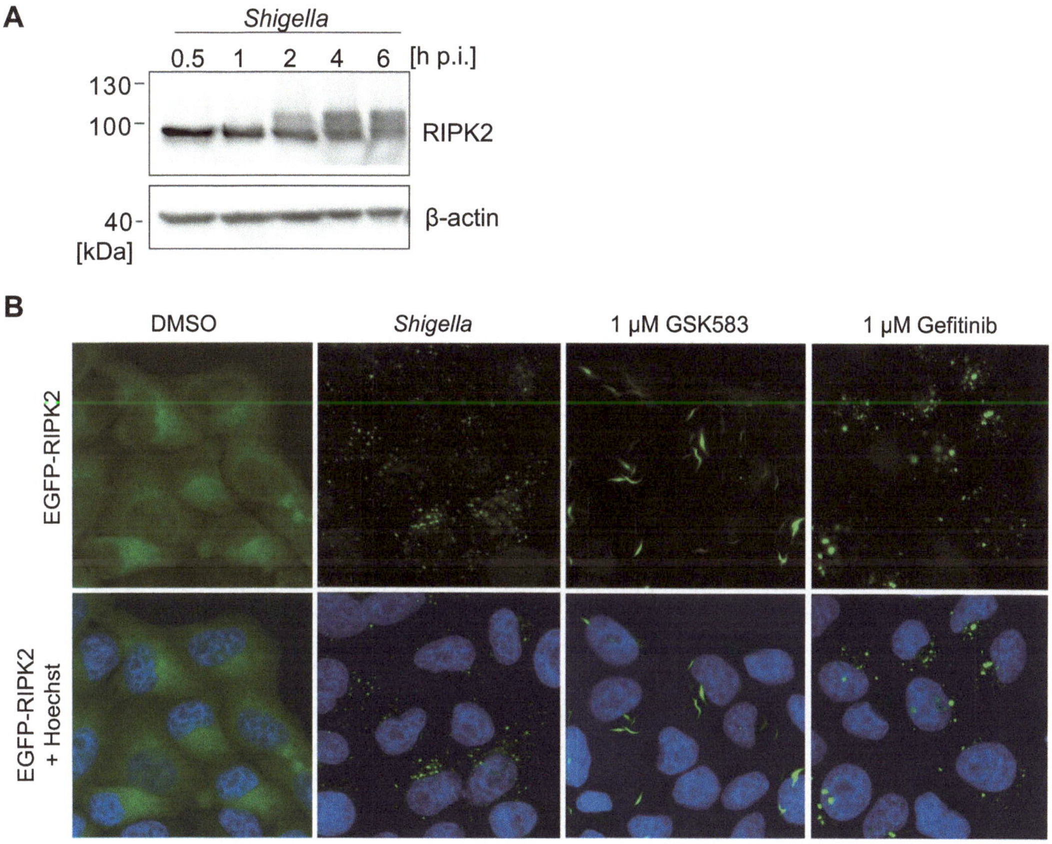

Fig. 2 Visualization of RIPosomes. (**a**) WB. Total cell lysates of HeLa eGFP-RIPK2 cells without and after infection with *Shigella* M90T cause an upshift of RIPK2 shown by the additional band at about 100–120 kDa. (**b**) Fluorescence microscopy. HeLa eGFP-RIPK2 cells were infected with *Shigella flexneri* M90T or treated with the RIPK2 inhibitors, GSK583 and gefitinib, for 2 h. RIPosomes' assembly shows a different structure for GSK treated cells (elongated) compared to *Shigella*-induced and gefitinib treated cells (punctuated). (Pictures are reproduced from Ref. [8])

3.5 Microscopic Analysis of RIPosomes

By tagging RIPK2 with eGFP, tracking is easily achieved by the detection with fluorescence microscopy. To preserve cellular structures in correlation to their native state, cells are fixed.

1. At the chosen time points p.i. with *Shigella* or inhibitor treatment (*see* Subheadings 3.2 and 3.3), the medium is removed from the cells on cover slides and 300 μL 4 % PFA in PBS is added per well for fixation.
2. After 15 min at room temperature, the slides are washed 3× times with PBS and dipped once in water to wash off salts that can cause artifacts.
3. For fluorescence imaging, slides are mounted with the cell side down on a 15 μL drop of Mowiol containing Hoechst33258 (for nuclear staining) and placed on a microscopic slide.
4. Slides are dried overnight at room temperature in the dark and can be analyzed by microscopy afterward.

For the detection of RIPosomes, fluorescence microscopy with 63× high-aperture oil immersion objective is recommended for getting the best resolution, but 40× is sufficient to observe these structures. To obtain three-dimensional reconstructions of cellular structures, optical sectioning by confocal microscopy can be applied. For wide-field microscopy, image stacks of adjacent z-sections can be captured, followed by software 3D-deconvolution. Deconvolution is an imaging processing technique, based on mathematical algorithms, to remove out-of-focus information and to generate sharper images with higher contrast.

For example, *Shigella* infection causes the assembly of eGFP-RIPK2 into multimeric structures after 4 h p.i., shown as puncta in the cells (Fig. 2b). Also the specific inhibition of RIPK2 by using the inhibitors GSK583 or gefitinib both induces RIPosome formation after treatment for 30 min and after an additional infection with *Shigella* at 4 h. Interestingly, RIPosomes induced in GSK583 treated cells appear in an elongated form rather than punctuated. The nuclear counterstaining provides information on the timing of *S. flexneri* M90T infection or RIPK2 inhibitor treatment and RIPosome formation.

Furthermore, co-localization of eGFP-RIPK2 or other cellular compartments with proteins, possibly involved in the RIPosome formation process, can be visualized by additional staining with fluorescently labeled proteins, specific antibodies, and fluorescent dyes of the infected or inhibitor treated and fixed cells.

However, the formation of RIPosomes after NOD1 activation by *S. flexneri* or by RIPK2 inhibitors is a dynamic process which can be visualized by live cell imaging in more detail. Single cells can be observed during the formation and disassembly of the RIPK2 complexes, and differences in RIPosome structure and their activation processes can be observed.

3.6 Microscopic Analysis of RIPosomes by Live Cell Imaging

For live cell imaging, a microscope with an incubation chamber is needed to incubate cells at 37 °C in a 5 % CO_2 atmosphere. The microscope and software should support time-lapse experiments with automatic image acquisition over a defined time frame. An automated stage to obtain different pre-defined position and z-stacks is of advantage. For tracking of differentially treated cells at the same time, the use of live cell imaging dishes with separated compartments is recommended.

1. HeLa cells, bacteria, and inhibitors are prepared as described in Subheadings 3.1, 3.2, and 3.3.
2. Culture medium is changed to medium without phenol red, not containing serum or antibiotics to reduce autofluorescence and to avoid interference with infection (*see* **Note 7**).
3. The microscope, stage, and objective are pre-warmed to 37 °C for at least 1 h before transferring the cells to the incubation chamber. Temperature fluctuations can occur and might interfere with infection efficacy and lead to problems with image focus.
4. Set live cell imaging parameters using the individual microscopy software and check the photosensitivity of your cells and dyes without infection (*see* **Note 17**). Chose 63× or 100× high-aperture oil or water immersion objective for getting the best resolution. To reduce light-induced cell damage and fluorochrome photobleaching, ensure to set the exposure time as short as possible. Do not use software-assisted autofocus functions. Hardware-controlled focus maintaining tools (e.g., Adaptive Focus Control (AFC)) are helpful to keep the cells in focus, even when performing long-term time lapse or multiple positioning, and do not contribute to phototoxicity or bleaching. Depending on the fluorescence intensity of your labeled protein, the exposition time must be adapted to obtain a sufficient signal-to-noise ratio.
5. According to the experimental design, time interval and total time of image acquisition need to be adapted. For long-term imaging up to 24 h, a time interval of 30 min to 1 h should be sufficient. For imaging of fast processes, short time intervals (one image at least each 15 s) should be used. For RIPosome formation, we typically take one image every 15 min over 6 h. For very weak signals, binning can be applied to increase signal-to-noise ratio and to reduce the risk of phototoxicity. However, it needs to be tested if the resulting reduced image resolution is acceptable.
6. Once the parameters are defined, the actual experiment can be started. To evaluate possible phototoxic or bleaching effects, compare morphology and condition of cells that were exposed to light during the experiment with cells in another

compartment of the same chamber that has not been illuminated. To rule out effects by insufficient environmental control during the experiment, you can furthermore compare your cells to cells incubated in the same medium in an independent cell culture incubator. If necessary, live cell imaging settings need to be adjusted and re-evaluated in consecutive experiments.

7. For *Shigella* infection and RIPK2 inhibitor experiments, use multi-compartment live cell imaging dishes. Treat cells as described in Subheadings 3.2 and 3.3 with the exception that FluoroBrite DMEM without supplements is used instead of DMEM containing phenol red.
   (a) Infect one compartment containing the cells with *S. flexneri*.
   (b) After sedimentation of bacteria (15 min at room temperature), the dish is transferred into the pre-warmed microscope incubation chamber at 37 °C and CO_2 is set to 5 %.
   (c) Two of the compartments are pre-treated for 30 min with the RIPK2 inhibitors, one with GKS582 and one with gefitinib, before the start of infection experiment.
   (d) One compartment serves as control and is left uninfected (DMEM only), or can be used as a solvent control for the inhibitors (DMSO diluted in DMEM respective to inhibitors).
8. The adjusted live cell imaging parameters are applied and positions and settings are defined as described above. Control the specificity of the effects seen upon infection or inhibitor treatment by simultaneous imaging of uninfected/untreated cells.
9. Optionally, *S. flexneri* can be visualized during imaging by adding appropriate fluorescent dyes to the bacterial culture medium. In any case, the labeling method should be compatible with the fluorescent tag of your protein of interest using fluorescently labeled bacteria.

4 Notes

1. As blocking solution, 3–5 % skimmed or dried milk powder solutions are commonly used. This is a broadly feasible and cheap option for many blocking applications, but can cause background staining. If this is the case, other blocking substances and concentrations (normally up to 10 %) should be tested, e.g., bovine serum albumin (BSA), fetal bovine serum (FBS), ovalbumin, or ready-to-use blocking concentrates/reagents. Additionally, 0.05–0.1 % Tween 20 can be added to the buffer to improve the results.

2. We use Roche blocking solution as antibody diluent. As for the blocking of WB membranes, BSA, FBS, or other different serum, like goat serum, can be used. Optimal blocking reagent is depending on the used antibody and blotting membranes and should be tested. For optimal performance, antibody concentrations should be adjusted. For reuse and storage of antibody solutions (at 4 °C), add 0.02 % sodium azide to the dilution to prevent contamination.
3. This represents a cheaper option compared to commercially available ECL substrates. However, some commercial solutions provide higher sensitivity for detection of proteins in a range of femtograms and are optimized for the detection of low antibody concentrations.
4. For live cell imaging, it is important to use a microscope equipped with a highly sensitive camera to obtain short exposure times and reduce light stress and an incubation chamber to ensure optimal bacterial invasion. For optimal temperature control, a heated stage insert is recommended, as *Shigella* needs 37 °C for invasion.
5. HeLa cells replicate approximately once every 24 h. This is important to keep in mind when estimating MOI for seeding and infection of the cells. On the day of infection, the actual cell number has to be estimated either by microscopic evaluation of cell densities or on the basis of extrapolation, e.g., 2.4×10^5 cells per well of a 24-well plate on day 2 (1.2×10^5 cells/well were plated on day 1). To obtain a MOI of 10, 250 μl DMEM with 2.4×10^6 bacteria and without supplements are needed for each well containing 2.4×10^5 cells.
6. The induction of a robust expression of the GOI (eGFP-RIPK2) in this system is performed by addition of 1 μg/ml doxycycline to the culture medium for 16–24 h before starting the experiment.
7. The use of medium without further supplements is important because serum can interfere with the infection and NOD activation and therefore influence RIPosome formation.
8. *Shigella* strains are Biosafety level 2 and have to be handled with care according to safety guidelines. Special regulations or permissions might be needed, dependent on local regulations. M90T *afaE*+ strains were kindly provided by Philippe Sansonetti, Institut Pasteur, Paris [18]. These strains harbor *afaE* on the pIL22 (spectinomycinR) plasmid, encoding the afimbrial adhesion factor afaE from uropathogenic *E. coli*. Alternatively, the invasive strain *S. flexneri* M90T Sm can be used. Congo Red is bound by virulent *Shigella* strains with an intact type III secretion system that enables to distinguish between virulent and non-virulent *Shigella* colonies by color [19].

9. For OD_{600} measurements, it is important to use the same photometer for repeated experiments to ensure reproducibility of the experiments, because of differences in OD_{600} between photometers. Determining the correlation of OD_{600} and bacterial counts typical an OD_{600} of 1 equals about 1×10^9 bacteria/ml. The exact value needs to be determined for each photometer by plating.
10. For infection, it is recommended to completely replace the medium with *Shigella* containing medium as dropwise addition might not lead to equal distribution of bacteria throughout the well.
11. The infection of cells begins at 37 °C as *Shigella* virulence gene expression is blocked at 30 °C or below [20]. When using bacteria without *afaE*, bacterial cells should be spun onto the HeLa cell for better synchrony of infection.
12. Typically, *S. flexneri* M90T invade cells around 15 min p. i. and start actin polymerization and movement at around 1.5 h p. i. [21, 22]. *Shigella* infection is ideally investigated over a time period up to 6 h to cover different aspects during infection. At later time points, *Shigella* infection induces cell death in the host cells. The optimal time points of *infection* should be assessed in a pilot experiment.
13. For analyzing RIPK2 complex formation, good starting time points are 2 h and 4 h p. i. to visualize the effect of *Shigella* infection and to ensure effective assembly of multimeric structures [8]. RIPosome formation can be detected usually starting from about 0.5–1 h p. i.
14. Inhibitors were dissolved in DMSO at 1 mM and diluted to a final concentration of 1 μM for the treatment of cells, which has shown to be effective in inhibition of RIPK2 activity before [10, 14]. After 30 min of incubation at 37 °C, RIPosomes are formed and inhibitors were tolerated by the cells up to 4 h of treatment. Longer incubation periods can harm the cells and cause toxicity. Therefore, the optimal time points for individual inhibitor treatments should be assessed in a pilot experiment.
15. The overlay with isopropanol prevents contact with atmospheric oxygen (which inhibits acrylamide polymerization) in addition to helping to level the resolving gel solution.
16. Prevent trapping of air bubbles in between the gel/filters and the membrane to ensure the complete transfer of proteins. Carefully, roll out air bubbles from the gel-membrane sandwich prior to placing in transfer cassette.
17. Cells are sensitive to photo damage, especially when fluorophores are present [23]. When fluorescent molecules are excited, they react with oxygen to produce free radicals that can lead to cell damage and death. Even without fluorophores, mammalian cells are sensitive to ultraviolet light.

References

1. Girardin SE, Tournebize R, Mavris M, Page AL, Li X, Stark GR, Bertin J, DiStefano PS, Yaniv M, Sansonetti PJ, Philpott DJ (2001) CARD4/Nod1 mediates NF-kappaB and JNK activation by invasive Shigella flexneri. EMBO Rep 2(8):736–742. https://doi.org/10.1093/embo-reports/kve155
2. Ogura Y, Inohara N, Benito A, Chen FF, Yamaoka S, Nunez G (2001) Nod2, a Nod1/Apaf-1 family member that is restricted to monocytes and activates NF-kappaB. J Biol Chem 276(7):4812–4818. https://doi.org/10.1074/jbc.M008072200
3. Philpott DJ, Sorbara MT, Robertson SJ, Croitoru K, Girardin SE (2014) NOD proteins: regulators of inflammation in health and disease. Nat Rev Immunol 14(1):9–23. https://doi.org/10.1038/nri3565
4. Topal Y, Gyrd-Hansen M (2020) RIPK2 NODs to XIAP and IBD. Semin Cell Dev Biol. https://doi.org/10.1016/j.semcdb.2020.07.001
5. Cooney R, Baker J, Brain O, Danis B, Pichulik T, Allan P, Ferguson DJP, Campbell BJ, Jewell D, Simmons A (2010) NOD2 stimulation induces autophagy in dendritic cells influencing bacterial handling and antigen presentation. Nat Med 16(1):90–97. https://doi.org/10.1038/nm.2069
6. Pellegrini E, Desfosses A, Wallmann A, Schulze WM, Rehbein K, Mas P, Signor L, Gaudon S, Zenkeviciute G, Hons M, Malet H, Gutsche I, Sachse C, Schoehn G, Oschkinat H, Cusack S (2018) RIP2 filament formation is required for NOD2 dependent NF-kappaB signalling. Nat Commun 9(1):4043. https://doi.org/10.1038/s41467-018-06451-3
7. Gong Q, Long Z, Zhong FL, Teo DET, Jin Y, Yin Z, Boo ZZ, Zhang Y, Zhang J, Yang R, Bhushan S, Reversade B, Li Z, Wu B (2018) Structural basis of RIP2 activation and signaling. Nat Commun 9(1):4993. https://doi.org/10.1038/s41467-018-07447-9
8. Ellwanger K, Briese S, Arnold C, Kienes I, Heim V, Nachbur U, Kufer TA (2019) XIAP controls RIPK2 signaling by preventing its deposition in speck-like structures. Life Sci Alliance 2(4). https://doi.org/10.26508/lsa.201900346
9. Pellegrini E, Signor L, Singh S, Boeri Erba E, Cusack S (2017) Structures of the inactive and active states of RIP2 kinase inform on the mechanism of activation. PLoS One 12(5): e0177161. https://doi.org/10.1371/journal.pone.0177161
10. Tigno-Aranjuez JT, Asara JM, Abbott DW (2010) Inhibition of RIP2's tyrosine kinase activity limits NOD2-driven cytokine responses. Genes Dev 24(23):2666–2677. https://doi.org/10.1101/gad.1964410
11. Krieg A, Correa RG, Garrison JB, Le Negrate G, Welsh K, Huang Z, Knoefel WT, Reed JC (2009) XIAP mediates NOD signaling via interaction with RIP2. Proc Natl Acad Sci U S A 106(34):14524–14529. https://doi.org/10.1073/pnas.0907131106
12. Goncharov T, Hedayati S, Mulvihill MM, Izrael-Tomasevic A, Zobel K, Jeet S, Fedorova AV, Eidenschenk C, deVoss J, Yu K, Shaw AS, Kirkpatrick DS, Fairbrother WJ, Deshayes K, Vucic D (2018) Disruption of XIAP-RIP2 association blocks NOD2-mediated inflammatory signaling. Mol Cell 69(4):551–565 e557. https://doi.org/10.1016/j.molcel.2018.01.016
13. Nachbur U, Stafford CA, Bankovacki A, Zhan Y, Lindqvist LM, Fiil BK, Khakham Y, Ko HJ, Sandow JJ, Falk H, Holien JK, Chau D, Hildebrand J, Vince JE, Sharp PP, Webb AI, Jackman KA, Muhlen S, Kennedy CL, Lowes KN, Murphy JM, Gyrd-Hansen M, Parker MW, Hartland EL, Lew AM, Huang DC, Lessene G, Silke J (2015) A RIPK2 inhibitor delays NOD signalling events yet prevents inflammatory cytokine production. Nat Commun 6:6442. https://doi.org/10.1038/ncomms7442
14. Haile PA, Votta BJ, Marquis RW, Bury MJ, Mehlmann JF, Singhaus R Jr, Charnley AK, Lakdawala AS, Convery MA, Lipshutz DB, Desai BM, Swift B, Capriotti CA, Berger SB, Mahajan MK, Reilly MA, Rivera EJ, Sun HH, Nagilla R, Beal AM, Finger JN, Cook MN, King BW, Ouellette MT, Totoritis RD, Pierdomenico M, Negroni A, Stronati L, Cucchiara S, Ziolkowski B, Vossenkamper A, MacDonald TT, Gough PJ, Bertin J, Casillas LN (2016) The identification and pharmacological characterization of 6-(tert-butylsulfonyl)-N-(5-fluoro-1H-indazol-3-yl)quinolin-4-amine (GSK583), a highly potent and selective inhibitor of RIP2 kinase. J Med Chem 59(10): 4867–4880. https://doi.org/10.1021/acs.jmedchem.6b00211
15. Steinle H, Ellwanger K, Mirza N, Briese S, Kienes I, Pfannstiel J, Kufer TA (2021) 14-3-3 and erlin proteins differentially interact with

RIPK2 complexes. J Cell Sci 134(12). https://doi.org/10.1242/jcs.258137

16. Moretti J, Blander JM (2017) Cell-autonomous stress responses in innate immunity. J Leukoc Biol 101(1):77–86. https://doi.org/10.1189/jlb.2MR0416-201R
17. Gutierrez O, Pipaon C, Inohara N, Fontalba A, Ogura Y, Prosper F, Nuñez G, Fernandez-Luna JL (2002) Induction of Nod2 in myelomonocytic and intestinal epithelial cells via nuclear factor-κB activation. J Biol Chem 277(44):41701–41705. https://doi.org/10.1074/jbc.M206473200
18. Clerc P, Sansonetti PJ (1987) Entry of Shigella flexneri into HeLa cells: evidence for directed phagocytosis involving actin polymerization and myosin accumulation. Infect Immun 55(11):2681–2688
19. Meitert T, Pencu E, Ciudin L, Tonciu M, Mihai I, Nicolescu S (1991) Correlation between Congo red binding as virulence marker in Shigella species and Sereny test. Roum Arch Microbiol Immunol 50(1):45–52
20. Maurelli AT (1989) Temperature regulation of virulence genes in pathogenic bacteria: a general strategy for human pathogens? Microb Pathog 7(1):1–10. https://doi.org/10.1016/0882-4010(89)90106-x
21. Kufer TA, Kremmer E, Adam AC, Philpott DJ, Sansonetti PJ (2008) The pattern-recognition molecule Nod1 is localized at the plasma membrane at sites of bacterial interaction. Cell Microbiol 10(2):477–486. https://doi.org/10.1111/j.1462-5822.2007.01062.x
22. Valencia-Gallardo CM, Carayol N, Tran Van Nhieu G (2015) Cytoskeletal mechanics during Shigella invasion and dissemination in epithelial cells. Cell Microbiol 17(2):174–182. https://doi.org/10.1111/cmi.12400
23. Stephens DJ, Allan VJ (2003) Light microscopy techniques for live cell imaging. Science 300(5616):82–86. https://doi.org/10.1126/science.1082160

Chapter 10

Identification of Novel Endogenous NOD Ligands: Quantitative Analysis of Binding Affinities of NOD1 or NOD2 with Sphingosine-1-Phosphate Using Microscale Thermophoresis

Gang Pei

Abstract

Nucleotide binding oligomerization domain-containing protein 1 (NOD1) and NOD2 have been identified as intracellular receptors for bacterial peptidoglycan for almost two decades; however, the direct binding with their respective ligands has only been recently demonstrated due to the difficulty of achieving large quantity of proteins with high purity. Here we describe a strategy combining immunoprecipitation of GFP-tagged proteins and microscale thermophoresis (MST) for efficient one-step purification of NOD1-GFP and NOD2-GFP and easy measurement of the binding affinities of NOD1 or NOD2 with sphingosine-1-phosphate (S1P) using small amount of proteins (nM range). This method will allow the identification of novel agonists/antagonists for NOD1/2.

Key words Nucleotide binding oligomerization domain (NOD)-like receptors, NOD1, NOD2, Dissociation constant, Sphingosine-1-phosphate (S1P), Microscale thermophoresis

1 Introduction

Nucleotide binding oligomerization domain (NOD)-like receptors (NLRs) are a group of intracellular pattern recognition receptors (PRRs) that play critical roles in microbial recognition and host defense. As two well-characterized immune receptors, NOD1 and NOD2 have been initially identified as NLR members triggering the activation of the nuclear factor kappa-light-chain-enhancer of activated B cells (NF-κB) upon bacterial peptidoglycan stimulation [1–3]. Subsequent structure-activity studies have elucidated γ-d-glutamyl-meso-diaminopimelic acid (iE-DAP) and muramyl dipeptide (MDP) as the minimal motifs activating NOD1 or NOD2, respectively [4–8]. The interactions with their respective ligands have been revealed by using streptavidin-bead pulldown [9],

Thomas A. Kufer and Maria Kaparakis-Liaskos (eds.), *Effector-Triggered Immunity: Methods and Protocols*, Methods in Molecular Biology, vol. 2523, https://doi.org/10.1007/978-1-0716-2449-4_10,

gel-sieve chromatography [10], or a cross-linking method [11]. Further, quantitative measurements of their binding dissociation constant (K_d) by surface plasmon resonance (SPR) have uncovered that the K_d values of NOD1 or NOD2 interacting with Tri-DAP or MDP are around 34.5 μM or 50 nM, respectively [12, 13]. Nevertheless, SPR and other commonly used methods for K_d measurements, such as isothermal titration calorimetry and fluorescence polarization, require large amount of highly purified proteins or an extra immobilization step, which are technically challenging and thereby limit their applications in NLRs-ligands/antagonists interaction studies [14]. Microscale thermophoresis (MST) is a rapidly developing method that detects the fluorescence change caused by the directed molecule movement in the temperature gradient. With hundreds of μL proteins at the concentration of nM range, MST allows for quantitative analysis of protein interactions with K_d range from pM to mM in solutions, even in cell lysates without protein purification [15, 16]. By using this method, we have successfully quantified the binding affinities of NOD1 or NOD2 with their specific peptidoglycan ligands and identified sphingosine-1-phosphate (S1P) as a novel activator for NOD1 and NOD2 by binding to their nucleotide binding domains (NBDs) [17].

Here we present our easy-to-follow protocol for the one-step purification of NOD1-GFP and NOD2-GFP using GFP-Trap agarose, and we also describe the detailed procedures for the quantitative analysis of the binding affinities of NOD1 or NOD2 with S1P by MST.

2 Materials

2.1 Cell Culture

1. T75 cell culture flask with vented cap.
2. 10-cm cell culture dish.
3. DPBS-Dulbecco's phosphate-buffered saline without calcium and magnesium.
4. Heat-inactivated fetal bovine serum (FBS): thaw FBS at 37 °C and then incubate at 56 °C for 30 min. Aliquot heat-inactivated serum into 50 ml tubes and store them at −20 °C.
5. Cell culture medium: Dulbecco's Modified Eagle Medium (DMEM) supplemented with 10% heat-inactivated FBS, 2-mM L-glutamine, 1-mM sodium pyruvate and 100-U/mL penicillin, 100-μg/mL streptomycin. Store at 4 °C.
6. Trypsin/EDTA solution.
7. Doxycycline: dissolve it in ultrapure water to the stock concentration at 5 μg/mL and pass it through the 0.2-μM filter. Store it at −20 °C.

8. HeLa doxycycline inducible NOD1-GFP or NOD2-GFP cells: they were generated and kindly provided by Prof. Thomas A. Kufer (University of Hohenheim, Germany). HeLa FlpIN T-REx cells were transfected with pcDNA5/FRT/TO-EGFP-NOD1 (Addgene, #131206) or pcDNA5/FRT/TO-EGFP-NOD2 (Addgene, #131207) together with pOG44 and selected with 10-μg/ml blasticidin (Invivogen) and 500-μg/ml hygromycin B (Invivogen) [17].
9. 0.4% trypan blue solution in water.
10. Neubauer counting chamber.

2.2 Protein Purification

All the solutions below are stored at 4 °C

1. GFP-Trap agarose (ChromoTek, #GTA-10): It is a GFP nanobody coupled to agarose beads with high specificity to GFP-fusion proteins.
2. Poly-Prep® gravity flow columns (Bio-Rad, #7311550).
3. Low protein binding micro tubes, 1.5 mL.
4. cOmplete® protease inhibitor cocktail tablets (Roche, #11697498001).
5. Lysis buffer: 50-mM Tris–HCl, pH 7.5, 150-mM NaCl, 1% NP-40, 10-mM DTT, 1-mM EDTA. Add one protease inhibitor cocktail tablet into 10-mL lysis buffer just before use.
6. Wash buffer I: 50-mM Tris–HCl, pH 7.5, 300-mM NaCl, 0.5% NP-40.
7. Wash buffer II: 50-mM Tris–HCl, pH 7.5, 150-mM NaCl, 5-mM $MgCl_2$, 1-mM ATP, 0.5% NP-40.
8. Elution buffer: 100-mM glycine, pH 2.5.
9. Neutralization buffer: 1-M Tris–HCl, pH 10.5.
10. Storage buffer: 50-mM Tris–HCl, pH 7.5, 150-mM NaCl, 0.5% NP-40, 5-mM DTT.
11. Zeba™ spin desalting columns, 7 K MWCO, 2 mL (Thermo Scientific, # 89890).
12. Ultrasonic homogenizer.
13. Centrifuges.

2.3 SDS-PAGE and Western Blotting

1. 4× Laemmli sample buffer: 0.2-M Tris–HCl, pH 6.8, 0.4-M DTT, 8% (w/v) SDS, 4% (v/v) β-mercaptoethanol, 40% (v/v) glycerol, 0.08% (w/v) bromophenol blue.
2. Coomassie staining solution: 0.1% Coomassie blue R-250, 10% acetic acid, 40% ethanol.
3. Destaining solution: 10% acetic acid, 40% ethanol.
4. Bio-Rad nitrocellulose membrane, 0.2 μM.

5. Heating block.
6. Bio-Rad Mini-PROTEAN tetra vertical chamber.
7. Bio-Rad semidry transfer cell.
8. Bio-Rad ChemiDoc touch system.

2.4 Protein Quantitation

1. Fluorescein isothiocyanate (FITC, Sigma-Aldrich, #F4274): dissolve it in DMSO to the stock concentration at 1 mM. Aliquot it in amber glass vials and store at −20 °C (*see* **Note 1**).
2. 96-well flat-bottom transparent microplates.
3. Amber glass vials, 2 mL (Sigma-Aldrich, # DWKW22468).
4. Multimode microplate reader.
5. GraphPad Prism software (version 9.0).

2.5 Microscale Thermophoresis Assay

1. S1P (Sigma-Aldrich, #73914): dissolve 1 mg of S1P in 1.28 mL of ultrapure water and add 20 μL of 1-N NaOH solution to increase the solubility. Aliquot it in amber glass vials and store at −20 °C (*see* **Note 2**).
2. Low protein binding micro tubes, 0.5 mL.
3. Monolith NT.115 standard capillaries.
4. Monolith NT.115 instrument.
5. MO.Affinity Analysis software (version 2.3).

3 Methods

3.1 Induction of NOD1-GFP and NOD2-GFP Expression

1. Culture HeLa inducible NOD1-GFP or NOD2-GFP cells in T75 flasks with the complete DMEM medium containing 10-μg/ml blasticidin and 500-μg/ml hygromycin B at 37 °C with 5% CO_2 atmosphere until reaching full confluency.
2. Wash twice with warm DPBS and incubate with 3-mL trypsin/EDTA solution at 37 °C for 5 min.
3. Add 10-mL DMEM complete medium to inactivate trypsin and resuspend cells with pipetting.
4. Mix same volume of cells and trypan blue solution and calculate the cell density with the counting chamber.
5. Seed 10× 10-cm cell culture dishes of HeLa inducible NOD1-GFP or NOD2-GFP cells in the complete DMEM medium and culture them overnight. For each dish, seed 2×10^6 cells.
6. Add doxycycline (1 μg/mL) into each dish and culture them for another 16 h to induce the expression of NOD1-GFP or NOD2-GFP. Afterward their expression is evaluated by Western blotting (Fig. 1).

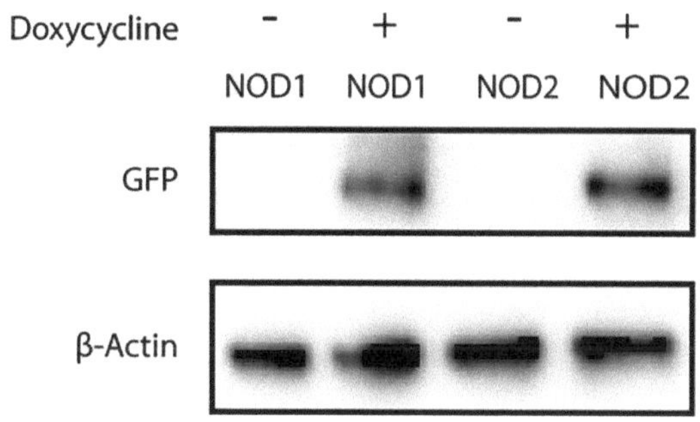

Fig. 1 Western blotting analysis of NOD1-GFP or NOD2-GFP expression. HeLa inducible NOD1 or NOD2 cells were stimulated with or without doxycycline for 16 h. Cells were collected and lysed with RIPA buffer for SDS-PAGE. β-Actin as the loading control

3.2 Purification of NOD1-GFP and NOD2-GFP

All procedures in this part are performed at 4 °C if not specified.

1. Wash the cells twice with cold DPBS and harvest cells of each dish with scrapers in 5-mL DPBS.
2. Collect cells by centrifuging at 300 × g for 10 min and discard the supernatants.
3. Lyse cells with 2-mL lysis buffer on ice for 20 min and centrifuge at 14,000 × g for 10 min to remove the pellets.
4. Add 200-μL GFP-Trap agarose into each Poly-Prep® gravity flow column and wash with 10-mL wash buffer I.
5. Seal the columns with caps and add supernatants from **step 3** into the columns. Mix with a rotator for 4 h.
6. Drain the columns and wash with 50 mL of wash buffer I followed by 50 mL of wash buffer II.
7. Elute each column with 300-μL elution buffer into low protein binding tubes and immediately add 30-μL neutralization buffer. Mix well by pipetting up and down.
8. Wash the agarose with 10-mL wash buffer I and store it in 20% ethanol (*see* **Note 3**).
9. Run SDS-PAGE to check the purified proteins (Fig. 2).
10. Change the buffer with storage buffer using Zeba™ spin desalting columns on the next day. Store the purified proteins at 4 °C (*see* **Note 4**).

3.3 Quantification of Purified NOD1-GFP and NOD2-GFP

Quantifying the precise concentration of purified NOD1 or NOD2 is critical to determine the binding affinity with their ligands. However, protein contaminants, possibly heat shock proteins (HSP70 and HSP90), are normally co-purified with NOD1 or NOD2 [18–20], which hinders the accurate measurement of the purified proteins by the Lowry or Bradford method. To solve this problem, we adopted a fluorescence-based method by normalizing

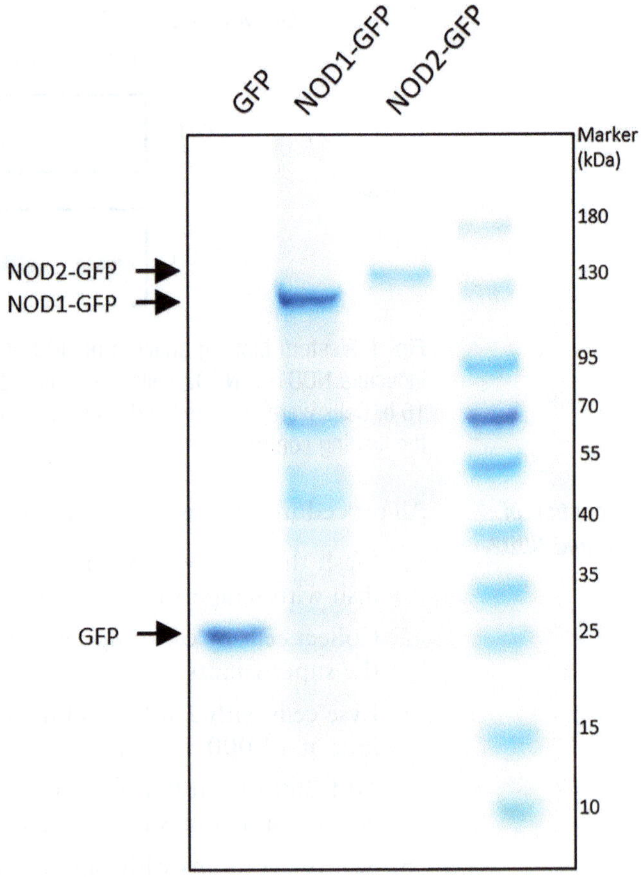

Fig. 2 Protein gel analysis of purified NOD1-GFP and NOD2-GFP. Purified proteins were boiled with 4× Laemmli sample buffer at 95 °C for 10 min and loaded into the 4–15% polyacrylamide gel. Gels were stained with the Coomassie staining solution

the fluorescence intensity of purified proteins to the FITC standard curves based on a previous report [21].

1. Dilute FITC stock solution with storage buffer and perform a twofold serial dilution from 25 μM to 48.8 nM. Add 100-μL FITC dilution and purified proteins into each well of microplates in triplicates.
2. Measure the fluorescence intensity of the FITC standard and the purified proteins using a plate reader with a 485/20 excitation, 530/25 emission filter set.
3. Fit the standard curve using the nonlinear regression with GraphPad and introduce the linear equation with the fitted slope and intercept. Calculate the fitted values of NOD1/2-GFP (X) according to their fluorescence intensity (Y) with the fitted linear equation (Fig. 3).

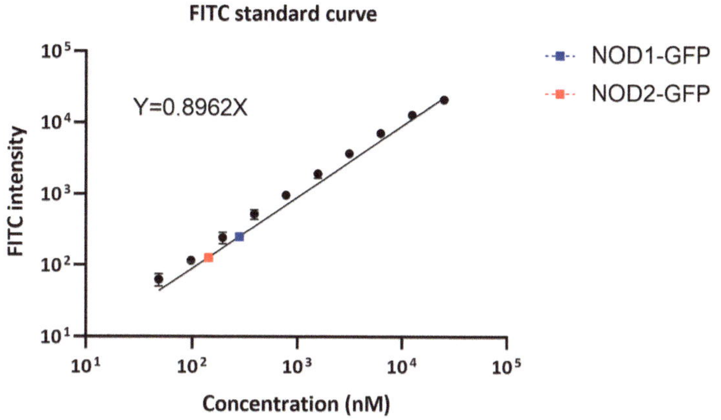

Fig. 3 An example of FITC standard curve. Data show means ±SD of three technical replicates from one representative experiment. Blue or red square indicates the fluorescence intensity of purified NOD1-GFP or NOD2-GFP from the same experiment, respectively

4. Calculate the real concentration of NOD1-GFP or NOD2-GFP using the ratio of fluorescein and GFP quantum yields, 0.85 and 0.60, respectively. The real concentration = (0.85/0.60) * fitted values of NOD1/2-GFP.

3.4 Quantification of Binding Affinity of NOD1-GFP or NOD2-GFP with S1P by MST

1. Dilute the purified NOD1-GFP or NOD2-GFP with storage buffer to the concentration of 40 nM. The final concentration in the binding assay is 20 nM (*see* **Note 5**).
2. Place 16 low protein binding tubes (0.5 mL) in the tube rack and add 10-μL diluted protein samples to the bottom of each tube (*see* **Note 6**).
3. Prepare a 16-step twofold serial dilution of S1P from 200 μM to 6.10 nM with storage buffer in low protein binding tubes. The final concentration of S1P is from 100 μM to 3.05 nM.
4. Transfer 10-μL S1P dilutions from tube 1 (200 μM) to tube 16 (6.10 nM) and sequentially add them into the 16 tubes with diluted proteins. Mix them well by pipetting up and down (*see* **Note 7**).
5. Incubate at room temperature for 10 min to reach the equilibrium binding.
6. Carefully put the Monolith NT.115 standard capillaries into the tubes underneath the solution. The samples will be aspirated due to capillary action (*see* **Note 8**).
7. Put the capillaries (1–16) in the slots of the sample tray. The capillary with the highest concentration of S1P is placed at position 1 of the tray, the lowest concentration at position 16.
8. Place the sample tray inside the Monolith NT.115 instrument and close the door.

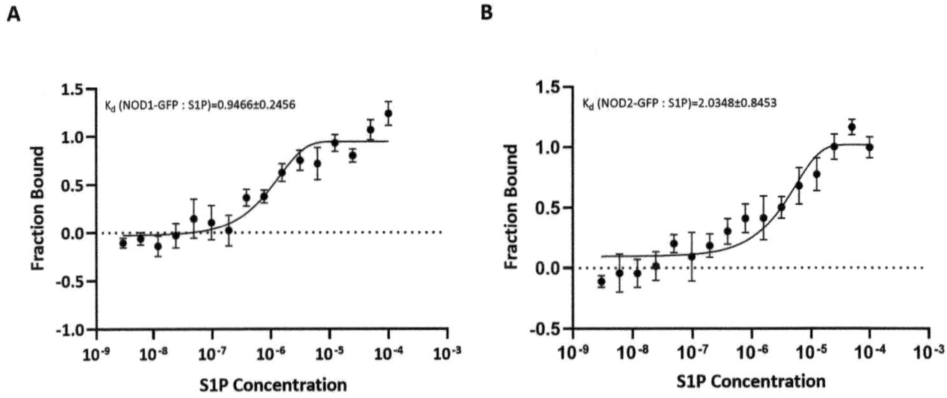

Fig. 4 Microscale thermophoresis measurements of the binding affinity of NOD1-GFP (**a**) or NOD2-GFP (**b**) with S1P. Data show means ±SD of three experiments using proteins from two different batches

9. Select the blue LED excitation source with $\lambda = 493$ nm and set the temperature of the instrument at 22 °C (*see* **Note 9**).
10. Perform MST experiments using 60% LED power and medium MST power. Set laser on and off times at 30 s and 5 s, respectively. Repeat the experiments three times with the same settings.
11. Select the experiments for analysis and fit curves with the K_d-binding model using the MO.Affinity Analysis software to calculate the K_d values.
12. Plot the binding response of NOD1-GFP or NOD2-GFP with S1P from 100 μM to 3.05 nM (Fig. 4).

4 Notes

1. FITC is light sensitive. Avoid prolonged exposure to light and avoid multiple freeze-thaw cycles.
2. Carefully check whether S1P is fully dissolved. If not, sonication on ice is applied until it is completely dissolved. Aliquot them in small volumes to avoid multiple freeze-thaw cycles.
3. The GFP-Trap agarose can be reused at least twice after elution with elution buffer.
4. Some proteins are unfolded after elution. Store the proteins at 4 °C overnight for refolding, which is indicated by the increase of fluorescence intensity.
5. Centrifuge the diluted proteins at 14,000 × g for 2 min to remove protein aggregates.
6. Pipette directly the proteins into the bottom of each tubes and avoid any residual solutions left in the tips to minimize the fluorescence variation.

7. Carefully mix the proteins and S1P dilutions by pipetting up and down. Do not vortex. Avoid bubble formation to reduce protein denaturation. Change the tip after every mixing.
8. Wear gloves and don't touch middle regions of the capillaries where the measurement will be performed.
9. Start the MST experiments until the target temperature of the machine is reached.

References

1. Bertin J, Nir WJ, Fischer CM, Tayber OV, Errada PR, Grant JR, Keilty JJ, Gosselin ML, Robison KE, Wong GH, Glucksmann MA, DiStefano PS (1999) Human CARD4 protein is a novel CED-4/Apaf-1 cell death family member that activates NF-kappaB. J Biol Chem 274(19):12955–12958. https://doi.org/10.1074/jbc.274.19.12955
2. Inohara N, Koseki T, del Peso L, Hu Y, Yee C, Chen S, Carrio R, Merino J, Liu D, Ni J, Nunez G (1999) Nod1, an Apaf-1-like activator of caspase-9 and nuclear factor-kappaB. J Biol Chem 274(21):14560–14567. https://doi.org/10.1074/jbc.274.21.14560
3. Ogura Y, Inohara N, Benito A, Chen FF, Yamaoka S, Nunez G (2001) Nod2, a Nod1/Apaf-1 family member that is restricted to monocytes and activates NF-kappaB. J Biol Chem 276(7):4812–4818. https://doi.org/10.1074/jbc.M008072200
4. Chamaillard M, Hashimoto M, Horie Y, Masumoto J, Qiu S, Saab L, Ogura Y, Kawasaki A, Fukase K, Kusumoto S, Valvano MA, Foster SJ, Mak TW, Nunez G, Inohara N (2003) An essential role for NOD1 in host recognition of bacterial peptidoglycan containing diaminopimelic acid. Nat Immunol 4(7): 702–707. https://doi.org/10.1038/ni945
5. Girardin SE, Boneca IG, Carneiro LA, Antignac A, Jehanno M, Viala J, Tedin K, Taha MK, Labigne A, Zahringer U, Coyle AJ, DiStefano PS, Bertin J, Sansonetti PJ, Philpott DJ (2003) Nod1 detects a unique muropeptide from gram-negative bacterial peptidoglycan. Science 300(5625):1584–1587. https://doi.org/10.1126/science.1084677
6. Girardin SE, Boneca IG, Viala J, Chamaillard M, Labigne A, Thomas G, Philpott DJ, Sansonetti PJ (2003) Nod2 is a general sensor of peptidoglycan through muramyl dipeptide (MDP) detection. J Biol Chem 278(11):8869–8872. https://doi.org/10.1074/jbc.C200651200
7. Girardin SE, Travassos LH, Herve M, Blanot D, Boneca IG, Philpott DJ, Sansonetti PJ, Mengin-Lecreulx D (2003) Peptidoglycan molecular requirements allowing detection by Nod1 and Nod2. J Biol Chem 278(43): 41702–41708. https://doi.org/10.1074/jbc.M307198200
8. Inohara N, Ogura Y, Fontalba A, Gutierrez O, Pons F, Crespo J, Fukase K, Inamura S, Kusumoto S, Hashimoto M, Foster SJ, Moran AP, Fernandez-Luna JL, Nunez G (2003) Host recognition of bacterial muramyl dipeptide mediated through NOD2. Implications for Crohn's disease. J Biol Chem 278(8): 5509–5512. https://doi.org/10.1074/jbc.C200673200
9. Mo J, Boyle JP, Howard CB, Monie TP, Davis BK, Duncan JA (2012) Pathogen sensing by nucleotide-binding oligomerization domain-containing protein 2 (NOD2) is mediated by direct binding to muramyl dipeptide and ATP. J Biol Chem 287(27):23057–23067. https://doi.org/10.1074/jbc.M112.344283
10. Askari N, Correa RG, Zhai D, Reed JC (2012) Expression, purification, and characterization of recombinant NOD1 (NLRC1): a NLR family member. J Biotechnol 157(1):75–81. https://doi.org/10.1016/j.jbiotec.2011.10.007
11. Wang YC, Westcott NP, Griffin ME, Hang HC (2019) Peptidoglycan metabolite photoaffinity reporters reveal direct binding to intracellular pattern recognition receptors and Arf GTPases. ACS Chem Biol 14(3):405–414. https://doi.org/10.1021/acschembio.8b01038
12. Laroui H, Yan Y, Narui Y, Ingersoll SA, Ayyadurai S, Charania MA, Zhou F, Wang B, Salaita K, Sitaraman SV, Merlin D (2011) L-Ala-gamma-D-Glu-meso-diaminopimelic acid (DAP) interacts directly with leucine-rich region domain of nucleotide-binding oligomerization domain 1, increasing phosphorylation activity of receptor-interacting serine/threonine-protein kinase 2 and its interaction with nucleotide-binding oligomerization domain 1. J Biol Chem 286(35):

31003–31013. https://doi.org/10.1074/jbc.M111.257501

13. Grimes CL, Ariyananda Lde Z, Melnyk JE, O'Shea EK (2012) The innate immune protein Nod2 binds directly to MDP, a bacterial cell wall fragment. J Am Chem Soc 134(33): 13535–13537. https://doi.org/10.1021/ja303883c
14. Podobnik M, Krasevec N, Bedina Zavec A, Naneh O, Flasker A, Caserman S, Hodnik V, Anderluh G (2016) How to study protein-protein interactions. Acta Chim Slov 63(3): 424–439. https://doi.org/10.17344/acsi.2016.2419
15. Jerabek-Willemsen M, Wienken CJ, Braun D, Baaske P, Duhr S (2011) Molecular interaction studies using microscale thermophoresis. Assay Drug Dev Technol 9(4):342–353. https://doi.org/10.1089/adt.2011.0380
16. Seidel SA, Dijkman PM, Lea WA, van den Bogaart G, Jerabek-Willemsen M, Lazic A, Joseph JS, Srinivasan P, Baaske P, Simeonov A, Katritch I, Melo FA, Ladbury JE, Schreiber G, Watts A, Braun D, Duhr S (2013) Microscale thermophoresis quantifies biomolecular interactions under previously challenging conditions. Methods 59(3): 301–315. https://doi.org/10.1016/j.ymeth.2012.12.005
17. Pei G, Zyla J, He L, Moura-Alves P, Steinle H, Saikali P, Lozza L, Nieuwenhuizen N, Weiner J, Mollenkopf HJ, Ellwanger K, Arnold C, Duan M, Dagil Y, Pashenkov M, Boneca IG, Kufer TA, Dorhoi A, Kaufmann SH (2021) Cellular stress promotes NOD1/2-dependent inflammation via the endogenous metabolite sphingosine-1-phosphate. EMBO J:e106272. https://doi.org/10.15252/embj.2020106272
18. Hahn JS (2005) Regulation of Nod1 by Hsp90 chaperone complex. FEBS Lett 579(20): 4513–4519. https://doi.org/10.1016/j.febslet.2005.07.024
19. Lee KH, Biswas A, Liu YJ, Kobayashi KS (2012) Proteasomal degradation of Nod2 protein mediates tolerance to bacterial cell wall components. J Biol Chem 287(47): 39800–39811. https://doi.org/10.1074/jbc.M112.410027
20. Mohanan V, Grimes CL (2014) The molecular chaperone HSP70 binds to and stabilizes NOD2, an important protein involved in Crohn disease. J Biol Chem 289(27): 18987–18998. https://doi.org/10.1074/jbc.M114.557686
21. Khavrutskii L, Yeh J, Timofeeva O, Tarasov SG, Pritt S, Stefanisko K, Tarasova N (2013) Protein purification-free method of binding affinity determination by microscale thermophoresis. J Vis Exp 78. https://doi.org/10.3791/50541

Chapter 11

A Surface Plasmon Resonance-Based Strategy to Characterize Interactions of NLR Proteins with Associated Factors

Jonas Moecking and Matthias Geyer

Abstract

NOD-like receptors (NLRs) are established as key regulators of the innate immune system. In recent years, an increasing number of interaction partners have been described that modulate receptor activity by direct binding. Characterizing these interactions can be challenging because these receptors tend to adopt different conformational states. We have developed a protocol that employs intracellular protein biotinylation to provide a straightforward immobilization strategy in surface plasmon resonance experiments. With this highly sensitive and label-free technique, the kinetics and affinities of NLR and co-factor interactions can be measured directly at the protein level.

Key words NOD-like receptor (NLR), Surface plasmon resonance (SPR), Biotinylation, Protein–-protein interaction, Kinetics, Affinity

1 Introduction

NOD-like receptors (NLRs) are key regulators of the innate immune system. As such, their activity is tightly controlled. There is an increasing amount of data describing an essential regulatory role of NLR binding partners. For instance, the activity of NLRP3 (NACHT, LRR, and PYD domains-containing protein 3) is regulated by the kinase NEK7 (never in mitosis A-related kinase 7) [1]. Similarly, activation of NLRC4 (NLR family CARD domain-containing protein 4) requires direct interaction with a single activated NAIP protein (neuronal apoptosis inhibitory protein) [2]. More recently, NLRP1 (NACHT, LRR, and PYD domains-containing protein 1) was shown to be kept in an autoinhibited conformation by direct binding to DPP9 (dipeptidyl peptidase 9) [3–5].

Surface plasmon resonance (SPR) is a widely used technique to characterize molecular interactions (e.g., protein–protein or

Thomas A. Kufer and Maria Kaparakis-Liaskos (eds.), *Effector-Triggered Immunity: Methods and Protocols*,
Methods in Molecular Biology, vol. 2523, https://doi.org/10.1007/978-1-0716-2449-4_11,

protein–small molecule) [6, 7]. However, purifying and measuring delicate proteins, like most NLRs are, with such a sensitive method can be difficult. In SPR experiments, one of the two interaction partners of interest is immobilized on the surface of a sensor chip. The immobilized protein is referred to as the ligand. Immobilization is commonly done by direct covalent coupling of the ligand to the sensor chip surface (e.g., by amine coupling). Alternative strategies include immobilization by binding of the ligand (or a fused tag) to a previously immobilized antibody (e.g., anti-GST or anti-FLAG) or by binding to a streptavidin functionalized surface. Using the streptavidin strategy requires biotinylation of the ligand. Protein biotinylation can be achieved with chemical and enzymatic approaches [8]. The enzymatic approach makes use of the enzyme BirA and allows site-specific intracellular biotinylation of a target peptide (e.g., Avi-tag). Different methods for intracellular biotinylation have been described previously for *Escherichia coli* and insect cell expression [9, 10]. Here we used a similar methodology to express the protein NLRP1 in baculovirus-infected *Sf9* insect cells. The protein was expressed from a pFastBac™ Dual vector as an N-terminal Avi-MBP fusion. Intracellular biotinylation of the Avi-tag was facilitated by co-expression of the BirA enzyme from the same vector. The protein was purified via the MBP-tag and subsequently immobilized on a streptavidin functionalized chip. This strategy enabled us to measure NLRP1/DPP9 interaction by SPR.

2 Materials

Prepare all solutions using ultrapure water (referred to as water in the protocol) and store at room temperature unless indicated otherwise. Analytical grade chemicals should be used for preparing buffers and stock solutions.

2.1 Cloning of Dual Expression Vector and Preparation of Bacmid

1. pFastBac™ Dual insect cell expression vector (Thermo Fisher Scientific).
2. Template DNA for BirA and NLRP1 (or the gene of interest).
3. Specific primers for use in PCR (restriction site highlighted in bold).
   - BirA (*Nco*I, fwd):

     5′CATG**CCATGG**ATGAAAGATAATACTGTTCCT C 3′
   - BirA (*Kpn*I, rev):

     5′ CGG**GGTACC**TCACTTCTCTGCGCTTCTC 3′

- Avi-MBP-tev (*Bgl*II, Avi, fwd):
  5′CGC**AGATCT**ATGGGCCTTAATGACATCTT
  CGAAGCTCAGAAAA
  TTGAGTGGCATGAAATGAAAATCGAAGAAG 3′
- Avi-MBP-tev (*Not*I, rev):
  5′ATAAGAAT**GCGGCCGC**GAATTCACCGGAT
  CCCTGAAAATACAG
  GTTTTC 3′ (*see* **Note 1**)
- GOI (e.g., NLRP1, *Bam*HI, fwd):
  5′CGC**GGATCC**ATGGCTGGTGGTGCTTGGG 3′
- GOI (e.g., NLRP1, *Eco*RI, rev):
  5′CG**GAATTC**AGGAGGACAGGGGCAGCAGG 3′

4. DNA polymerase and the corresponding reaction buffer for amplification of specific genes of interest in PCR.
5. PCR reagents: nuclease-free water and 2.5-mM dNTP mix.
6. Restriction enzymes: *Nco*I, *Kpn*I, *Bgl*II, *Bam*HI, *Eco*RI, *Not*I (New England Biolabs). Preferably restriction enzymes should be HF versions (High Fidelity).
7. 50× TAE buffer: 242-g Tris–HCl, 57.1-mL acetic acid (100%), and 100 mL of 0.5 M EDTA (pH 8.0) to 700 mL of water. Adjust volume to 1000 mL. The pH of the buffer is not adjusted and should be around 8.3.
8. Agarose gel electrophoresis equipment: electrophoresis chamber, power pack, agarose gel tray.
9. Agarose gel electrophoresis reagents: agarose, loading dye (usually available from plasmid preparation kit), DNA ladder, DNA dye such as peqGREEN.
10. T4 DNA Ligase and T4 DNA Ligase reaction buffer.
11. Chemically competent *Escherichia coli* (DH5α or NEB 10-beta). Chemically competent DH10Bac *E. coli*.
12. Water bath (42 °C).
13. 50-mg/mL kanamycin in water. Stored at −20 °C for long-term storage.
14. 100-mg/mL ampicillin in water. Stored at −20 °C for long-term storage.
15. 10-mg/mL gentamycin in water. Stored at −20 °C for long-term storage.
16. 10-mg/mL tetracycline in 2-propanol. Stored at −20 °C for long-term storage.
17. 1-M IPTG in water. Store at −20 °C.

18. 20-mg/mL Bluo-Gal in dimethylformamide (*see* **Note 2**). Store at −20 °C.
19. LB agar plates containing only ampicillin (100 μg/mL) or ampicillin (100 μg/mL), kanamycin (50 μg/mL), gentamycin (10 μg/mL), tetracycline (10 μg/mL), IPTG (1 mM), and Bluo-Gal (100 μg/mL). LB media containing the same antibiotics as described for the LB agar plates.
20. Plasmid preparation kit or DNA purification columns.
21. Agarose gel DNA purification kit.
22. UV light transilluminator.
23. 100% 2-propanol and 70% ethanol.
24. Facility to sequence plasmids.
25. Blue ice and an ice bucket.

2.2 Transfection of Insect Cells, Protein Expression, and Purification

1. Insect cells (e.g., *Sf9*) and compatible growth medium.
2. Syringes with Luer lock and compatible syringe filters (0.22 μm, 0.45 μm).
3. Cell counter and compatible slides (alternatively a Neubauer chamber and a microscope work as well).
4. 6-well tissue culture plate.
5. Transfection reagent suitable for insect cell transfection.
6. Conical glass flasks (50, 500, and 5000 mL).
7. Sonicator for cell lysis.
8. 15-mL and 50-mL tubes.
9. Table top centrifuge for spinning (suitable for 15-mL and 50-mL tubes).
10. 1× PBS for washing cells after harvest.
11. Purification buffer A: 20-mM HEPES/NaOH (pH 7.8), 150-mM NaCl, 5-mM β-mercaptoethanol.
12. Purification buffer B: Buffer A containing 10-mM maltose.
13. Purification buffer C: Buffer A containing 0.5-mM Tris(2-carboxyethyl)phosphin (TCEP) instead of β-mercaptoethanol.
14. 0.1-M PMSF in 2-propanol. Store at −20 °C.
15. 1-mg/mL DNase I in 50% glycerol, 20-mM Tris/HCl (pH 7.5), 1-mM $MgCl_2$. Store at −20 °C.
16. Amylose resin (New England Biolabs) and gravity columns (e.g., 15 mL).
17. NanoDrop (Thermo Fisher Scientific).
18. Blue ice and an ice bucket.

19. SDS-PAGE equipment (electrophoresis chamber, SDS gel (12% acrylamide), SDS-PAGE running buffer, SDS sample buffer, protein ladder, Coomassie stain and destain solutions, heat block (95 °C)).
20. Protein concentration filter with 100-kDa MWCO.

2.3 SPR Measurement

Solutions and buffers used for SPR must be filtered (0.2-µm filter) and degassed (unless the system used contains a built-in degasser). This is of utmost importance to avoid any particulate matter from entering the system, which can potentially clog the microfluidics or damage the membrane valves.

1. SPR device (e.g., Biacore 8 K, Cytiva) and suitable control and evaluation software.
2. Maintenance Chip (Cytiva).
3. Series S Sensor Chip SA (Cytiva).
4. Purified recombinant proteins (Biotin-NLR and interaction partner).
5. Desorb solution 1: 5-g/L SDS.
6. Desorb solution 2: 50-mM Glycine/NaOH (pH 9.5).
7. Running buffer: 10-mM HEPES/NaOH (pH 7.8), 150-mM NaCl, 0.5-mM TCEP, 0.05% TWEEN20.
8. Polypropylene U-bottom 96-well plate (*see* **Note 3**).
9. 96 deep-well plate (1 mL).
10. Normalizing solution: 70% glycerol.
11. Sensor chip activation solution: 1-M NaCl in 50-mM NaOH.
12. Wash solution: 50% 2-propanol in 1-M NaCl and 50-mM NaOH (*see* **Note 4**).
13. Regeneration solution: 10-mM glycine, pH 3.0 (*see* **Note 5**).

3 Methods

3.1 Cloning of pFastBac™ Dual with BirA and Avi-MBP-tev Expression Cassettes

1. Set up the PCR for the BirA, Avi-MBP-tev, and GOI insert in PCR tubes. Mix stock solutions thoroughly before use. For a 25 µL reaction using Q5 polymerase, combine 5 µL of 5× reaction buffer, 4 µL of dNTPs, 1.25 µL of the corresponding forward and reverse primer and template DNA (~100 ng). Finally, adjust the volume to 24.5 µL with nuclease-free water and add 0.5 µL of polymerase. Run the PCR according to the instructions provided by the polymerase manufacturer.
2. While the PCR is running, set up the restriction enzyme digest of the pFastBac™ Dual vector to make it compatible for ligation with the BirA insert. Mix 2–10 µg of the vector with 1 µL

of each restriction enzyme (*Kpn*I HF and *Nco*I HF) and 2 μL of 10× CutSmart buffer. Adjust volume to 20 μL with nuclease-free water. Incubate at 37 °C for 1 h (*see* **Note 6**).

3. Prepare an agarose gel by dissolving 1-g agarose in 50-mL 1× TAE buffer in a glass flask or beaker. Heat up the mixture (e.g., in a microwave) to completely dissolve the agarose. Let the solution cool down for 5–10 min at RT before adding 2.5-μL DNA dye. Swirl to mix and pour an agarose gel in a tray previously sealed at its front and back end with tape. Place a comb at the top of the gel with an appropriate number of wells. Let the agarose gel solidify at RT for at least 20 min.
4. Fill the electrophoresis chamber with 1× TAE buffer. Remove the comb of the gel and place it (still in the tray) in the chamber.
5. Add 5 μL of 6× sample dye to the PCR reactions and 4 μL of loading dye to the restriction digest. Mix the dye and sample by pipetting and load each sample into a well of the agarose gel.
6. Run the gel for 35 min at 100 V.
7. Place the gel on a UV-table and excise the bands corresponding to the PCR products and the cleaved vector. Use a clean scalpel and wipe it with 80% ethanol after cutting out each band.
8. Purify the DNA fragments using the agarose DNA purification kit according to the protocol of the manufacturer. Only use 16 μL of water to elute the PCR product. Use 30 μL of water to elute the digested vector. Measure the concentration of the purified vector using a NanoDrop (note: yield is typically low but should be between 15 and 30 μg/μL at least).
9. Set up the restriction digest of the BirA PCR product to make it compatible for ligation into the digested vector. Add 2 μL 10× CutSmart buffer and 1 μL of each enzyme (*Kpn*I HF and *Nco*I HF) directly to the purified PCR product and incubate at 37 °C for 1 h.
10. Purify the digested PCR product again using the kit (*see* **Note 7**).
11. Combine 9 μL of digested and purified PCR product and 3 μL of the digested and purified vector together with 2 μL of 10 x T4 DNA Ligase buffer, 0.5 μL of T4 DNA Ligase, and 5.5 μL water.
12. Incubate at 16 °C overnight (*see* **Note 8**).
13. Thaw a tube with 60 μL of chemically competent *E. coli* on ice. Add 10 μL of the ligation reaction and incubate for 10 min on ice.

14. Heat-shock the bacteria in a water bath at 42 °C for 30 s. Immediately place the tube on ice again for 2 min. Add 800 μL of LB media to the tube and incubate at 37 °C shaking at 800 rpm.
15. Spin down the bacteria at 1500 × g for 2 min. Discard 800 μL of the supernatant and resuspend the pellet in the remaining liquid.
16. Plate the resuspended bacteria on an LB agar plate with ampicillin and incubate at 37 °C overnight.
17. Add 4 mL of LB with ampicillin in a 15-mL tube. Pick a single colony of the plate using a small pipette tip and drop the tip into the 15-mL tube. Pick a total of five colonies and incubate o/n in a tilted shaking incubator at 37 °C.
18. Purify the plasmid from the bacteria using the plasmid mini kit following the instructions of the manufacturer.
19. Send the purified plasmids for sequencing to confirm insertion (*see* **Note 9**).
20. A purified plasmid containing the BirA insert can now be used for entering the Avi-MBP tag (*Bgl*II + *Not*I restriction sites) (note: digest vector *Bam*HI/*Not*I to fuse *Bam*HI with *Bgl*II site → results in removal of the initial *Bam*HI site and reverse primer introduces new *Bam*HI and *Eco*RI sites) following the above-described procedure.
21. Repeat the above-described procedure with a positively sequenced plasmid containing the Avi-MBP insert (*Bam*HI and *Eco*RI restriction sites) to insert the NLR gene of interest (e.g., NLRP1).
22. Thaw DH10Bac competent *E. coli* cells on ice and add 100 ng of purified positively sequenced plasmid containing the BirA and Avi-MBP-NLR inserts.
23. Follow the transformation procedure described above (*see* **steps 13–16**). Use plates containing ampicillin, kanamycin, tetracycline, gentamycin, IPTG, and X-gal. Incubate the plates at 37 °C for 36–48 h after streaking the bacteria.
24. For bacmid production, pick a single white colony following the procedure described in **step 17** (using LB media containing ampicillin, kanamycin, gentamycin, tetracycline) (*see* **Note 10**).
25. To isolate the bacmid from the bacterial cultures, the plasmid mini kit can be used following the kit protocol up to the point of spinning down the neutralized cell lysate. After adding neutralization buffer and spinning down the lysate, continue with **step 26**.

26. Prepare a tube containing 750 μL of ice cold 2-propanol. Add 750 μL of the neutralized and clarified lysate containing the bacmid.
27. Centrifuge at 21,000 g and 4 °C for 30 min.
28. Carefully remove the supernatant using a pipette (*see* **Note 11**).
29. Wash with 70% ethanol. Centrifuge at 21,000 g and 4 °C for 10 min. Carefully remove the supernatant with a pipette and repeat this wash step once.
30. Remove the ethanol using a pipette and air-dry the bacmid at RT for 30 min. To avoid disturbing the pellet, a smaller pipette should be used to remove the last remainder of the supernatant.
31. Resuspend the bacmid in 50 μL nuclease-free water. Determine the concentration using a NanoDrop. The purified resuspended bacmid can be stored at 4 °C for up to 4 weeks.

3.2 Transfection of Insect Cells and Expression of Protein

All steps with insect cells should be carried out in a laminar flow safety cabinet to ensure sterile conditions. Equipment and materials used for handling insect cells should be thoroughly sterilized by autoclaving (if possible) and wiping with 80% ethanol (*see* **Note 12**).

1. All following steps up to the harvest of the protein expression culture should be carried out under sterile conditions.
2. Dilute a continuous culture of *Sf9* (or *Sf21*) to 0.3×10^6 cells/mL and transfer 3 mL into a well of a 6-well tissue culture plate. Make sure the viability of the cells is at least 95%.
3. To allow the cells to settle down and adhere, incubate the plate at 27 °C in a stationary incubator for 20–30 min.
4. Prepare the transfection mix (2-μg purified bacmid, 200 μL of serum-free insect cell medium, and 5-μL transfection reagent) in a sterile 1.5-mL reaction tube. Incubate at RT for 30 min.
5. Add the entire transfection mix to the well containing the cells. Try to distribute by adding small drops to different areas of the well.
6. Incubate at 27 °C for 72 h.
7. Harvest the supernatant containing virus using a sterile 3-mL syringe. Attach a sterile syringe top filter (0.22 μm) to the top and filter and transfer virus to a sterile tube. The filtered V_0 virus can be stored at 4 °C for up to 4 weeks.
8. Set up a fresh 50-mL flask containing 10 mL of insect cells at a density of 0.6×10^6 cells/mL and add 1 mL of V_0 virus.
9. Incubate the virus infected culture at 27 °C and 80 rpm (shaking throw: 25 mm). Monitor cell growth and viability. Each day take a 1-mL sample from the flask using a sterile 5-mL pipette.

Take 10 μL of that sample, mix with 10-μL trypan blue, and analyze cell viability using the cell counter. After 3–4 days, the cells should stop dividing, the viability should decrease significantly (~80–90%), and the cell size should increase (14–15 μm).

10. Transfer the culture to a 50-mL tube and centrifuge for 5 min at 400 g. Harvest the V_1 virus from the supernatant using a 20-mL syringe and a sterile syringe filter (0.22 μm) (*see* **Note 13**). The filtered V_1 virus can be stored at 4 °C for up to 8 weeks.
11. Set up a second virus amplification in a 500-mL flask with 100 mL of cells at 1.0×10^6 cells/mL and add 1 mL of the V_1 virus.
12. Incubate and monitor the cells as described in **step 9**.
13. Harvest the virus using a 50-mL syringe and sterile syringe filters (0.22 μm) as described in **step 10** (*see* **Note 14**). The V_2 virus should be prepared within a week before setting up an expression culture.
14. To set up the expression culture, prepare a 5-L flask containing 450 mL of cells at 2.0×10^6 cells/mL and add 50 mL of V_2 virus. Incubate at 27 °C and 80 rpm for 50 h.
15. Harvest the expression culture by centrifugation at 750 g for 20 min in 50-mL tubes (or larger centrifuge buckets if available). Wash the cell pellet with 30-mL cold 1× PBS and pool the pellet in a single 50-mL tube. Repeat the centrifugation step and discard the supernatant. The pellet can be directly used for the purification procedure or snap frozen in liquid N_2 and stored at −80 °C until further use.

3.3 Purification of Biotinylated Protein

Protein purification should be carried out at 4 °C or on ice and purified protein stored at −80 °C in small aliquots.

1. Resuspend the insect cell pellet in the 50-mL tube by adding 40-mL buffer A supplemented with PMSF (1 mM) and DNase I (10 μg/mL). Resuspending can be achieved by adding a small stir bar to the 50-mL tube and placing it on a magnetic stirrer or by manually pipetting the pellet up and down.
2. Lyse the cells by sonication for 5 min (pulse 5 s ON/5 s OFF, 30% amplitude). Place the 50-mL tube on ice for sonication. Make sure that the sonicator probe is well emerged in the lysate before starting the sonication.
3. Spin down the lysate by centrifugation at 75,000 g for 45 min. Filter the supernatant using a syringe and syringe filters (0.45 μm). If the filter is clogged, use a new filter. When the filters clog immediately, diluting the supernatant in buffer A or adding more DNase I can help.

4. Transfer 5 mL of amylose resin into a 50-mL tube and wash 3× with 15-mL buffer A. To pellet the resin, centrifuge for 1 min at 500 g (set the break to the lowest value possible). Take a 6-μL sample of the supernatant and directly add it to 2 μL of 4× SDS sample buffer for analysis by SDS-PAGE.
5. Add the supernatant of the centrifuged lysate (*see* **step 3**) to the equilibrated amylose resin. Incubate at 4 °C for 1 h while rotating.
6. Prepare a Protino column to collect the resin (e.g., using a laboratory support stand) and 50-mL tubes to collect the different fractions of the purification.
7. Pour the entire lysate with the resin into the Protino column in a stepwise manner and collect the flow-through in a 50-mL tube. Take a 6-μL sample of the flow-through and directly add it to 2 μL of 4× SDS sample buffer for analysis by SDS-PAGE.
8. Wash the resin 3× by pouring 50 resin volumes of buffer A into the column. Collect each wash in a separate 50-mL tube. Take a 10-μL sample of each wash fraction for analysis by SDS-PAGE.
9. Elute the protein by adding two resin volumes of buffer B and collect in a 15-mL tube. Repeat the elution step five times. Take 5 μL of each elution fraction for analysis by SDS-PAGE.
10. Check the different fractions of the affinity purification in SDS-PAGE to confirm the presence and purity of the protein of interest.
11. Determine the protein concentration using a NanoDrop photometer. Depending on the concentration, it may be required to concentrate using an Amicon Ultra Centrifugal Filter (for the described NLRP1 construct, a 100-kDa molecular weight cutoff was used). Reaching a concentration of 1 μM is recommended.
12. After eluting (and concentrating), immediately prepare small aliquots (e.g., 20 μL), snap-freeze in liquid N_2, and store at −80 °C until further use. Wear appropriate protective equipment when handling liquid nitrogen (cryo gloves and a face shield).
13. Purification of His_6-DPP9 was previously described [11]. Note that higher analyte concentrations (up to 10 μM) may be required for kinetic measurements.

3.4 Surface Plasmon Resonance Measurement

All steps are described for a Biacore 8 K SPR system. Make sure the loading and positioning of different solutions in different plates outlined below are in accordance with the corresponding scheme provided by the software. For other SPR spectrometers, some parts of the protocol may be different (e.g., required volumes). Always pay attention to the specific requirements of your SPR device.

1. Set up and save the method for the immobilization procedure in the Biacore™ 8 K Control Software.

Startup cycle:	Contact time, 60 s; dissociation time, 60 s; flow rate, 30 μL/min (run 2× cycles).
Surface activation:	Contact time, 60 s; flow rate, 10 μL/min, 3×.
Ligand injection:	Flow cell 1 only; contact time, 1200 s (*see* **Note 15**); flow rate, 2 μL/min.
Wash:	Use "Wash" command.
Wait:	Use "Wait" command, 1200 s (baseline stabilization).

(Surface activation steps, ligand injection, wash and wait can be run within the same cycle.)

2. Set up and save the method for the kinetic measurement in the Biacore™ 8 K Control Software. For the kinetic measurement, both a single-cycle and a multi-cycle approach can be used. Start with a single-cycle approach with a dilution series of three to five concentrations (also *see* **step 11**) to screen for a suitable concentration range for kinetic measurements. Iterative repeats of this step may be required to find the ideal concentration range. For a robust measurement, the dilution series should ideally have at least five concentrations and cover a range of 0.1–10 × K_D.

Startup cycle:	Contact time, 60 s; dissociation time, 60 s; flow rate, 30 μL/min (run 5× cycles).
Single cycle kinetics:	Contact time, 120 s; dissociation time, 240 s; flow rate, 30 μL/min (run 2 cycles, the first cycle being the blank cycle with only buffer injections and the second cycle being the actual measurement cycle).
Regeneration:	Contact time, 30 s; flow rate, 30 μL/min (*see* **Note 6**).

3. Thaw the ligand (protein to be immobilized) and analyte (interaction partner to be measured) and keep them on ice. Prepare the running buffer (*see* **Note 16**).
4. Clean the SPR device using the "Desorb" function directly before mounting the sensor chip. Transfer 900 μL of Desorb solution 1 into the first row of a 1-mL deepwell plate and 900 μL of Desorb solution 2 into the second row. Place the deepwell plate in corresponding position of the hotel and start the procedure.
5. Let the sensor chip equilibrate at room temperature for 15 min before opening the packaging. Dock the Series S Sensor Chip SA. Connect your buffer and equilibrate the system using the "Change solution" function. Allow the system to run in standby flow for 15 min.

6. Run the "Normalize" function with normalizing solution. Transfer 110 μL into a row of a 96-well plate. Place the plate in the corresponding position of the hotel. Run the procedure and let the system run in standby flow for 15 min after it is done.
7. Based on the molecular weight of your ligand and analyte, calculate the level of protein you need to immobilize to achieve a theoretical R_{max} (maximum expected response of an analyte of interest) of 100 RU using the following formula:

$$R_{ligand} = \frac{R_{max} \cdot MW_{ligand}}{MW_{analyte} \cdot Valency_{ligand}} \text{ (RU)} \quad (\textit{see } \textbf{Note 17})$$

R_{ligand}: Response obtained after ligand immobilization

MW_{ligand}: Molecular weight of ligand (Da)

$MW_{analyte}$: Molecular weight of analyte (Da)

$Valency_{ligand}$: Binding sites for analyte on ligand (e.g., 1 for a 1:1 binding)

8. Prepare 5 mL of sensor chip activation solution and wash solution in 15-mL tubes (*see* **Note 5**).
9. Prepare your ligand by diluting it in running buffer in a 1.5-mL reaction tube. Typically, a ligand concentration of 100 nM is a good starting point (*see* **Note 18**).
10. Prepare a 96-well plate with all the solutions needed according to the scheme provided by the control software in the immobilization method and start the procedure. Only run the procedure for a single channel on the sensor chip. Note the resulting immobilization level (Fig. 1) and calculate the theoretical R_{max} using the equation in 7. The resulting theoretical R_{max} should be at least 20 for an initial single-cycle kinetics measurement. If it is below this threshold, increase the concentration of the ligand or the contact time and repeat the immobilization.
11. Prepare the dilution series of the analyte in running buffer. Start with a concentration of 5000 nM and set up a series of five dilutions (1:5 → 5000, 1000, 250, 50, 10 nM).
12. Transfer the dilution series of the analyte to a 96-well plate according to the scheme provided by the control software in the kinetic measurement method and start the procedure.
13. If a regeneration step is included in the kinetic measurement, it is advisable to repeat the kinetic measurement with the same

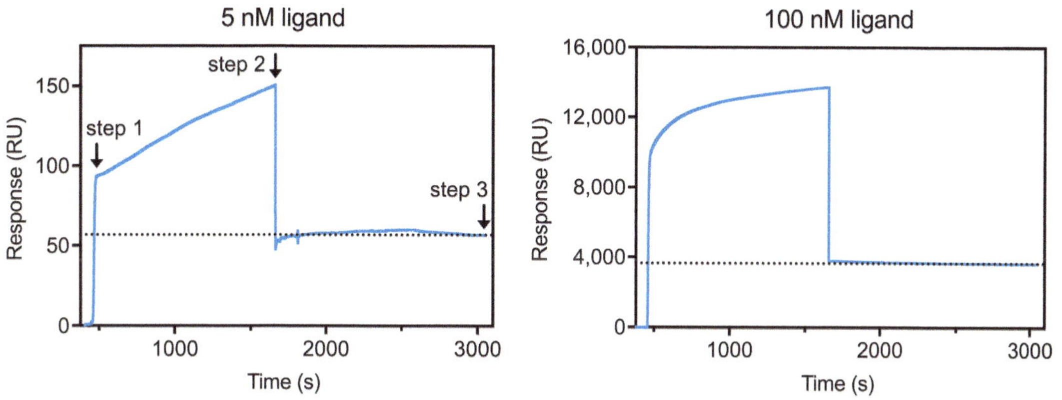

Fig. 1 Biotin-MBP-NLRP1 immobilization at different concentrations. The sensorgrams visualize the increase of the response upon injection of the Biotin-MBP-NLRP1 (step 1). The drop of RU at the end of the ligand injection (step 2) is something we frequently observe when immobilizing NLR proteins. After a stabilization of the ligand on the chip (steps 2–3), the final response can be observed from the sensorgram (step 3)

dilution series. If the regeneration was successful, the results will be comparable with the previous measurement (*see* **Note 19**).

14. Carry out iterative cycles of the immobilization and kinetic measurement procedures to identify the ideal concentrations for the ligand and analyte (*see* **Note 20**).
15. Run a final experiment using the previously identified concentrations for the ligand and analyte to determine kinetic constants of the interaction (Fig. 2) (*see* **Note 21**).
16. Choose a suitable model to fit the kinetic measurement. Typically, a 1:1 model is used. However, different models may be used for more complex interactions. In addition, plot the steady-state affinity in the evaluation software (Fig. 2, right panel) (*see* **Note 22**).
17. Compare the previously calculated to the observed R_{max}. If the observed R_{max} is too high, this may indicate unspecific binding. If it is much lower than the calculated R_{max}, this may indicate a high ratio of inactive (i.e., not able to interact) immobilized ligand (*see* **Note 23**).
18. As a quality control of the obtained data, repeat the final experiment with different levels of immobilized protein. The kinetic constants should be comparable between experiments independently of the immobilization level, provided that the signal is still sufficient at each analyte concentration.

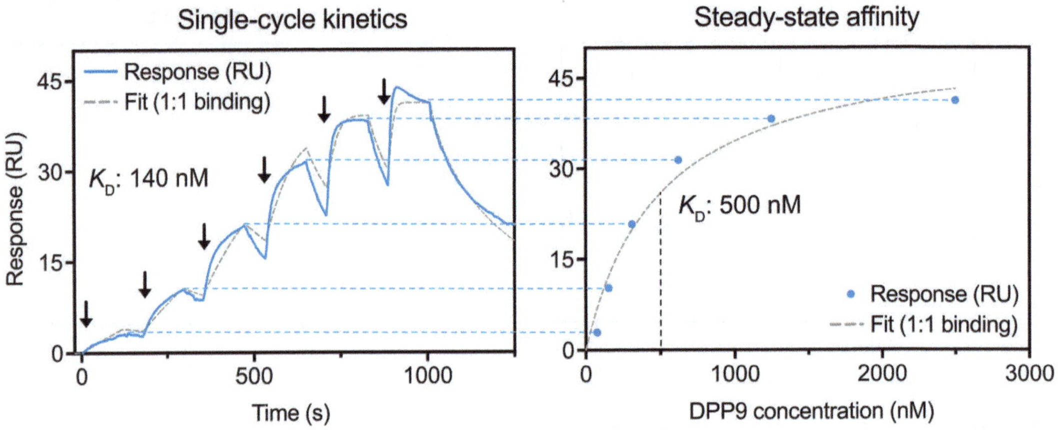

Fig. 2 Single-cycle kinetics of Biotin-MBP-NLRP1 and His$_6$-DPP9. The sensorgram (left panel) shows the double-referenced signal (Fc2–Fc1, analysis cycle–blank cycle) of a single-cycle kinetics measurement of His$_6$-DPP9 (analyte) on Biotin-MBP-NLRP1 (ligand). Each arrow indicates the injection of increasing concentrations of analyte (78.125, 156.25, 312.5, 625, 1250, 2500 nM). Kinetic constants such as the K_D (140 nM) can be derived from the 1:1 binding fit. This can be done using the Biacore Insight Evaluation software. The steady-state affinity (right panel) can be derived directly from the kinetic measurement. In the displayed plot, the last 5 s of each injection were used to determine the steady-state response. The resulting values were then plotted over the concentration of the corresponding injection. If the signal reaches a steady state for each injection, the steady-state affinity can also be used to calculate a K_D (here 500 nM) (*see* **Note 22**)

4 Notes

1. This primer binds to a region coding for a TEV protease cleavage site and is used as a universal primer in our lab to generate vectors for fusion constructs with a tag and a TEV cleavage site.
2. The Bluo-Gal stock solution needs to be prepared in a glass tube. Since DMF is a solvent, it should be handled with appropriate protective equipment under a fume hood.
3. We recommend using polypropylene U-bottom plates for any experiment. Some experimental setups may require low percentages of DMSO in the samples and running buffer, which is incompatible with polystyrene.
4. Prepare a stock solution of 2-M NaCl in 100-mM NaOH. Dilute 1:1 in water to obtain activation solution and 1:1 in 2-propanol to obtain wash solution. The stock solution can be stored at RT. 2-propanol is a solvent, and it should be handled with appropriate protective equipment under a fume hood.
5. The ideal regeneration solution can vary significantly for different ligands. Sometimes regeneration is not possible at all without affecting stability of the immobilized ligand. To find ideal conditions for regeneration, the regeneration scouting protocol from Cytiva is a useful guide.

6. According to the manual, the reaction time of a digestion with HF enzymes can be as short as 5–15 min. We typically do 1 h digests, since it results in complete cleavage without seeing any unspecific fragments.
7. This purification step does not require running a gel. Simply use the same kit and directly apply digested PCR product to the column after mixing it with the appropriate buffer. Note that this buffer usually differs from the buffer used to dissolve the agarose gel.
8. Ligation may also work with an incubation for 2 h at RT. However, from experience, ligation at 16 °C o/n is more efficient.
9. A test digest (with the respective restriction enzymes used for cloning) or colony PCR (with specific primers) can be carried out as a control prior to sequencing possible clones.
10. The manual of the bac-to-bac system describes the entire process including the handling in detail. However, a slightly different protocol is used in our laboratory. For further reference, the bac-to-bac manual can still be used [12].
11. No pellet might be visible after this step. Continue with procedure until resuspending the DNA in water and use a NanoDrop to confirm the presence of bacmid DNA.
12. Since no antibiotics are used in insect cell culture, it is important to be extra careful to avoid contaminations.
13. For easy loading of the supernatant into the syringe, remove the plunger of the syringe and attach the sterile filter. Pour the supernatant into the syringe and apply the plunger while placing the bottom of the syringe with the filter over a collection tube.
14. For larger volumes, the sterile filter might get clogged. Simply replace the filter with a new one, if this happens. Alternatively, a sterile bottle top filter can be used instead of a syringe filter.
15. Depending on the required immobilization level, longer contact times up to 60 min are possible. This can be useful when measuring low molecular weight analytes. If a sufficient amount of immobilization is achieved before the end of the injection time set in the method, the immobilization can be manually stopped by using the stop injection button. However, we noticed a drop in the RU for NLRP1 at the end of the injection (Fig. 1). This has to be taken into account when manually stopping the injection.
16. Always prepare the running buffer freshly just before starting an experiment. Even minor contaminations or precipitate can significantly affect the signal-to-noise ratio of the measurement.

17. Exemplary calculation for required immobilization level:
    Desired R_{max}: 100 RU
    MW_{ligand}: 209,000 (Da)
    $MW_{analyte}$: 100,000 (Da) (DPP9 forms dimers in solution → 200,000 Da)
    $Valency_{ligand}$: 1 (one dimer of DPP9 binds one NLRP1 molecule)

$$\frac{100\,(\mathrm{RU})\cdot 209{,}000\,(\mathrm{Da})}{200{,}000\,(\mathrm{Da})\cdot 1} = 104.5\,(\mathrm{RU})$$

18. Similar to the contact time, this concentration may be adjusted, if higher or lower immobilization levels are required.
19. Regeneration scouting is advised if the protein seems to be difficult to regenerate. For NLRP1 we did not find suitable regeneration conditions. In such a case, no regeneration is advised. However, multiple measurements on the same channel without regeneration should be analyzed with care. For instance, if the analyte does not completely dissociate from the sensor chip after the final dissociation, there will be less ligand available for binding in subsequent measurements, most likely resulting in a lower response.
20. To obtain robust data, it is generally best to use the lowest immobilization level possible while maintaining a sufficient signal-to-noise level. Thus, the ideal immobilization level depends on the interaction you are investigating, especially the molecular weight of the analyte, and the sensitivity of the SPR system you are using.
21. Since we did not find suitable regeneration conditions for NLRP1, we chose to run the final experiment using a single-cycle kinetics approach. With suitable regeneration conditions, running an experiment using a multi-cycle kinetics approach can have certain advantages such as easy overlay of different concentrations.
22. It can be difficult to reach a steady state (i.e., a stable response) for each injection in single-cycle kinetics experiments especially at lower concentrations. This will affect the accuracy of the steady-state data (Fig. 2). In that case, kinetic constants should be derived from the single-cycle kinetic data and corresponding fit. Running multi-cycle kinetics is often more suitable to reach a steady-state even at low analyte concentrations.
23. In both cases, buffer optimization can improve the performance. A different pH and even a change of the buffer substance (e.g., Tris-HCl instead of HEPES) can have significant effects. Another strategy to reduce unspecific binding is to add carboxymethyl-dextran to the buffer at a final concentration of 1 mg/mL.

Acknowledgment

This work is supported by the Deutsche Forschungsgemeinschaft (DFG) under Germany's Excellence Strategy – EXC2151–390873048.

References

1. Schmid-Burgk JL, Chauhan D, Schmidt T et al (2016) A genome-wide CRISPR (clustered regularly interspaced short palindromic repeats) screen identifies NEK7 as an essential component of NLRP3 inflammasome activation. J Biol Chem 291:103–109
2. Lightfield KL, Persson J, Brubaker SW et al (2008) Critical function for Naip5 in inflammasome activation by a conserved carboxy-terminal domain of flagellin. Nature Immunol 9:1171–1178
3. Zhong FL, Robinson K, Teo DET et al (2018) Human DPP9 represses NLRP1 inflammasome and protects against autoinflammatory diseases via both peptidase activity and FIIND domain binding. J Biol Chem 293: 18864–18878
4. Robert Hollingsworth L, David L, Li Y et al (2021) Mechanism of filament formation in UPA-promoted CARD8 and NLRP1 inflammasomes. Nat Commun 12:189
5. Huang M, Zhang X, Toh GA et al (2021) Structural and biochemical mechanisms of NLRP1 inhibition by DPP9. Nature 592: 773–777
6. Karlsson R, Michaelsson A, Mattsson L (1991) Kinetic analysis of monoclonal antibody-antigen interactions with a new biosensor based analytical system. J Immunol Methods 145:229–240
7. Karlsson R (2004) SPR for molecular interaction analysis: a review of emerging application areas. J Mol Recognit 17:151–161
8. Fairhead M, Howarth M (2015) Site-specific biotinylation of purified proteins using BirA. Methods Mol Biol 1266:171–184
9. Duffy S, Tsao K, Waugh DS (1998) Site-specific, enzymatic biotinylation of recombinant proteins in Spodoptera frugiperda cells using biotin acceptor peptides. Anal Biochem 1(128):122–128
10. Tsao K, Debarbieri B, Michel H et al (1996) A versatile plasmid expression vector for the production of biotinylated proteins by site-specific, enzymatic modification in Escherichia coli. Gene 169:59–64
11. Ross B, Krapp S, Augustin M et al (2018) Structures and mechanism of dipeptidyl peptidases 8 and 9, important players in cellular homeostasis and cancer. Proc Natl Acad Sci 115:E1437–E1445
12. Invitrogen (2015), Bac-to-Bac® Baculovirus Expression System. http://tools.thermofisher.com/content/sfs/manuals/bactobac_man.pdf

Chapter 12

Functional Assessment of Disease-Associated Pyrin Variants

Daria Chirita, Yvan Jamilloux, Thomas Henry, and Flora Magnotti

Abstract

The pyrin inflammasome detects effectors and toxins that inhibit RhoA GTPases and triggers inflammatory cytokines release and a fast cell death termed pyroptosis. Ancient plague pandemics in the Mediterranean basin have selected in the human population pyrin variants that can trigger an autoinflammatory disease termed familial Mediterranean fever (FMF). In addition, distinct mutations in *MEFV*, the gene encoding pyrin, cause a different rare autoinflammatory disease termed pyrin-associated autoinflammation with neutrophilic dermatosis (PAAND). As of today, more than 385 *MEFV* variants have been described although for most of them, whether they are pathogenic variant or benign polymorphism is unknown.

Here, we describe different methods using primary human monocytes or engineered monocytic cell lines to functionally characterize *MEFV* variants, determine their potential pathogenicity, and classify them as either FMF-like or PAAND-like variants.

Key words Pyrin, Inflammasome, Familial Mediterranean fever, Pyrin-associated autoinflammation with neutrophilic dermatosis, MEFV, Pyroptosis, Real-time cell death assay, IL-1β

1 Introduction

MEFV encodes the pyrin protein, an important player of the antibacterial innate immune responses [1–5]. Mutations in this gene are associated with two auto-inflammatory disorders: familial Mediterranean fever (FMF), the most frequent hereditary autoinflammatory disease [6–8], and a rare disease termed pyrin-associated autoinflammation with neutrophilic dermatosis (PAAND) [9]. In these patients, *MEFV* mutations lead to a deregulation in the pyrin-associated inflammatory responses causing recurrent inflammation [10, 11].

Pyrin is an inflammasome sensor on the front line of the innate immune response against pathogens. Inflammasomes work as multi-protein complexes formed in response to cytosolic pathogen-associated molecular patterns (PAMPs), endogenous damage-associated molecular patterns (DAMPs), or homeostasis-

Thomas A. Kufer and Maria Kaparakis-Liaskos (eds.), *Effector-Triggered Immunity: Methods and Protocols*, Methods in Molecular Biology, vol. 2523, https://doi.org/10.1007/978-1-0716-2449-4_12,

altering molecular processes (HAMPs) [12]. The following three major events characterize pyrin inflammasome formation and can be used as readouts to monitor inflammasome activation:

1. ASC (apoptosis-associated speck-like protein containing a CARD) speck formation. ASC is the inflammasome adaptor. It oligomerizes to form a "speck" in the cytosol as the inflammasome complex is assembled [13]. ASC speck constitutes a platform onto which the inflammasome effector, caspase-1, is recruited and activated.
2. Pyroptosis. Pyroptosis is an inflammatory form of cell death due to the caspase-1-mediated cleavage of gasdermin D. Gasdermin D is a pore-forming protein that, upon cleavage, oligomerizes in the plasma membrane and mediates a fast necrotic cell death [14, 15].
3. The release of pro-inflammatory cytokines (interleukin-1β (IL-1β) and interleukin-18 (IL-18)) into the extracellular space [12, 16].

Pyrin acts as a guard mechanism for RhoA GTPases inhibition [2–4]. Numerous bacterial effectors such as YopE and YopT from *Yersinia* spp. [3, 4] and TecA from *Burkholderia* [5] and numerous toxins such as *Clostridioides difficile* toxins A and B (TcdA/TcdB) inhibit RhoA by covalent modifications. Pyrin does not recognize neither the bacterial effectors/toxins nor the covalent modifications but senses RhoA inhibition [2]. Indeed, at steady state, RhoA maintains pyrin inactive via the dynamic activation of the two RhoA effectors, PKN1 and PKN2. PKN1 and PKN2 are kinases from the PKC superfamily that phosphorylate pyrin on two serine residues (S208 and S242). The phosphorylated pyrin binds 14-3-3 chaperone proteins, which sequester pyrin away from an active inflammasome [17, 18]. When RhoA GTPases are inhibited, PKN1 and PKN2 are no longer activated, and pyrin is no longer phosphorylated. In the dephosphorylated state, pyrin is able to assemble the inflammasome complex and triggers the inflammatory response [10, 17–19]. *MEFV* mutations affecting one of the phosphorylation sites or 14-3-3 binding sites alleviate this regulatory mechanism and are causing PAAND [9, 10, 20].

The pyrin inflammasome activation mechanism has not been fully deciphered, but we demonstrated that pyrin dephosphorylation is insufficient to activate the inflammasome. Indeed, blocking pyrin phosphorylation through PKN1/PKN2 inhibition (using the PKC superfamily inhibitor UCN-01) or genetic approaches (by mutating the S208/S242 residues) does not activate pyrin in primary monocytes from healthy donors. On the contrary, dephosphorylation of pyrin is sufficient to activate the pyrin inflammasome in monocytes from FMF patients [21]. Pyrin activation is thus controlled by two mechanisms, the first one depending on (de)phosphorylation which is constitutively activated in PAAND patients and a second poorly understood mechanism, which is constitutively activated in FMF patients. This differential response to pyrin dephosphorylation (i.e., to UCN-01 treatment) is at the basis of the assays described here to discriminate FMF-like *MEFV* variants patients from WT-like variants. Of note, another technique based on the in vitro resistance to colchicine inhibition has been previously described [19]. Furthermore, combining a PAAND mutation (constitutive activation of signal 1) with a FMF mutation (constitutive activation of signal 2) results in a constitutively active pyrin inflammasome. We thus present a technique based on genetic engineering of a monocytic cell line to classify *MEFV* variants for their potential pathogenicity (FMF-like or PAAND-like variant). Finally, we detail methods to investigate the impact of *MEFV* mutation on pyrin phosphorylation level. Altogether, the techniques presented here allow to monitor activation of the pyrin inflammasome using several stimuli and several readouts, in primary human monocytes and in *MEFV* variant-expressing monocytic cell lines to be able to assess the functional impact of *MEFV* mutations and determine their potential pathogenic impact.

2 Materials

2.1 Pyrin-Expressing U937 Cell Lines

Generation of pyrin-expressing U937 cell lines is not described as a specific method since these cells are generated by standard transduction methods [21]. 3xFLAG-*MEFV*-expressing lentiviral plasmids are available at Addgene (e.g., #134702, 134703, 134706 for 3xFLAG-WT, p.M694V, and p.S208C/S242R/M694V *MEFV*-expressing plasmids, respectively). Site-directed mutagenesis of *MEFV* gene to generate novel sequence variant of interest can be performed either on the lentiviral plasmid or on the parental entry plasmid (Addgene #167018, 167118, 167020 for pENTR1A-WT, p.M694V, p.S242R *MEFV*) (*see* **Note 1**). pINDUCER21 vector [22] derivatives are used and contain GFP for FACS sorting-based selection and a doxycycline-inducible promoter driving *MEFV* expression.

1. Cell culture medium: RPMI 1640 with GlutaMAX, 10% fetal calf serum (FCS).
2. Doxycycline (Sigma, D9891): stock at 1 mg/mL in water, filter sterilize, aliquot, and store at −20 °C. Final concentration: 1 μg/mL (*see* **Note 2**).

Table 1
Cell number required per sample for the different experiments

Experiment	Number of U937 cells/ condition	U937 cells/ mL	Number of primary cells/ condition	Primary cells/mL
IP	2×10^6	1×10^6	–	–
IF	2×10^5	2×10^6	12×10^4	1×10^6
PI incorporation	5×10^4	0.5×10^6	2×10^4	0.2×10^6
ELISA IL-1β	5×10^4	0.5×10^6	5×10^3	0.05×10^6
ELISA IL-18	2×10^5	2×10^6	5×10^3	0.05×10^6

2.2 Primary Monocytes

Isolation of primary monocytes is not described as a specific method since standard procedures are used. Experimental work on primary human cells requires ethical approval. Procedures have been optimized to limit the cell number [23] (Table 1), and we routinely perform several inflammasome stimuli and readouts with 4 mL of blood drawn in heparin-coated tubes. Tubes are kept overnight at room temperature (RT) before isolation the next day using density gradient centrifugation and CD14-based positive selections [21, 24].

2.3 Pyrin Phosphorylation Level

1. Lysis buffer: 25-mM Tris HCl, pH 7.0, 150-mM NaCl, 1-mM EDTA, 0.1% Igepal CA-630 (NP40) supplemented with 10-mM sodium fluoride (phosphatase inhibitor) and cOmplete™ EDTA-free Protease Inhibitor Cocktail (Roche, 05056489001).
2. Anti-FLAG beads: Anti-FLAG M2 Affinity Gel (Sigma, A2220) or Anti-FLAG® M2 Magnetic Beads (Sigma, M8823).
3. WB blocking buffer: Tris-buffered saline buffer (25-mM Tris-base, 137-mM NaCl, 2.7-mM KCl, pH 7.5), 1% (v/v) Tween, 5% (w/v) skimmed milk.
4. Antibodies: Anti-Pyrin (phospho S241) antibody [EPR19570] (Abcam, ab200420) diluted 1:1000, anti-Pyrin (phospho S205) antibody [EPR19567] (Abcam, ab201784) diluted 1: 1000 (*see* **Note 3**), anti-Pyrin (human), pAb (AL196) (AdipoGen, AG-25B-0020-C100) diluted 1:2000, Anti-Actin Antibody clone C4 (Millipore, MAB1501) diluted 1:5000.

2.4 ASC Speck Immunofluorescence

1. Fixative: 2% (w/v) paraformaldehyde in PBS.
2. Cytocentrifuge: Cytospin 3 (Shandon), cytofunnels, cytoclips, disposable filter cards (Thermo Scientific, 5991022).
3. Permeabilization buffer (PBS-T): PBS, 0.1% (v/v) Triton X100.

Table 2
Inflammasome activators used and working and stock concentrations

Compound	Working concentration in U937 assays	Working concentration in primary monocytes assays	Stock concentration; solvent; reference
LPS	0.5 μg/mL for IL-1β 1 μg/mL for IL-18 and RT cell death assay (in the case of the LPS + nigericin stimulus)	10 ng/mL	5 mg/mL; water; InvivoGen (tlrl-3pelps)
Nigericin	50 μg/mL	5 μg/mL	5 mg/mL; ethanol; InvivoGen (tlrl-nig)
TcdA	1 μg/mL	1 μg/mL	380 μg/mL; water; purified from *Clostridioides difficile* VPI10463 strain, as previously described [27]
TcdB	–	125 ng/mL	200 μg/mL; water; Abcam (ab124001)
UCN-01	12.5 μM	12.5 μM	10.4 mM; DMSO; Sigma (U6508)

4. Blocking buffer: PBS, 0.1% (v/v) Triton X100, 3% (w/v) BSA, 3-mM NaN_3.
5. PAP thick pen for immunostaining (Dutscher, 490001).
6. Antibodies: Anti-ASC (Santa Cruz, sc22514R, 4 μg/mL); Alexa 488-goat anti-rabbit antibodies (Invitrogen, A-110088, 10 μg/mL).
7. DAPI: Stock at 1 mg/mL in water, use at 100 ng/mL in PBS-T.
8. Fluoromount-G Mounting Medium (Invitrogen, 00-4958-02).

2.5 Inflammasome Stimuli

Inflammasome stimuli are presented in Table 2.

2.6 Real-Time Cell Death Assay

1. 96-well plates black flat bottom.
2. Supplemented CO_2-independent medium: CO_2-independent medium supplemented with 10% FCS, 2-mM glutamine.
3. Propidium iodide: Stock 1 mg/mL in water, filter sterilize, use at 5 μg/mL final.
4. Triton X100 (Sigma, X100): Stock at 10% (v/v) in water, use at 1% final.
5. Required apparatus: Plate fluorimeter with thermoregulation (CO_2 chamber dispensable).

2.7 ELISA

1. Nunc MaxiSorp™ flat bottom, for ELISA assay.
2. Phorbol 12-myristate 13-acetate (PMA) (InvivoGen, tlrl-pma): Stock at 5 mg/mL in water, use at 100 ng/mL final.
3. ELISA IL-1β: Human IL-1 beta/IL-1F2 DuoSet ELISA (R&D, DY201).
4. ELISA IL-18: Capture antibody, anti-human IL18 (R&D, D044-3) at 4 μg/mL; detection antibody, biotin-labeled anti-human IL18 (R&D, D045-6) at 500 μg/mL; standard protein, recombinant human IL18 (R&D, B001-5).

3 Methods

To functionally assess the role of *MEFV* mutations, we evaluate the activity of the corresponding pyrin inflammasome in response to various stimuli (e.g., UCN-01 for FMF-like variants) or by combining them with the well-known FMF mutation (p.M694V) or PAAND mutation (p.S242R). The decision tree to classify novel pyrin variants as FMF-like, PAAND-like, or likely benign based on the techniques described here and the current knowledge in the field is presented in Fig. 1. Pyrin phosphorylation level can be investigated as an additional mechanistic insight especially for PAAND-like mutations. Three readouts for pyrin inflammasome activation are routinely examined: (i) ASC speck formation, (ii) pyroptosis, and (iii) the release of pro-inflammatory cytokines. All these analyses can be performed using *MEFV*-expressing U937 cell lines (*see* Subheading 2.1) or primary human monocytes, with the exception of pyrin phosphorylation monitoring by Western blot analysis, which has been optimized for cell lines.

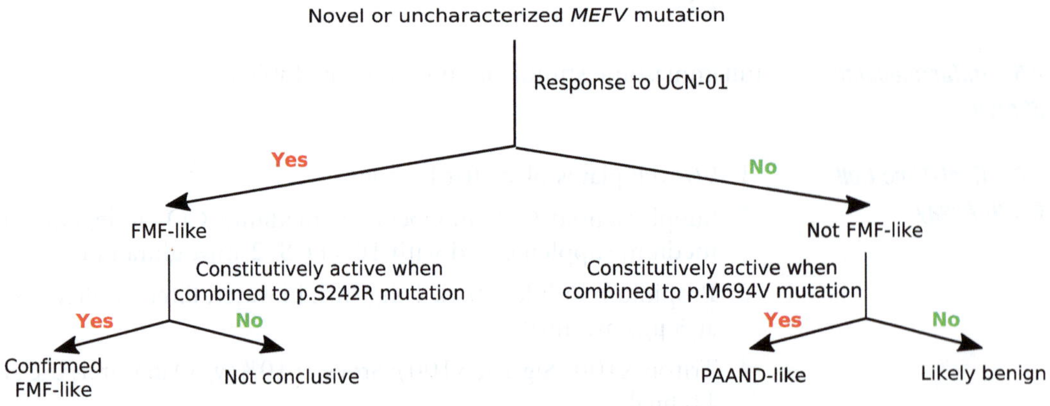

Fig. 1 Decision tree to classify a novel or uncharacterized *MEFV* variant based on the techniques presented here and the current knowledge in the field. Only the first step is applicable to primary human monocytes

3.1 Determination of Pyrin Phosphorylation Level by Immunoprecipitation Followed by Western Blot

This protocol aims at determining the phosphorylation level of serine 208 and 242 residues of pyrin using 3xFlag-*MEFV*-expressing U937 cell lines. Typical results are presented Fig. 2.

1. Incubate the cells in cell culture medium supplemented with 1-μg/mL doxycycline overnight at 37 °C, 5% CO_2, to induce *MEFV* expression. Keep 2×10^6 U937 cells without doxycycline (or a control U937 cell line not expressing a flag-tagged protein) to saturate the anti-FLAG beads.
2. Count the cells and seed the appropriate amount (Table 1) in fresh medium. Proceed to the treatment (e.g., UCN-01, 12.5 μM, 15 min).
3. Collect cells in 1.5-mL tubes. Wash the cells once with 1-mL ice-cold PBS (centrifuge 5 min 120 rcf 4 °C). Resuspend the

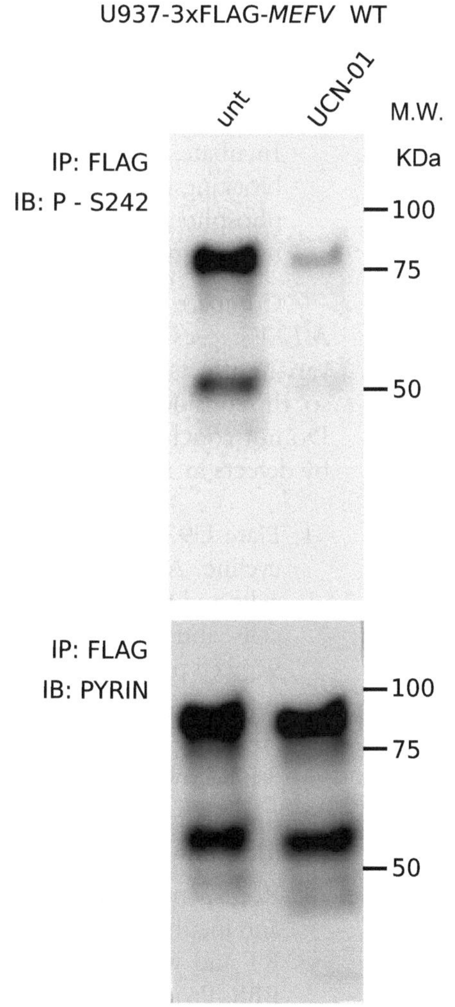

Fig. 2 Phosphorylation level of pyrin S242 as determined by immunoblotting (IB) following immunoprecipitation (IP). Treatment with UCN-01 (12.5 μM, 15 min) triggers pyrin S242 dephosphorylation. Unt: untreated. MW: molecular weight

cell pellet in 400 μL of ice-cold lysis buffer. Lyse cells during 1 h at 4 °C on a rotating wheel or store at −80 °C. Clarify the lysate by centrifuging 10 min at 14,000 rcf 4 °C. Set aside 30 μL for input control (add 10 μL of loading dye, boil for 5 min at 95 °C, and store at −20 °C).

4. Take 10 μL of beads slurry per sample. Wash three times with 1-mL lysis buffer. Add the clarified U937 lysates (400 μL, without doxycycline) to the beads to saturate the non-specific binding sites. Incubate on a rotating wheel at 4 °C for 30 min. Wash once in lysis buffer and resuspend in 100 μL of lysis buffer/sample.
5. Add the clarified lysate sample to the 100-μL beads mix. Incubate on a rotating wheel at 4 °C for 2 h. Wash the beads three times in ice-cold lysis buffer. Resuspend beads in 20-μL loading dye, boil for 5 min at 95 °C, and proceed to Western blot analysis or store at −20 °C.
6. Perform a classic SDS-PAGE followed by transfer on PVDF membrane. Block membranes in WB blocking buffer for 1 h. Incubate membranes with primary antibodies diluted in WB blocking buffer for 1 h at RT or overnight at 4 °C using anti-phospho pyrin, anti-pyrin, and anti-actin antibodies. Proceed using regular Western blotting protocol.

Compare phosphorylation level to WT, p.S208C, and p.S242R *MEFV*-expressing cells to evaluate whether the considered *MEFV* variant affects serine residue 208 or 242 phosphorylation level. If so, the considered variant can be classified as a PAAND-like variant. Do not conclude on negative results since PAAND can be caused by defects in 14-3-3 binding [9].

3.2 ASC Speck Visualization by Immunofluorescence (IF)

1. Plate U937 cells and incubate overnight with 1 μg/mL doxycycline. Alternatively, for IF on primary monocytes, seed the cells and proceed immediately with the treatments. Count the cells and seed the appropriate cell number in a V-bottom 96-well plate.
2. Proceed with the treatments as applies in your experiment (typically, 1-untreated, 2-UCN-01 12.5 μM 30 min, 3-LPS 3 h followed by nigericin during 90 min).
3. Following the indicated incubation period, centrifuge the plate for 5 min 120 rcf at RT and wash with PBS.
4. Resuspend the cells in 100 μL of paraformaldehyde 2% in PBS and incubate for 15 min at 37 °C. Centrifuge for 5 min 120 rcf RT and wash the cells once with PBS. Resuspend in 100 μL PBS. Proceed with the slide preparation or store at 4 °C.
5. Use cytospin to transfer cells onto glass slides. Place the slide in the cytoclip, place the filter card (absorbent side up) and cytofunnel on top, close the cytoclip, and make sure that the circles

in the filter card and the cytofunnel align by looking at the bottom. Place the assembled device in the cytospin and deposit the sample through the funnel. Spin for 5 min at 450 rpm RT. Undo the cytoclip and carefully remove the filter card, without disturbing the cells now stuck onto the slide.

6. Delineate a circle around the samples using a hydrophobic marker and place the slide in PBS-T—the slides should be covered with liquid—to permeabilize the cells. Incubate for 10 min at RT. Replace the PBS-T with blocking buffer and incubate slides for 1 h at RT or store at 4 °C overnight.
7. Prepare a humidity chamber to accommodate all your slides. Prepare 50 μL per sample of anti-ASC antibody in blocking buffer. Refresh the hydrophobic marker circle around the samples. Carefully place 50 μL of the antibody mix over the samples. Incubate in the humidity chamber for 1 h at RT.
8. Wash the slides three times by dipping them in three consecutive 50-mL tubes of PBS-T. Refresh the hydrophobic marker around the samples. Prepare 50 μL per sample of the secondary antibody (e.g., Goat anti-Rabbit-Alexa 594) in blocking buffer and depose it over the samples. Incubate 1 h at RT.
9. Wash three times as described above. Stain with DAPI for 5 min at RT.
10. Wash the slides twice and add one final wash with Milli-Ro water to remove salts. Place 2 μL of the mounting solution in the center of the cover glass, spread it with the tip of the pipette, and carefully place the cover glass on top of the sample, avoiding bubbles. Let the slide dry away from daylight and preferably at 4 °C.
11. Acquire images with a fluorescence microscope or confocal microscope (Fig. 3). If you observe specks formation in response to UCN-01, your cells likely harbor a FMF-like *MEFV* variant (*see* **Note 4**). Compare with primary monocytes from healthy donors (HD) and p.M694V homozygous FMF patient or the corresponding U937 cell lines.

3.3 Real-Time Cell Death Analysis by Propidium Iodide (PI) Incorporation

1. Seed the cells in triplicates for each condition and plan an untreated condition and a "Triton X-100" condition, which will provide the fluorescence value corresponding to 100% of cell death.
2. Count the cells and seed cells in a black flat bottom 96-well plate in supplemented CO_2-independent medium (*see* **Note 5**) in the presence or absence of doxycycline at 1 μg/mL.
3. Add Triton X-100 1% final in the "total cell death" wells. Add propidium iodide (PI) to each well (5 μg/mL final).
4. Immediately start the acquisition of the PI fluorescence, corresponding to the PI incorporation into dead cells, using a

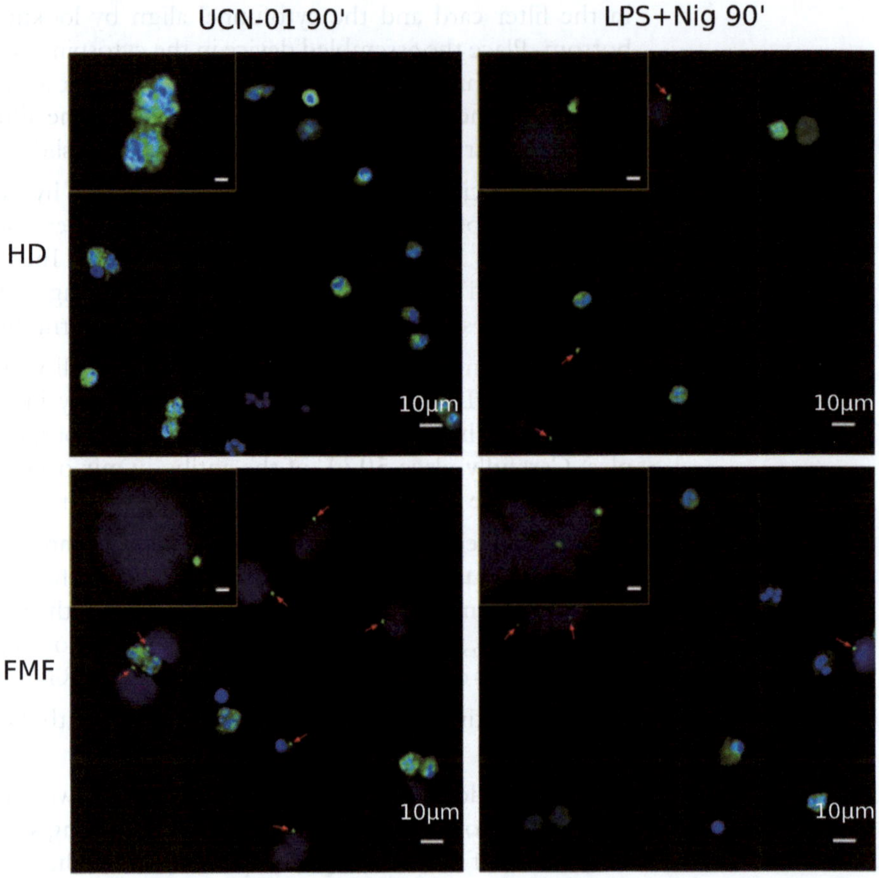

Fig. 3 ASC specks formation in response to UCN-01 discriminates FMF patient's monocytes from healthy donor (HD) monocytes. ASC immunofluorescence on monocytes from one HD and one FMF patient (FMF) treated with the indicated stimuli for the indicated time points. The red arrows indicate the ASC specks. Left: cells were treated with UCN-01 for 90 min. Only FMF monocytes displayed ASC specks. Right: cells were treated with LPS for 3 h and nigericin for 90 min to activate the NLRP3 inflammasome as a positive control (*see* **Note 5**) for ASC specks formation occurring in monocytes from both HD and FMF patient

thermostatic fluorimeter (e.g., Tecan M200) at 37 °C with the following wavelengths: excitation 535 nm (bandwidth 15 nm) and emission 635 nm (bandwidth 15 nm) [25, 26]. To evaluate the cell death associated with *MEFV* variant expression (constitutive activation), measure PI incorporation every 15 min for 16 h.

5. 16 h later, stop the fluorimeter and add inflammasome stimuli (e.g., UCN-01 at 12.5 μM; *see* **Note 6**). Start a novel acquisition with readings every 5 min for 3 h (*see* **Note 7**).

   Alternatively, for primary monocytes, proceed immediately to inflammasome stimuli treatment and short kinetics readings every 5 min for 3 h.

6. To determine the percentage of dead cells for each condition, analyze the data as follows: Calculate the average of the untreated triplicates for each time point. Calculate the average of the Triton X-100 triplicates for each time point. Select the maximum Triton X-100 mean value (this value corresponds to the PI fluorescence corresponding to 100% of dead cells). Determine the percentage of dead cells for each condition (each well) and time point using the following calculation: Cell $death_{well\ X}$ = ((fluorescence value well X − average untreated)/(Max Triton X-100 mean value − average untreated) × 100).
7. Plot the cell death results along time using the appropriate graphing software (e.g., GraphPad Prism) to obtain the cell death curves in relation with time (Figs. 4 and 5).

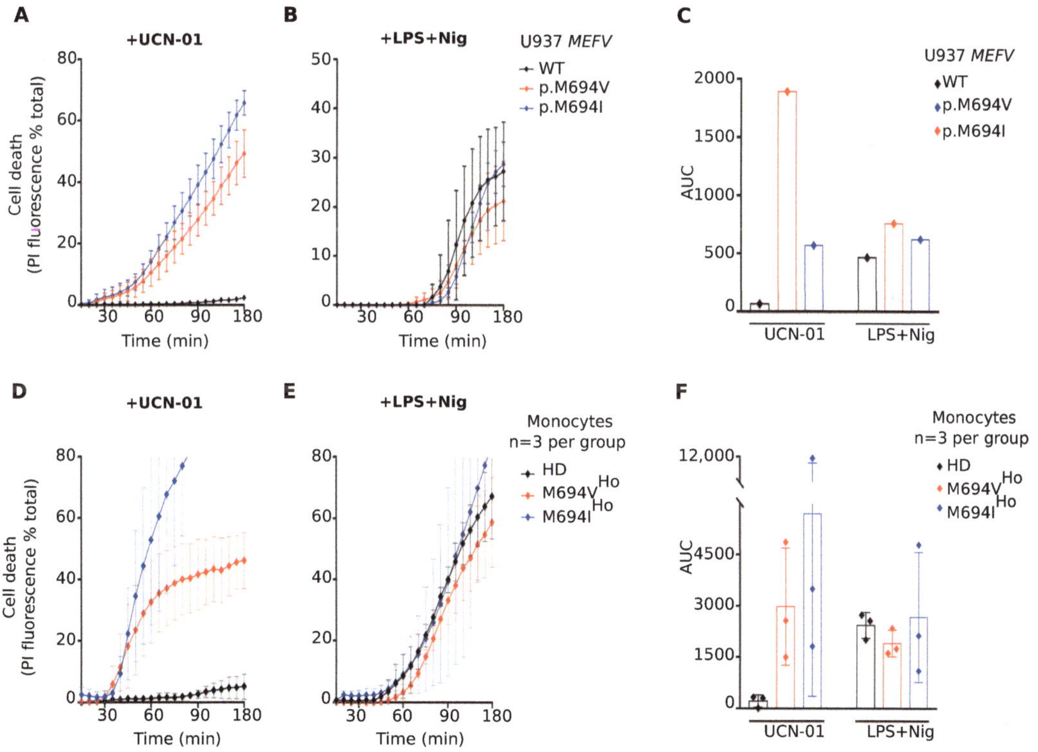

Fig. 4 Characterization of *MEFV* variant by real-time cell death assays (short kinetics). (**a**, **b**) Real-time cell death analysis in U937 cells expressing WT *MEFV* and two prototypical FMF-associated *MEFV* variants p. M694V and p.M694I. (**d**, **e**) Real-time cell death analysis in monocytes from three HD and three FMF patients homozygous for the same variants. UCN-01 treatment induces a fast cell death (after 30–40 min) when cells express a FMF-associated *MEFV* variant. No difference is detected after LPS + nigericin treatment. (**c**, **f**) To quantify these experiments, the Area Under the Curve (AUC) was computed. After UCN-01 treatment, the AUC value is higher when cells express a FMF-associated *MEFV* variant than when cells express the WT *MEFV*. On the contrary, the AUC values after to the LPS + nigericin treatment are similar

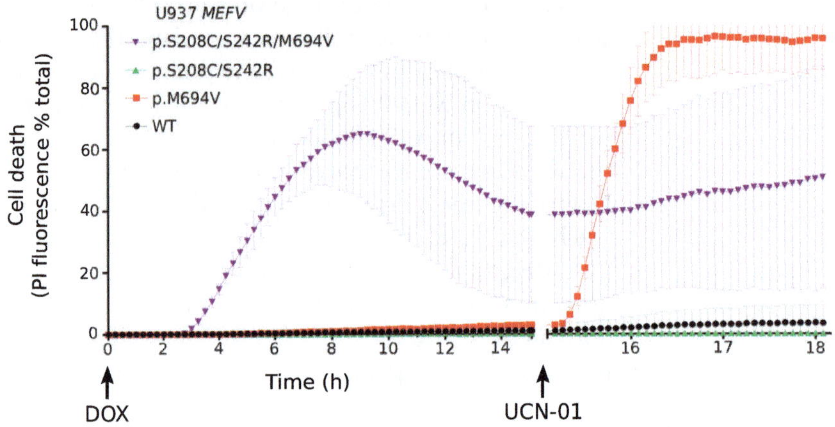

Fig. 5 Characterization of *MEFV* variant by real-time cell death assay (long kinetics). Real-time cell death analysis in U937 cells expressing WT; p.M694V; p.S208C/S242R; p.S208C/S242R/M694V *MEFV* variants. Cell death was monitored for 15 h following doxycycline treatment (DOX). Cells expressing the *MEFV* p.M694V variant do not substantially die in response to doxycycline, indicating that the corresponding pyrin variant is not constitutively active. Yet, these cells die after UCN-01 addition (red line). The p.S208C/S242R *MEFV* variant includes the two prototypical "PAAND" mutations. When the p.S208C/S242R mutations are associated to the "FMF" p.M694V mutation (purple line), the corresponding pyrin variant is constitutively active, and cell death is observed starting at 3 h post-doxycycline addition

Quantifying the corresponding area under the curves (AUC) allows to compare statistically the real-time cell death curves. The AUC can be calculated with specific software such as the GraphPad prism feature: Analyze/XY Analyses/Area under curve (Fig. 4c, f).

If, when combined with p.M694V mutation, the studied *MEFV* mutation triggers constitutive cell death (i.e., in the presence of doxycycline), the corresponding mutation is a PAAND-like mutation. If, when combined with the p.S242R mutation, the studied *MEFV* mutation triggers constitutive cell death (i.e., in the presence of doxycycline), the corresponding mutation is a FMF-like mutation. If the studied *MEFV* mutation triggers UCN-01-mediated cell death, the mutation classifies as a FMF-like mutation (Fig. 1) (*see* **Note 8**).

3.4 Quantification of IL-1β and IL-18 Release by ELISA

1. Seed the cells in triplicates for each condition. To assess IL-1β release, U937 cells require differentiation in macrophages with PMA treatment (*see* **Note 9**). IL-18 quantification does not require PMA differentiation but requires a greater cell number (Table 2). Primary monocytes do not require PMA neither for IL-1β nor for IL-18. Include a "LPS alone" control.
2. Count and plate the cells in a flat bottom 96-well plate at the required concentration in cell culture medium. For U937 cells and IL-1β ELISA, seed cells at day 0 and perform a 3 h pulse

with PMA (100 ng/mL). Wash cells and incubate for 24 h. On day 1, treat with doxycycline (1 μg/mL). 12–16 h later (day 2), add the relevant stimuli. For U937 cells and IL-18 ELISA, start at the doxycycline step. For primary monocytes, proceed directly with the treatments.

3. Spin the plate for 5 min and 120 rcf at RT and replace medium with fresh cell culture medium.
4. Prime with LPS during 3 h at 37 °C, 5% CO_2.
5. Add the signal 2 (e.g., nigericin or UCN-01). Incubate at 37 °C, 5% CO_2 for 3 h (1.5 h for primary monocytes).
6. Spin plates for 5 min and 120 rcf at RT and collect supernatant in a separate 96-well plate. Store at −80 °C until the ELISA is performed according to manufacturer's protocol (Fig. 6).

   Analyze the results in comparison to HD monocytes/WT *MEFV* and p.M694V *MEFV*-expressing U937 cells and conclude on the nature of the studied *MEFV* variant using Fig. 1 decision tree.

4 Notes

1. Site-directed mutagenesis to introduce novel mutation in the *MEFV* sequence can be performed in the lentiviral plasmid (e.g., pINDUCER21-WT MEFV) using a proofreading polymerase with high processivity (e.g., Pfu Ultra II, Agilent 600672). This can be challenging due to the large size of the vector (≈13 kb). We routinely perform site-directed mutagenesis on the parental entry plasmid (e.g., pENTR1A-WT, p.M694V, p.S242R *MEFV*, ≈5 kb) and clone the mutated *MEFV* variant in pInducer 21 using Gateway cloning (Gateway® LR Clonase® Enzyme mix, Invitrogen 11791-019).
2. Doxycycline-mediated induction of *MEFV* is observed at the protein level as soon as 3–6 h post-doxycycline addition. We typically used overnight induction, but shorter duration can be used. The doxycycline concentration routinely used (1 μg/mL) is saturating. To titrate pyrin protein level, decrease the concentration to (50 ng/mL).
3. Antibodies against phospho-pyrin have been raised against murine pyrin [18]. Yet, they cross-react against human phospho-pyrin. Serine 205 and 241 in murine pyrin correspond to serine 208 and 242 in human pyrin.
4. Counting the percentage of ASC speck-containing cells is strongly suggested. Yet, it can be challenging since ASC speck formation is associated with pyroptosis and cell loss.

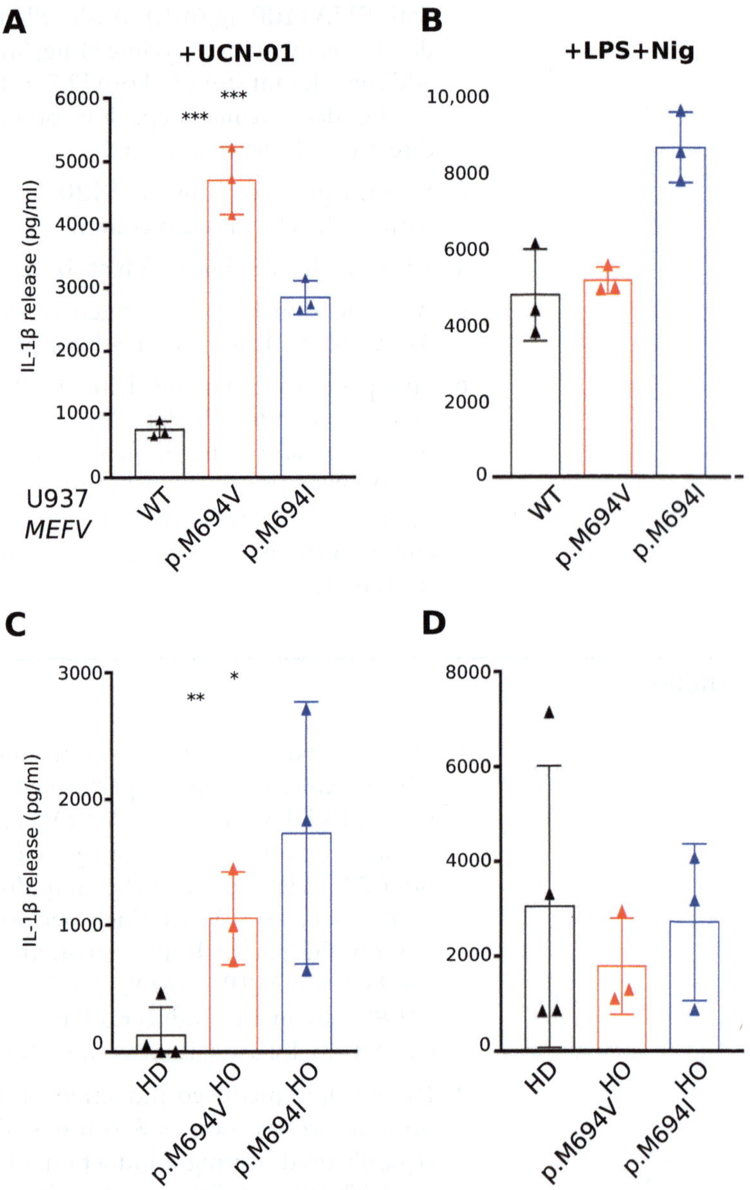

Fig. 6 Characterization of *MEFV* variant by IL-1β ELISA. IL-1β quantification by ELISA in (**a**, **b**) U937 expressing WT (black), p.M694V (red), and p.M694I (blue) *MEFV* after PMA differentiation and (**c**, **d**) primary monocytes from three HD (black), three patients homozygous for the p.M694V mutation (red), or the p. M694I mutation (blue). Higher levels of IL-1β have been detected in the supernatant of cells expressing FMF *MEFV* variants than in the supernatant of cells expressing WT *MEFV*. (**a**, **b**) Each point represents the value of one well; the bar represents the mean and SD of the triplicate. (**c**, **d**) Each point represents the mean value of a triplicate for one individual; the bar represents the mean and SD of the values of three individuals

5. For real-time cell death analysis, cells can be resuspended in supplemented RPMI if the assay is meant to last for <5 h. If the plate stays longer inside the fluorimeter in the absence of CO_2 supply, it is recommended to seed the cells in supplemented CO_2-independent medium.
6. As positive control to monitor inflammasome activation, the NLRP3 inflammasome is primed with LPS during 3 h followed by addition of the NLRP3 signal 2, nigericin.
7. The real-time cell death analysis can be performed on U937 cells or primary monocytes. Regarding the U937 cells, there are two possibilities, as shown in Figs. 4 and 5. Cells can be treated with doxycycline overnight, seeded, and analyzed for cell death only upon inflammasome stimulation the day after (3 h kinetics). This short kinetics aims at measuring the effects of the treatments when pyrin is already produced. To investigate the potential constitutive activation of a particular pyrin variant, cell death monitoring can be started at the time of doxycycline addition. 16 h later, inflammasome activators (e.g., UCN-01) can be added before returning the plate to the fluorimeter.
8. Long-term UCN-01 treatment can trigger apoptosis [21]. To conclude on pyroptosis, only the first 3 h post-UCN-01 addition should be considered, and, in the case of U937 cells, the dependency on *MEFV* expression (i.e., the presence of doxycycline) should be checked.
9. PMA is a PKC activator. Since pyrin inflammasome is under the control of the PKC superfamily PKN1/PKN2, we recommend a 3 h PMA pulse followed by one wash and 36 h of incubation.

Acknowledgment

This work has received funding from the European Union's Horizon 2020 research and innovation program under grant agreement #779295 (ImmunAID). We are grateful to Michel Popoff, Institut Pasteur, Paris, for providing TcdA toxin.

References

1. Jamilloux Y, Magnotti F, Belot A, Henry T (2018) The pyrin inflammasome: from sensing RhoA GTPases-inhibiting toxins to triggering autoinflammatory syndromes. Pathog Dis 76(3). https://doi.org/10.1093/femspd/fty020
2. Xu H, Yang J, Gao W et al (2014) Innate immune sensing of bacterial modifications of Rho GTPases by the pyrin inflammasome. Nature 513:237–241. https://doi.org/10.1038/nature13449
3. Ratner D, Orning MPA, Proulx MK et al (2016) The Yersinia pestis effector YopM inhibits pyrin inflammasome activation. PLoS Pathog 12:e1006035. https://doi.org/10.1371/journal.ppat.1006035

4. Chung LK, Park YH, Zheng Y et al (2016) The Yersinia virulence factor YopM hijacks host kinases to inhibit type III effector-triggered activation of the pyrin inflammasome. Cell Host Microbe 20:296–306. https://doi.org/10.1016/j.chom.2016.07.018
5. Aubert DF, Xu H, Yang J et al (2016) A Burkholderia type VI effector deamidates rho GTPases to activate the pyrin inflammasome and trigger inflammation. Cell Host Microbe 19:664–674. https://doi.org/10.1016/j.chom.2016.04.004
6. Manukyan G, Aminov R (2016) Update on pyrin functions and mechanisms of familial Mediterranean fever. Front Microbiol 7:456. https://doi.org/10.3389/fmicb.2016.00456
7. The International FMF Consortium (1997) Ancient missense mutations in a new member of the RoRet gene family are likely to cause familial Mediterranean fever. The International FMF Consortium. Cell 90:797–807. https://doi.org/10.1016/s0092-8674(00)80539-5
8. French FMF Consortium (1997) A candidate gene for familial Mediterranean fever. Nat Genet 17:25–31. https://doi.org/10.1038/ng0997-25
9. Moghaddas F, Llamas R, De Nardo D et al (2017) A novel pyrin-associated autoinflammation with neutrophilic dermatosis mutation further defines 14-3-3 binding of pyrin and distinction to familial Mediterranean fever. Ann Rheum Dis 76:2085–2094. https://doi.org/10.1136/annrheumdis-2017-211473
10. Masters SL, Lagou V, Jéru I et al (2016) Familial autoinflammation with neutrophilic dermatosis reveals a regulatory mechanism of pyrin activation. Sci Transl Med 8:332ra45. https://doi.org/10.1126/scitranslmed.aaf1471
11. Ozen S, Bakkaloglu A, Yilmaz E et al (2003) Mutations in the gene for familial Mediterranean fever: do they predispose to inflammation? J Rheumatol 30:2014–2018
12. Martinon F, Burns K, Tschopp J (2002) The inflammasome: a molecular platform triggering activation of inflammatory caspases and processing of proIL-beta. Mol Cell 10:417–426. https://doi.org/10.1016/s1097-2765(02)00599-3
13. Fernandes-Alnemri T, Wu J, Yu JW et al (2007) The pyroptosome: a supramolecular assembly of ASC dimers mediating inflammatory cell death via caspase-1 activation. Cell Death Differ 14:1590–1604
14. Liu X, Zhang Z, Ruan J et al (2016) Inflammasome-activated gasdermin D causes pyroptosis by forming membrane pores. Nature 535:153–158. https://doi.org/10.1038/nature18629
15. Kayagaki N, Stowe IB, Lee BL et al (2015) Caspase-11 cleaves gasdermin D for non-canonical inflammasome signalling. Nature 526:666–671. https://doi.org/10.1038/nature15541
16. Cookson BT, Brennan MA (2001) Pro-inflammatory programmed cell death. Trends Microbiol 9:113–114. https://doi.org/10.1016/s0966-842x(00)01936-3
17. Park YH, Wood G, Kastner DL, Chae JJ (2016) Pyrin inflammasome activation and RhoA signaling in the autoinflammatory diseases FMF and HIDS. Nat Immunol 17:914–921. https://doi.org/10.1038/ni.3457
18. Gao W, Yang J, Liu W et al (2016) Site-specific phosphorylation and microtubule dynamics control pyrin inflammasome activation. Proc Natl Acad Sci U S A 113:E4857–E4866. https://doi.org/10.1073/pnas.1601700113
19. Van Gorp H, Saavedra PHV, de Vasconcelos NM et al (2016) Familial Mediterranean fever mutations lift the obligatory requirement for microtubules in pyrin inflammasome activation. Proc Natl Acad Sci U S A 113:14384–14389. https://doi.org/10.1073/pnas.1613156113
20. Hong Y, Standing ASI, Nanthapisal S et al (2019) Autoinflammation due to homozygous S208 MEFV mutation. Ann Rheum Dis 78:571. https://doi.org/10.1136/annrheumdis-2018-214102
21. Magnotti F, Lefeuvre L, Benezech S et al (2019) Pyrin dephosphorylation is sufficient to trigger inflammasome activation in familial Mediterranean fever patients. EMBO Mol Med 11:e10547. https://doi.org/10.15252/emmm.201910547
22. Meerbrey KL, Hu G, Kessler JD et al (2011) The pINDUCER lentiviral toolkit for inducible RNA interference in vitro and in vivo. Proc Natl Acad Sci U S A 108:3665–3670. https://doi.org/10.1073/pnas.1019736108
23. Jamilloux Y, Lefeuvre L, Magnotti F et al (2018) Familial Mediterranean fever mutations are hypermorphic mutations that specifically decrease the activation threshold of the pyrin inflammasome. Rheumatology (Oxford) 57:100–111. https://doi.org/10.1093/rheumatology/kex373
24. Magnotti F, Malsot T, Georgin-Lavialle S et al (2020) Fast diagnostic test for familial Mediterranean fever based on a kinase inhibitor. Ann Rheum Dis 80:128–132. https://doi.org/10.1136/annrheumdis-2020-218366

25. Case CL, Roy CR (2011) Asc modulates the function of NLRC4 in response to infection of macrophages by Legionella pneumophila. mBio 2:e00117-11. https://doi.org/10.1128/mBio.00117-11
26. Pierini R, Juruj C, Perret M et al (2012) AIM2/ASC triggers caspase-8-dependent apoptosis in Francisella-infected caspase-1-deficient macrophages. Cell Death Differ 19:1709–1721. https://doi.org/10.1038/cdd.2012.51
27. Popoff MR (1987) Purification and characterization of Clostridium sordellii lethal toxin and cross-reactivity with Clostridium difficile cytotoxin. Infect Immun 55:35–43

Chapter 13

An Assay for the Seeding of Homotypic Pyrin Domain Filament Transitions

Inga V. Hochheiser and Matthias Geyer

Abstract

Pattern recognition receptors of innate immune cells allow the recognition of invariant microbial structures. The nucleotide-binding oligomerization domain-like receptors (NLRs) comprise 22 members, divided into 3 subfamilies. Homotypic pyrin domain (PYD) interactions were shown to mediate the interaction of inflammasome forming NLRPs with the adaptor protein ASC, bridging the interaction to caspase-1 and resulting in caspase-1-induced cytokine maturation and pyroptotic cell death. Here we describe a $NLRP3^{PYD}$-mediated ASC polymerization assay that reconstitutes the transition from the $NLRP3^{PYD}$ nucleation seed to ASC adaptor filament elongation with recombinant proteins.

Key words NLRP3, ASC, Pyrin domain, PYD, Inflammasome, Innate immunity, Filament nucleation, Filament elongation, Electron microscopy

1 Introduction

Inflammasomes are supramolecular complexes that regulate the activation of highly pro-inflammatory cytokines and pyroptotic cell death [1, 2]. Canonical NACHT, LRR, and PYD domain-containing protein 3 (NLRP3) inflammasome formation involves the activation-induced oligomerization of the NLRP3 sensor protein that in turn recruits the adaptor protein apoptosis-associated speck-like protein (ASC) via homotypic pyrin domain (PYD) interactions. Subsequently, ASC-CARD domains assemble, constituting a nucleation seed for homotypic domain interactions with caspase activation and recruitment domains (CARDs) of the caspase-1 zymogen. Proximity-induced autoproteolytic cleavage yields active caspase-1 that mediates activation and release of highly pro-inflammatory cytokines of the IL-1 family and triggers a pro-inflammatory form of cell death called pyroptosis [3, 4].

In cellular systems, the N-terminal NLRP3 pyrin domain ($NLRP3^{PYD}$) has been shown to form filamentous assemblies that

Thomas A. Kufer and Maria Kaparakis-Liaskos (eds.), *Effector-Triggered Immunity: Methods and Protocols*, Methods in Molecular Biology, vol. 2523, https://doi.org/10.1007/978-1-0716-2449-4_13,

ultimately induce the formation of ASC specks [5, 6]. These specks are large puncta of polymerized ASC protein that mark the point of full inflammasome formation [7]. In vitro assays that reconstitute ASC-dependent inflammasome formation are rare and laborious to achieve, despite their importance for understanding the molecular basis of NLRP-mediated ASC specking.

We developed a NLRP3^{PYD} seeding ASC polymerization assay that allows to reconstitute the transition from the NLRP3^{PYD} nucleation seed to ASC adaptor filament elongation in vitro [8]. This protocol describes the preparation of monomeric soluble forms of recombinant human NLRP3^{PYD} and ASC-mCherry proteins that are subsequently used in a fine-tuned titration assay to generate NLRP3^{PYD}–ASC-mCherry transitions in a test tube, the latter being ultimately analyzed by negative stain electron microscopy (EM). The described system can be potentially applied to other filament forming NLRPs and can be used to study the directionality of ASC elongation on sensor protein seeds, as well as the molecular interactions taking place at the interface site. Blocking inflammasome formation at the NLRP3^{PYD}–ASC transition site could be an effective mechanism to inhibit these higher-order signalosomes and lead to new therapeutic intervention in inflammation.

2 Materials

2.1 Cloning, Expression, and Purification of Monomeric NLRP3^{PYD}

1. pGEX-4T1 bacterial expression vector (Addgene catalog number: 27458001).
2. Heat competent BL21(DE3) *E. coli* cells.
3. Water bath (42 °C).
4. Lysogeny broth (LB) medium, dissolve in ddH_2O.
5. Table top centrifuge (suitable for 2-mL safe-lock tubes; maximum speed, 19,000 g; maximum capacity, 48 mL).
6. LB agar plate containing ampicillin (100 μg/mL).
7. 100 mg/mL ampicillin in ddH_2O.
8. Photometer and cuvettes to determine the optical density (OD) at 600 nm of the liquid microbial culture.
9. 1-M isopropyl β-D-1-thiogalactopyranoside (IPTG) in ddH_2O.
10. Floor-standing high-speed centrifuge (maximum speed, 81,771 g; maximum capacity, 6 L).
11. Lysis buffer, 20-mM HEPES (pH 7.5), 150-mM NaCl, and 0.5-mM Tris(2-carboxyethyl)-phosphin (TCEP).

12. 100-mM phenylmethylsulfonyl fluoride (PMSF) in 2-propanol.
13. Sonicator for cell lysis.
14. Elution buffer, 20-mM HEPES (pH 7.5), 150-mM NaCl, 0.5-mM TCEP, 15-mM L-glutathione.
15. Tobacco etch virus (TEV) protease, purified according to Tropea et al. [9].
16. Dialysis membrane with 6–8-kDa cutoff.
17. Dialysis buffer, 50-mM glycine (pH 3.8), 150-mM NaCl, and 2-mM dithiothreitol (DTT).
18. Gel filtration buffer, 50-mM glycine (pH 3.8), 150-mM NaCl, 0.5-mM TCEP.
19. 5 mL GSTrap column (GE Healthcare).
20. Superdex 75 gel filtration column 16/600 (GE Healthcare).
21. FPLC system.
22. SDS-PAGE equipment (electrophoresis chamber, SDS gel (15% acrylamide), SDS sample buffer, SDS-PAGE running buffer, protein marker, Coomassie stain and destaining solutions, heating block (95 °C)).

2.2 Cloning, Expression, and Purification of Monomeric ASC-mCherry and Filament Polymerization

1. pET-23a bacterial expression vector (Addgene catalog number: 69745-3).
2. Heat competent BL21(DE3) *E. coli* cells.
3. Water bath (42 °C).
4. Lysogeny broth (LB) medium, dissolve in ddH_2O.
5. Table top centrifuge (suitable for 2-mL safe-lock tubes; maximum speed, 19,000 g; maximum capacity, 48 mL).
6. LB agar plate containing ampicillin (100 μg/mL).
7. 100 mg/mL ampicillin in ddH_2O.
8. 1-M isopropyl β-D-1-thiogalactopyranoside (IPTG) in ddH_2O.
9. Floor-standing high-speed centrifuge (maximum speed, 81,771 g; maximum capacity, 6 L).
10. Lysis buffer, 20-mM Tris (pH 8.0), 500-mM NaCl, and 5-mM imidazole.
11. 100-mM phenylmethylsulfonyl fluoride (PMSF) in 2-propanol.
12. Sonicator for cell lysis.
13. Solubilization buffer, 20-mM Tris (pH 8.0), 500-mM NaCl, 5-mM imidazole, 4-M guanidine hydrochloride (Gdn-HCl).
14. 5-mL HisTrap column (GE Healthcare).

15. Wash buffer, 20-mM Tris (pH 8.0), 500-mM NaCl, 20-mM imidazole, 4-M Gdn-HCl.
16. Elution buffer, 20-mM Tris (pH 8.0), 500-mM NaCl, 300-mM imidazole, 4-M Gdn-HCl.
17. Dialysis membrane with 14-kDa cutoff.
18. Dialysis buffer, 50-mM glycine (pH 3.8), 150-mM NaCl, 2-mM DTT.
19. Gel filtration buffer, 50-mM glycine (pH 3.8), 150-mM NaCl, 0.5-mM TCEP.
20. Superdex 75 gel filtration column 16/600 (GE Healthcare).
21. FPLC system.
22. SDS-PAGE equipment (electrophoresis chamber, SDS gel (12% acrylamide), SDS sample buffer, SDS-PAGE running buffer, protein marker, Coomassie stain and destaining solutions, heating block (95 °C)).
23. 3-M Tris-HCL in ddH_2O.

2.3 Negative Stain EM

1. Carbon-coated copper grids.
2. PELCO easiGlow™ Glow Discharge Cleaning System.
3. 2% uranyl formate solution in ddH_2O (*see* **Note 1**).
4. Cellulose filter paper.
5. Transmission electron microscope (TEM) operating at 200 kV.

3 Methods

3.1 Cloning, Expression, and Purification of Monomeric NLRP3^{PYD}

The human NLRP3^{PYD} (UniProt accession number Q96P20, residues 3–110) flanked by 5′ *Nco*I and 3′ *Eco*RI restriction sites, respectively, was cloned into a pGEX-4T1 expression vector, providing a N-terminal glutathione S-transferase (GST) tag followed by a tobacco etch virus (TEV) protease cleavage site.

1. Prior to transformation, add 100 ng of the expression plasmid to a 60-μL aliquot of heat competent BL21(DE3) *E. coli* cells and incubate on ice for 10 min.
2. For transformation, incubate the cell/DNA suspension at 42 °C for 1 min in a water bath.
3. After transformation, incubate the sample on ice for 2 min and subsequently add 1 mL of LB medium.
4. Incubate the sample at 37 °C at 1000 rpm shaking for 1 h.
5. Spin down the cell suspension at 1500 g for 5 min, remove the supernatant, and plate out remaining cells on an ampicillin-containing LB agar plate, followed by incubation of the plate at 37 °C overnight.

6. For recombinant protein expression, generate a pre-culture by picking a single clone of the transformed BL21(DE3) *E. coli* cells and adding it to a 250-mL chicane flask containing 50-mL LB medium supplemented with ampicillin (final conc.: 100 μg/mL). Incubate the pre-culture at 37 °C shaking at 140 rpm overnight.
7. On the next day, determine the OD_{600} of the pre-culture using a photometer.
8. For setting up the high-scale expression, prepare a 5-L chicane flask containing 1-L LB medium supplemented with ampicillin (final conc.: 100 μg/mL) and add an appropriate amount of pre-culture to reach an OD_{600nm} of 0.1.
9. Incubate the expression culture at 37 °C at 110 rpm until an OD_{600nm} of 0.8 is reached.
10. Cool down the expression culture to roughly 20 °C by incubation on ice before inducing expression by adding 300 μL of 1 M IPTG to reach a final IPTG concentration of 0.3 mM.
11. Incubate the induced expression culture at 20 °C at 110 rpm overnight.
12. The next day, harvest the expression culture by centrifugation at 4000 g for 20 min and use the cell pellet for subsequent protein purification or shock-freeze cells in liquid nitrogen and store at −80 °C.
13. For protein purification, add 40-mL lysis buffer, supplemented with 1-mM PMSF to 1 L of pelleted cells, and resuspend at 4 °C stirring.
14. Lyse the cells on ice by sonication at 40% intensity for 4 min with on and off cycle of 10 s and 5 s, respectively.
15. Centrifuge the cell lysate at 49,000 g for 30 min at 4 °C to spin down cellular debris.
16. Filter the supernatant through a 0.45-μm syringe filter and apply it onto a 5-mL GSTrap column equilibrated with lysis buffer and connected to an ÄKTA prime FPLC system. Wash the column with ten column volumes (CV) of lysis buffer and elute the protein with five CV of elution buffer. Collect the eluate in 1-mL fractions and pool protein fractions containing recombinant GST-NLRP3^{PYD} fusion protein. Determine the protein concentration of the fusion protein using a spectrophotometer stating the molecular weight and the extinction coefficient (MW, 39,682 Da; extinction coefficient, 65,320 L/mol/cm).
17. For removal of the N-terminal GST-tag, adjust the fusion protein concentration to 3 mg/mL and add TEV protease to

reach a concentration ratio of TEV protease and fusion protein of 1:25 and incubate at 4 °C overnight.

18. On the following day, centrifuge the TEV-digested protein sample at 11,000 g for 10 min and 4 °C and dialyze the supernatant against dialysis buffer using a dialysis membrane with 6–8-kDa cutoff for 8 h at 4 °C under gentle stirring (*see* **Note 2**).
19. Prior to gel filtration, centrifuge the dialyzed protein at 11,000 g for 10 min to spin down protein aggregates (*see* **Note 3**). Transfer the supernatant into a new 1.5-mL test tube and load it onto a Superdex 75 gel filtration column 16/600 (GE Healthcare) connected in series with a 5-mL GSTrap column equilibrated with gel filtration buffer (*see* **Note 4**). Collect the eluate in 1-mL fractions and analyze protein containing fractions via SDS-PAGE. Pool the elution fractions containing the monomeric $NLRP3^{PYD}$ protein and concentrate the sample to a final concentration of 0.8 mg/mL (parameters for concentration determination: MW, 12,595 Da; extinction coefficient, 20,970 L/mol/cm). Snap-freeze protein aliquots in liquid nitrogen to store them at −80 °C until further use.

3.2 Cloning, Expression, and Purification of Monomeric ASC-mCherry

The cDNA encoding full-length human ASC (UniProt accession number Q9ULZ3, residues 1–195) followed by a TEV protease cleavage site and mCherry flanked by *Nde*I/*Xho*I restriction sites at the 5′ and 3′ ends, respectively, was cloned into a pET-23a expression vector providing a C-terminal hexa-histidine tag according to Venegas et al. [10]. All steps from transformation to pre-culture generation were performed in accordance to the previously described protocol for the expression of the recombinant GST-$NLRP3^{PYD}$ protein.

1. For setting up the high-scale expression culture, prepare a 5-L chicane flask containing 1-L LB medium supplemented with ampicillin (final conc.: 100 μg/mL) and add an appropriate amount of pre-culture to reach an OD_{600nm} of 0.1.
2. Incubate the expression culture at 37 °C at 110 rpm until an OD_{600nm} of 0.8 is reached.
3. Induce expression by adding IPTG (final concentration: 0.1 mM).
4. Incubate the induced expression culture at 37 °C at 110 rpm for 4 h and subsequently harvest the cells by centrifugation at 4000 g for 20 min. Use the cell pellet immediately or shock-freeze cells in liquid nitrogen and store at −80 °C.

5. For protein purification, add 40-mL lysis buffer, supplemented with 1-mM PMSF to 1 L of pelleted cells, and stir at 4 °C for pellet resuspension.
6. Lyse the cells on ice by sonication at 40% intensity for 4 min with on and off cycles of 10 s and 5 s, respectively.
7. Centrifuge the cell lysate at 49,000 g for 30 min at 4 °C to spin down cellular debris also containing *E. coli* inclusion bodies, from which the recombinant ASC-mCherry protein is going to be purified.
8. Discard the supernatant and dissolve the pellet in 5-mL solubilization buffer. Incubate the sample on ice for 1 h under continuous stirring to solubilize the protein containing inclusion bodies.
9. Centrifuge the suspension at 49,000 g for 30 min at 4 °C and collect the supernatant, containing the recombinant ASC-mCherry protein.
10. Load the protein suspension onto a HisTrap column pre-equilibrated with solubilization buffer using a high-performance FPLC system.
11. Wash the column with one column volume (CV) of solubilization buffer followed by ten CV of wash buffer.
12. Elute the protein in 1-mL fractions with elution buffer.
13. Dialyze the pooled elution fractions containing the ASC-mCherry fusion protein against dialysis buffer at 4 °C overnight under continuous, gentle stirring (*see* **Note 5**).
14. On the following day, spin down visible protein aggregates at 11,000 g for 10 min and concentrate the dialyzed protein sample if needed (*see* **Note 6**). Prior to gel filtration, again centrifuge the sample at 11,000 g for 10 min and load the supernatant onto a Superdex 75 gel filtration column (GE Healthcare) equilibrated with gel filtration buffer.
15. Analyze protein containing fractions by SDS-PAGE and pool the elution fractions containing monomeric, soluble ASC-mCherry protein.
16. Concentrate the pooled protein fractions to 1 mg/mL (parameters for concentration determination: MW, 50,406 Da; extinction coefficient, 61,310 L/mol/cm) and snap-freeze protein aliquots in liquid nitrogen for storage at −80 °C.

3.3 NLRP3^{PYD} Seeded ASC-mCherry Filament Polymerization

1. Thaw aliquots of monomeric NLRP3^{PYD} and monomeric ASC-mCherry proteins stored at −80 °C quickly in the hands.
2. Spin down thawed aliquots at 11,000 g for 10 min to pellet protein aggregates and transfer the supernatant containing the

soluble monomeric protein fraction into a new 1.5-mL test tube.

3. Determine the protein concentrations using a spectrophotometer stating the MW and extinction coefficient of the NLRP3^{PYD} and ASC-mCherry, respectively.
4. Dilute the monomeric NLRP3^{PYD} and monomeric ASC-mCherry proteins to concentrations of 50 μM and 16 μM, respectively.
5. Mix 50 μL of monomeric NLRP3^{PYD} protein at a concentration of 50 μM with 1 μL of 3-M Tris-HCl (pH 8.0) in a 1.5-mL test tube by gently pipetting up and down and incubate for 3 min at 25 °C in a heating block to induce NLRP3^{PYD} filament formation (Fig. 1).
6. To the same tube, add 1.6-μL monomeric ASC-mCherry at a concentration of 16 μM to reach a molar ratio of ASC-mCherry:NLRP3^{PYD} of 1:100. Gently mix by pipetting and incubate for another 5 min at 25 °C in a heating block (Fig. 1).
7. To the same tube, add 1.6-μL monomeric ASC-mCherry at a concentration of 16 μM to reach a molar ratio of ASC-mCherry:NLRP3^{PYD} of 1:50, mix gently, and incubate for another 5 min at 25 °C in a heating block (Fig. 1).
8. Again to the same tube, add 3-μL monomeric ASC-mCherry at a concentration of 16 μM to reach a molar ratio of ASC-mCherry:NLRP3^{PYD} of 1:25, mix gently, and incubate for another 5 min at 25 °C in a heating block (Fig. 1).
9. Check the last consecutive incubation **steps 3** and **4** by negative stain EM as described in the respective paragraph.

3.4 Negative Stain EM

Incubation **steps 3** and **4** of the NLRP3^{PYD} seeded ASC-mCherry polymerization assay are examined by negative stain EM to evaluate the formation of NLRP3^{PYD} to ASC-mCherry filament transitions (Fig. 2).

1. Glow discharge a carbon-coated copper grid (PLANO) using the PELCO easiGlow™ Glow Discharge Cleaning System (or other system) to hydrophilize the carbon support film of the grid using an argon gas plasma at 0.4 mbar, 15 mA for 45 s.
2. Add 4 μL of the protein suspension of incubation **step 3** or **4**, respectively, onto the glow-discharged carbon-coated copper grid and incubate for 1 min.
3. Blot away excess sample with a filter paper and wash the grid sample side by dipping it into three individual 20 μL drops of protein buffer with a blotting step in between.

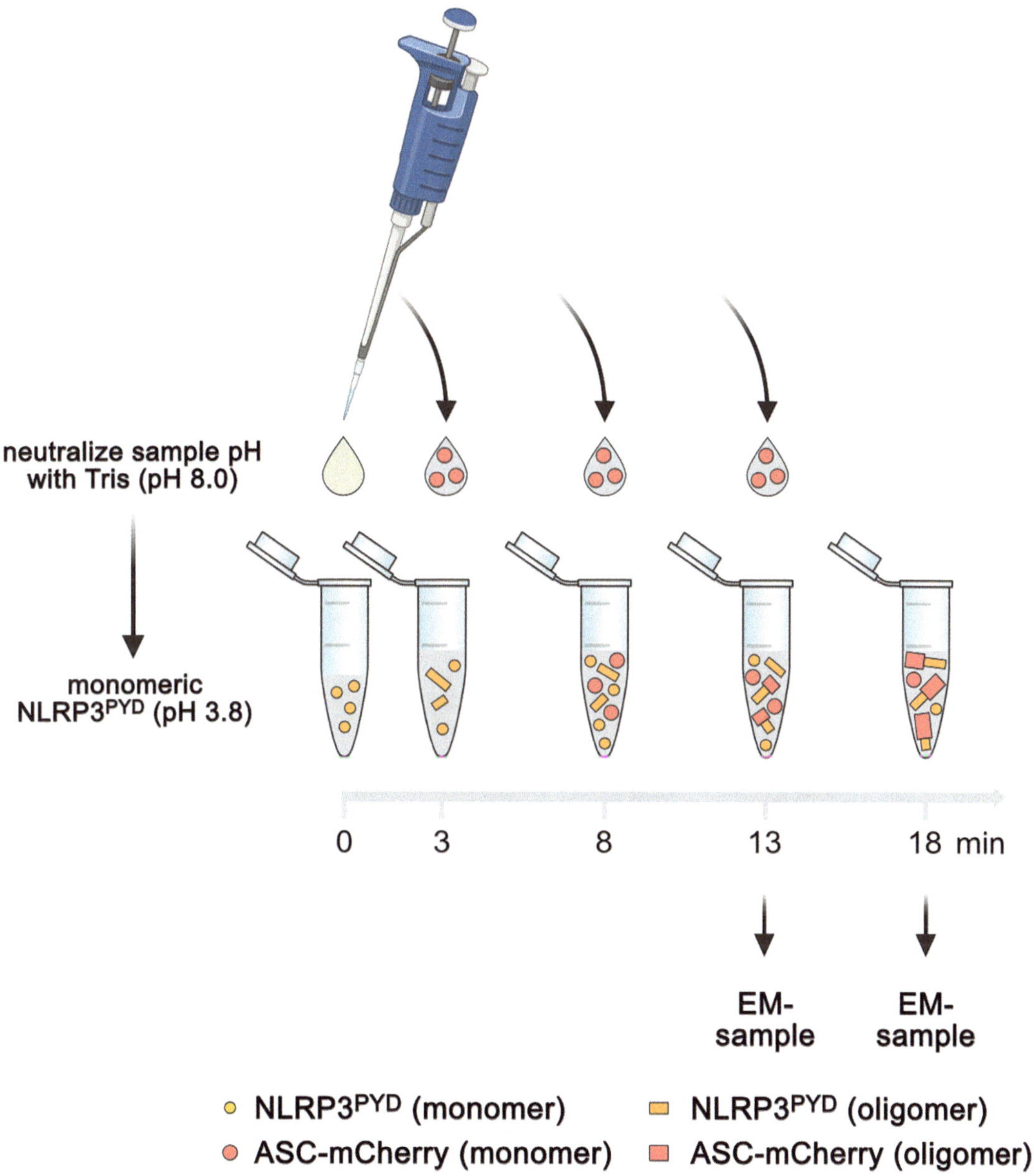

Fig. 1 Schematic of the NLRP3^{PYD}-induced ASC filament polymerization assay. Formation of NLRP3^{PYD} filament seeds is induced by neutralization to pH 8.0. Stepwise titration of highly diluted ASC-mCherry leads to the elongation of ASC-mCherry on NLRP3 filament seeds. (The figure was in part created using BioRender)

4. Negative stain the sample by dipping the sample side into 20 μL of 2% uranyl formate for 30 s and blot away excess stain using a stain-drained filter paper to prevent too harsh blotting.
5. Let the grid air-dry for 2 min.
6. Image the negative-stained sample grids using a transmission electron microscope (TEM).

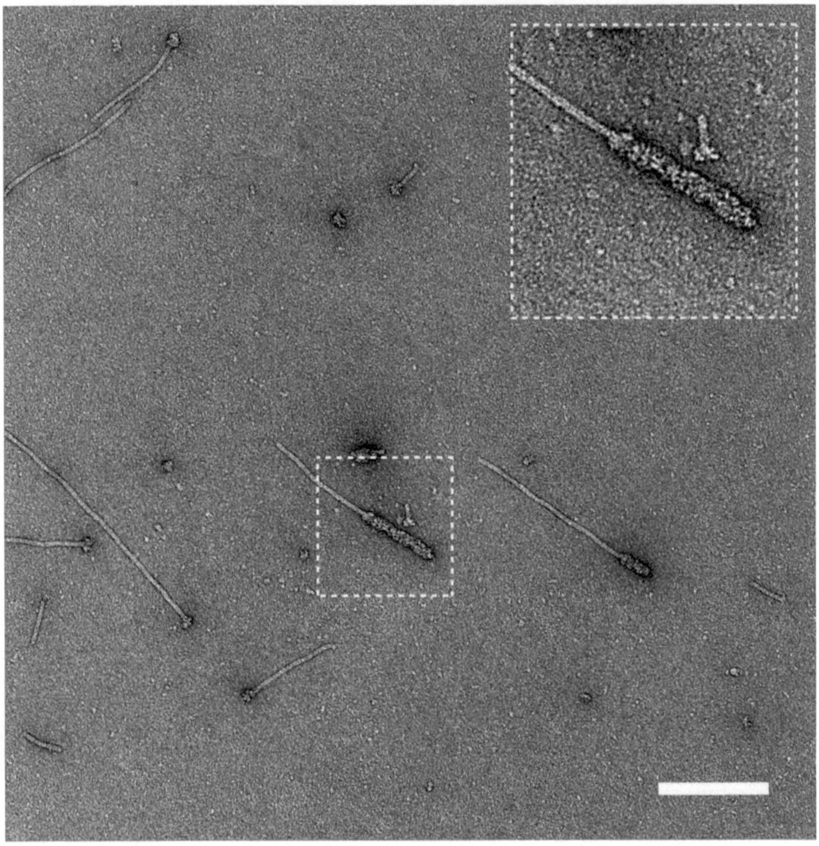

Fig. 2 Negative-stain electron micrograph of NLRP3^{PYD} to ASC-mCherry transitions after the first two consecutive titration steps. Several thin NLRP3^{PYD} filaments elongate into thicker ASC-mCherry filaments. Interestingly, the transition is exclusively observed at only one end of the NLRP3^{PYD} filament, suggesting a directionality of filament growth. The scale bar equals 250 nm

4 Notes

1. Uranyl formate is a harmful substance. Special handling precautions have to be obeyed.
2. For dialysis of the TEV-digested NLRP3^{PYD} into the acidic buffer, it is recommended to dialyze 5-mL sample against 4 L of buffer solution. If less dialysis buffer volume is used, two dialysis steps are recommended.
3. Dialysis of the TEV-digested NLRP3^{PYD} against the acidic buffer causes a lot of protein to aggregate. Centrifugation prior to gel filtration is therefore absolutely required in order to prevent blockage of the gel filtration column.
4. Doing a reverse GSTrap purification by connecting a GSTrap column in series with the gel filtration column is highly

recommended as this allows a good separation of the TEV-digested $NLRP3^{PYD}$ from the cleaved off GST-tag.

5. For dialysis of the ASC-mCherry protein into the acidic buffer, it is recommended to dialyze 5-mL sample against 4 L of buffer solution. If less dialysis buffer volume is used, two dialysis steps are recommended.
6. Dialysis of the ASC-mCherry protein into the acidic buffer causes some protein aggregation. It is therefore recommended to spin down protein aggregates prior to concentrating the sample and subsequent loading onto the gel filtration column.

Acknowledgment

This work was supported by the Deutsche Forschungsgemeinschaft (DFG) under Germany's Excellence Strategy—EXC2151–390873048.

References

1. Martinon F, Burns K, Tschopp J (2002) The inflammasome: a molecular platform triggering activation of inflammatory caspases and processing of proIL-beta. Mol Cell 10:417–426. https://doi.org/10.1016/s1097-2765(02)00599-3
2. Kufer TA, Sansonetti PJ (2011) NLR functions beyond pathogen recognition. Nat Immunol 12:121–128. https://doi.org/10.1038/ni.1985
3. Franchi L, Muñoz-Planillo R, Núñez G (2012) Sensing and reacting to microbes via the inflammasomes. Nat Immunol 13:325–332. https://doi.org/10.1038/ni.2231
4. Latz E, Xiao TS, Stutz A (2013) Activation and regulation of the inflammasomes. Nat Rev Immunol 13:397–411. https://doi.org/10.1038/nri3452
5. Stutz A, Kolbe C-C, Stahl R et al (2017) NLRP3 inflammasome assembly is regulated by phosphorylation of the pyrin domain. J Exp Med 214:1725–1736. https://doi.org/10.1084/jem.20160933
6. Marleaux M, Anand K, Latz E, Geyer M (2020) Crystal structure of the human NLRP9 pyrin domain suggests a distinct mode of inflammasome assembly. FEBS Lett 594:2383–2395. https://doi.org/10.1002/1873-3468.13865
7. Fernandes-Alnemri T, Wu J, Yu J-W et al (2007) The pyroptosome: a supramolecular assembly of ASC dimers mediating inflammatory cell death via caspase-1 activation. Cell Death Differ 14:1590–1604. https://doi.org/10.1038/sj.cdd.4402194
8. Hochheiser IV, Behrmann H, Hagelueken G et al (2022) Directionality of PYD filament growth determined by the transition of NLRP3 nucleation seeds to ASC elongation. Sci Adv 8:eabn7583. https://doi.org/10.1126/sciadv.abn7583
9. Tropea JE, Cherry S, Waugh DS (2009) Expression and purification of soluble His6-tagged TEV protease. In: Doyle SA (ed) High throughput protein expression and purification: methods and protocols. Humana Press, Totowa, pp 297–307
10. Venegas C, Kumar S, Franklin BS et al (2017) Microglia-derived ASC specks cross-seed amyloid-β in Alzheimer's disease. Nature 552:355–361. https://doi.org/10.1038/nature25158

Chapter 14

Detection of Gasdermin Activation and Lytic Cell Death During Pyroptosis and Apoptosis

Benjamin Demarco, Saray Ramos, and Petr Broz

Abstract

Cytosolic pattern recognition receptors trigger pyroptosis by detection of danger- or pathogen-associated molecular patterns. These receptors initiate the assembly of inflammasomes, multimeric protein complexes that drive caspase-1 activation. Active caspase-1 cleaves the proinflammatory cytokines IL-1β and IL-18 and the pore-forming protein gasdermin-D (GSDMD) thereby liberating its N-terminal domain. The GSDMD N-termini form multimeric pores at the plasma membrane that allow leakage of intracellular content and ultimately trigger a type of cell death called "pyroptosis." Emerging studies have revealed that GSDMD is also processed by apoptotic caspases-8/-3/-7. In this chapter, we aim to describe methods to monitor lytic cell death and to distinguish between GSDMD processing events and the GSDMD fragments that are generated after pyroptosis or apoptosis induction. We also illustrate the difference between GSDMD pore formation, and final cell lysis, and how this affects to the release of intracellular content. Finally, we show that the activation of another pore-forming protein, gasdermin-E, does not exclusively translate into lytic cell death in bone marrow-derived macrophages.

Key words Inflammasomes, Cell death, Pyroptosis, Apoptosis, Gasdermin, Immunoblotting, IL-1β quantification, LDH detection

1 Introduction

1.1 Inflammasomes and Pyroptosis

The proinflammatory cell death pyroptosis, derived from the Greek "pyro" (fire, fever) and "ptosis" (falling) [1], is triggered when specialized cytosolic pattern recognition receptors (PRRs) detect danger- or pathogen-associated molecular patterns, which initiate the assembly of multimeric protein complexes called "inflammasomes" [2]. Inflammasomes were discovered in 2002, when the Tschopp laboratory reported that PRR nucleotide-binding oligomerization domain, leucine-rich repeat (NLR), and pyrin domain containing 1 (NLRP1) interact with a protein called apoptosis-associated speck-like protein containing a caspase recruitment domain (ASC) to recruit the effector enzyme caspase-1 [2]. In the last years, diverse studies have discovered other inflammasomes

Thomas A. Kufer and Maria Kaparakis-Liaskos (eds.), *Effector-Triggered Immunity: Methods and Protocols*, Methods in Molecular Biology, vol. 2523, https://doi.org/10.1007/978-1-0716-2449-4_14,

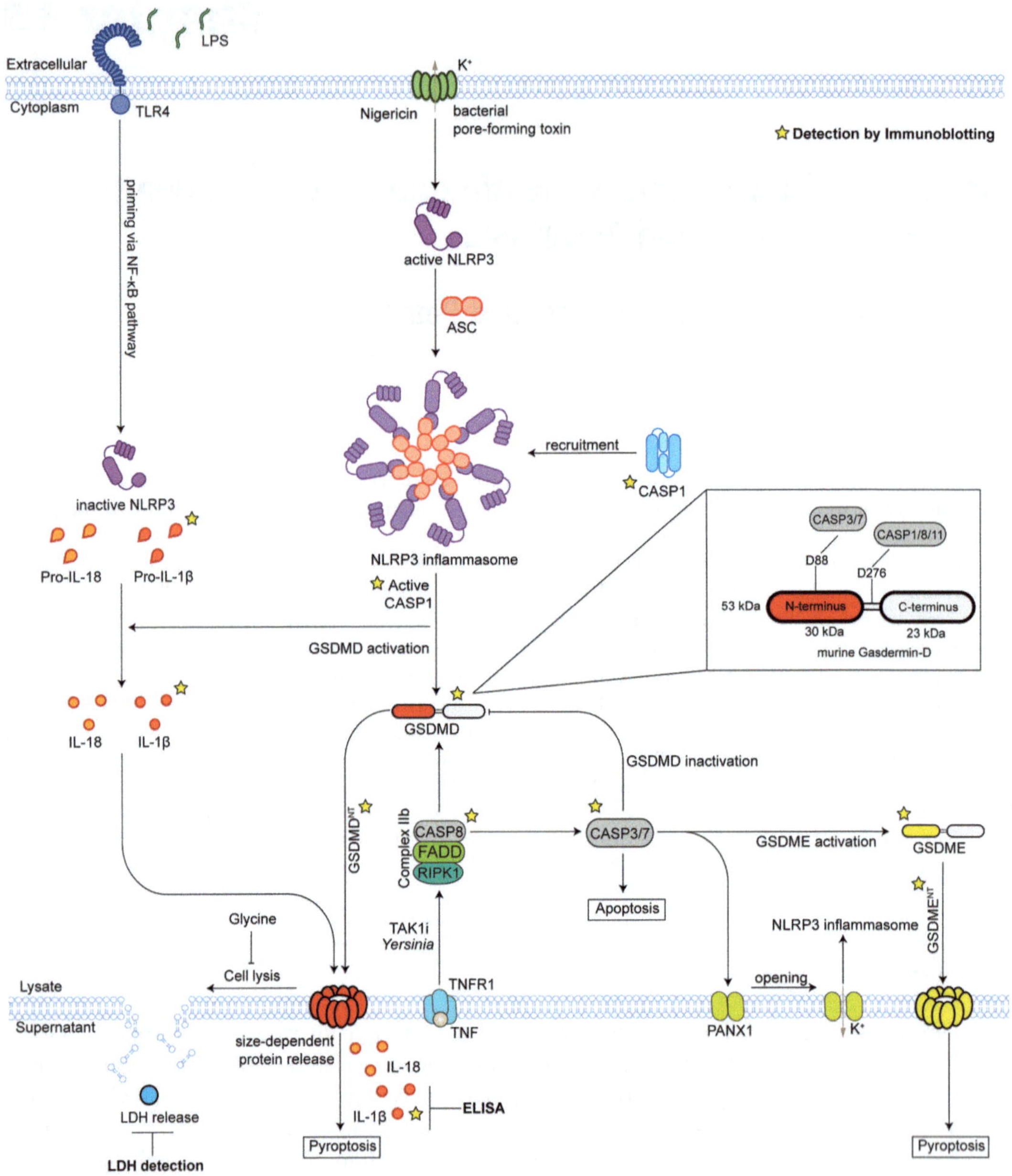

Fig. 1 Schematical overview of inflammasome assembly, pyroptosis, extrinsic apoptosis, and gasdermin activation. Upon LPS detection in the extracellular space by TLR4, the pattern recognition receptor NLRP3 and the proinflammatory cytokines IL-1β and IL-18 are upregulated via the NF-κB pathway, a signaling step known as "priming" or "signal 1." NLRP3 is however only activated by a second signal ("signal 2") such as the loss of intracellular potassium triggered by diverse pores or channels (as an example, the bacterial pore-forming toxin nigericin is depicted). Active NLRP3 binds to the adaptor protein ASC and forms multiprotein complexes known as "inflammasomes," where caspase-1 is recruited and activated. Active caspase-1 proteolytically activates IL-1β and IL-18 and also claves the pore-forming protein GSDMD in its linker region at aspartic acid (D) 276 in mouse (D277 in human). The GSDMD N-terminal domain ($GSDMD^{NT}$) targets the plasma membrane where it oligomerizes and forms pores. Caspase-8 activated in complex IIb also cleaves

which are named according to the involved PRR. For instance, the absent in melanoma 2 (AIM2) inflammasome harbors the PRR AIM2 that recognizes cytoplasmic double-stranded DNA and plays a crucial role in host defense against bacterial and viral infections [3–5]. The neuronal apoptosis inhibitory protein/NOD-like receptor containing a caspase activating and recruitment domain 4 (NAIP/NLRC4) inflammasome detects bacterial proteins and flagellin [6–15], while the pyrin inflammasome is activated by bacterial toxins that target the small GTPase RhoA [16, 17]. The PRR NLR family pyrin domain containing 3 (NLRP3) does not recognize pathogens directly but rather detects the loss of cellular homeostasis induced by pathogens or sterile cell damage [18]. Upon inflammasome assembly, caspase-1 undergoes autoproteolysis that leads to its activation. Active caspase-1 cleaves proinflammatory cytokines such as pro-interleukin (IL)-1β and pro-IL-18 to its biologically active form and the pore-forming protein gasdermin-D (GSDMD) in its linker region (aspartic acid (D) 276 in mouse, D277 in human). Caspase-1-mediated GSDMD processing liberates the GSDMD N-terminal domain ($GSDMD^{NT}$) from the autoinhibition by its C-terminal domain [19, 20] (Fig. 1). Once released by cleavage, $GSDMD^{NT}$ targets the plasma membrane, where it oligomerizes and forms large multimeric pores [21–24]. Similarly, active murine caspase-11 (caspase-4/-5 in human) cleaves GSDMD as the same site as caspase-1 after detection of cytoplasmic LPS from Gram-negative bacteria [19]. Caspase-11 activation does not require a PRR or the adaptor protein ASC [25]. Thus, this pathway is also known as "non-canonical inflammasome pathway." However, interferon-inducible proteins such as guanylate-binding proteins are needed for full caspase-11 activity [26–34]. Recently, it has also been

Fig. 1 (continued) GSDMD at the same site as caspase-1 and murine caspase-11 (caspases-4/-5 in human) and executor caspases-3/-7 to drive apoptosis. Active caspases-3/-7 inactivate GSDMD by cleaving at D88 in mouse (D87 in human) and induce PANX1 opening that triggers potassium efflux and subsequent NLRP3 inflammasome activation. GSDME is also a caspases-3/-7 substrate that triggers pyroptosis in certain cell lines when activated. GSDMD pore formation allows the release of small proteins such as cleaved IL-1β (17 kDa) which can be detected and quantified in the supernatant by ELISA. Besides, lactate dehydrogenase, a larger protein (147 kDa), is only released when cells lyse and can be detected in the supernatant by an LDH detection kit. The osmoprotectant glycine blocks cell lysis, but not GSDMD pore formation, and allows the study of intracellular protein leakage through GSDMD pores exclusively. Proteins detected by immunoblotting described in this chapter are marked with a yellow star. ASC: Apoptosis-associated speck-like protein containing a caspase recruitment domain. CASP1: Caspase-1. CASP8: Caspase-8. CASP3/7: Caspase-3/-7. ELISA: Enzyme-linked immunosorbent assay. GSDMD: Gasdermin D. $GSDMD^{NT}$: GSDMD N-terminal domain. GSDME: Gasdermin E. FADD: Fas-associated protein with death domain. IL: Interleukin. NLRP3: NOD-, LRR-, and pyrin domain-containing protein 3. PANX1: Pannexin-1. RIPK1: Receptor interacting serine/threonine kinase 1. TAK1i: Transforming growth factor-β-activated kinase 1 inhibitor. TLR4: Toll-like receptor 4. TNF: Tumor necrosis factor. TNFR1: TNF receptor superfamily member 1A

demonstrated that caspase-11 activity is regulated by IFN-inducible Igrm2 and the non-canonical autophagy effector Gate-16 [35–37]. While caspase-11 cannot process the proinflammatory cytokines IL-1β and IL-18, GSDMD activation by caspase-11 triggers NLRP3 inflammasome activation in a potassium efflux-dependent manner and subsequent cytokine processing by caspase-1 [38]. Importantly, the intracellular content is released through GSDMD pores in a size-dependent manner [39]. Small proteins such as cleaved IL-1β and IL-18 are released through GSDMD pores, but larger proteins (>50 kDa) are only detected in the extracellular space after cell lysis, where the intracellular content is indiscriminately released. Interestingly, a recent report demonstrated that pyroptotic cell lysis requires nerve injury-induced protein 1 (NINJ1) [40]; however, the activation mechanisms and regulation of NINJ1 during cell death are still unknown. Since pyroptosis is highly proinflammatory, it is tempting to speculate that cells possess intracellular mechanisms to dampen cell lysis and avoid hyperinflammation. Indeed, we have demonstrated that, similar to mixed lineage kinase domain like pseudokinase (MLKL) pores during necroptosis [41], the endosomal sorting complexes required for transport (ESCRT) machinery negatively regulate GSDMD-induced cell death by removing GSDMD pores from the plasma membrane in form of ectosomes [42].

1.2 Apoptosis and Inflammation

Opposite to pyroptosis, the programmed cell death apoptosis is traditionally viewed as immunologically silent, and it is crucial for embryonic development and the removal of superfluous cells such as autoreactive lymphocytes [43]. Depending on which initiator caspase is activated, apoptosis is mainly categorized in intrinsic apoptosis (caspase-9-driven, also called "mitochondrial apoptosis") or extrinsic apoptosis (caspase-8-driven, initiated by death-containing cell surface receptors). Both caspases activate executor caspases-3/-7 to drive the sequential cellular breakdown [44]. In physiological conditions, apoptotic cells release nucleotides such as ATP and UTP ("find-me" signal) and expose the inner membrane lipid phosphatidylserine ("eat-me" signal) signals to facilitate a rapid apoptotic cell clearance by phagocytes before the intracellular content can be released, thus avoiding any inflammatory response [45–48]. However, in some scenarios such as infections, tissue injury, autoimmune syndromes, and cancer, the cellular homeostasis is dysregulated, where the phagocytic capacity is overwhelmed, and apoptotic cells can promote an inflammatory response (reviewed in detail in [49–51]). In agreement with this dogma, emerging studies have shown that apoptotic myeloid cells such as macrophages, dendritic cells, and neutrophils can trigger an inflammatory cell death [52–61]. For instance, inhibition of cellular inhibitor of apoptosis (cIAPs) or transforming growth factor β-activated kinase 1 (TAK1) by chemotherapeutical drugs trigger caspase-8-driven apoptosis accompanied by IL-1β processing and secretion. Interestingly, IL-1β is cleaved not only by caspase-1 but also by caspase-8 directly, highlighting a proinflammatory function of caspase-8 [52]. Remarkably, cleaved IL-1β release has also been reported in vivo upon Fas ligand challenge, a stimulation known to trigger caspase-8-driven extrinsic apoptosis [62]. Similarly, caspase-1-independent IL-1β processing and release has been reported in other studies in mice in an acute arthritis model [63, 64]. Another caspase-8 proinflammatory function is its involvement in endotoxic shock with LPS or *E. coli* in mice, where caspase-8 collaborates with caspase-11 to amplify tissue damage [65]. Overall, these studies have demonstrated that apoptosis is not immunologically silent under certain circumstances and highlight diverse proinflammatory functions of caspase-8.

1.3 Cross Talk Between Apoptosis and Pyroptosis

Consistent with the dogma that apoptosis is not exclusively a silent form of cell death, two recent studies have proved that the NLRP3 inflammasome is activated during apoptosis in a potassium efflux-dependent manner. A later study from us provided the activation mechanism of NLRP3 inflammasome activation where we demonstrated that the plasma membrane channel pannexin-1 (PNX1), a channel known to be cleaved and activated by caspase-3/-7 [45–48], triggers potassium efflux and NLRP3 inflammasome activation that contributes to GSDMD-driven lytic cell death [66, 67] (Fig. 1). While PNX1 is required for NLRP3 inflammasome activation during apoptosis, it is dispensable during canonical or non-canonical inflammasome pathway [68, 69]. Paradoxically, caspase-3/-7 downregulates pyroptosis by cleaving GSDMD within its cytotoxic N-terminal domain at D88 in mouse (D87 in human) [66]. We generated a novel $Gsdmd^{D88A}$ knock-in mice, where the aspartic acid at position 88 was exchanged for an alanine ($GSDMD^{D88A}$) to generate a caspase-3/-7 cleavage-resistant GSDMD version [66, 67]. Remarkably, macrophages harboring $GSDMD^{D88A}$ displayed enhanced lytic cell death upon extrinsic apoptosis induction compared to wild-type (WT) cells, which correlates with an accumulation of the cytotoxic $GSDMD^{NT}$. Opposite that during apoptosis, GSDMD is not inactivated during pyroptosis, since cells rapidly undergo cell lysis without activating apoptotic caspase-3/-7 [70]. Since caspases-3/-7 inactivates GSDMD, it was tempting to speculate that GSDMD might be activated during apoptosis, and this inactivation is a mechanism to avoid lytic cell death. Indeed, we and others have independently shown that caspase-8 directly cleaves GSDMD downstream of TNF receptor at the same site as caspase-1 in cells infected with *Yersinia* or treated with certain chemotherapeutical drugs, known to trigger extrinsic apoptosis [67, 71–73] (Fig. 1). Interestingly, caspase-8 recruited in the RIPK1-containing complex-IIb exclusively, but not in complex-IIa or death-inducing signaling complex (DISC), is strong enough activated to directly cleave GSDMD in macrophages [70]. Remarkably, TNF-induced lethality, an in vivo model where mice lethality is dependent on caspase-8 and RIPK1 [70, 74–76], also required GSDMD, but not caspase-1, suggesting that caspase-8-mediated GSDMD activation and subsequent cell lysis and tissue damage lead to mice mortality in TNF-induced shock [70, 74–76].

The pore-forming protein gasdermin-E (GSDME, originally identified as DFNA5) is another member of the gasdermin protein family [77]. GSDME is cleaved by caspase-3/-7 in its linker region that liberates its N-terminal domain to induce lytic cell death [77] (Fig. 1). Thus, GSDME has been proposed to mediate secondary necrosis in apoptotic cells, switching apoptosis to lytic cell death, especially in certain cancer cell lines where GSDME is highly expressed [77–79]. Remarkably, chemotherapy drug-induced toxicity is also triggered by GSDME-induced pyroptosis in vivo, where mice lacking GSDME are more resistant to tissue damage in the small intestine and weight loss upon cisplatin treatment compared to WT counterparts [78]. Interestingly, it has also been proposed that GSDME is responsible for secondary necrosis in primary bone marrow-derived macrophages (BMDMs) [77]. However, we and others have not observed a GSDME-mediated cell lysis upon apoptosis stimulation in myeloid cells including BMDMs despite the detection of GSDME processing in its cytotoxic fragment [53, 67, 72, 80–82]. It is conceivable to speculate that GSDME-mediated cell lysis strongly depends on its endogenous expression in cells and the magnitude of caspase-3/-7 activation triggered by the different treatments.

1.4 Objectives of This Chapter

In this chapter, we present our routinely used methodology to detect GSDMD activation as a readout of pyroptotic cell death, focusing on three well-established methods: the quantification of lactate dehydrogenase (LDH) release (*see* Subheading 3.2), IL-1β detection by ELISA (*see* Subheading 3.3), and immunoblotting (*see* Subheading 3.8) (Fig. 1). As an example, we present a protocol that describes the procedure of stimulation with either classical pyroptosis or extrinsic apoptosis inducers in bone marrow-derived macrophages (BMDMs) (*see* Subheading 3.1). For pyroptosis, we treat LPS-primed wild-type (WT) and *Gsdmd*$^{-/-}$ BMDMs with the pore-forming toxin nigericin, a well-known NLRP3-inflammasome activator, in the presence or absence of exogenous glycine, a compound known to block cell lysis [83].

1.4.1 LDH Release as a Measure of Lytic Cell Death, But Not GSDMD Pore Formation

Since exogenous glycine blocks cell lysis, the usage of this compound allows to investigate the release of intracellular content that occurs in absence of membrane disruption and exclusively through GSDMD pores. Consistent to a previous study from our laboratory [39], glycine pretreatment abolished LDH detection but does not influence IL-1β levels in the supernatant (Fig. 2a, b),

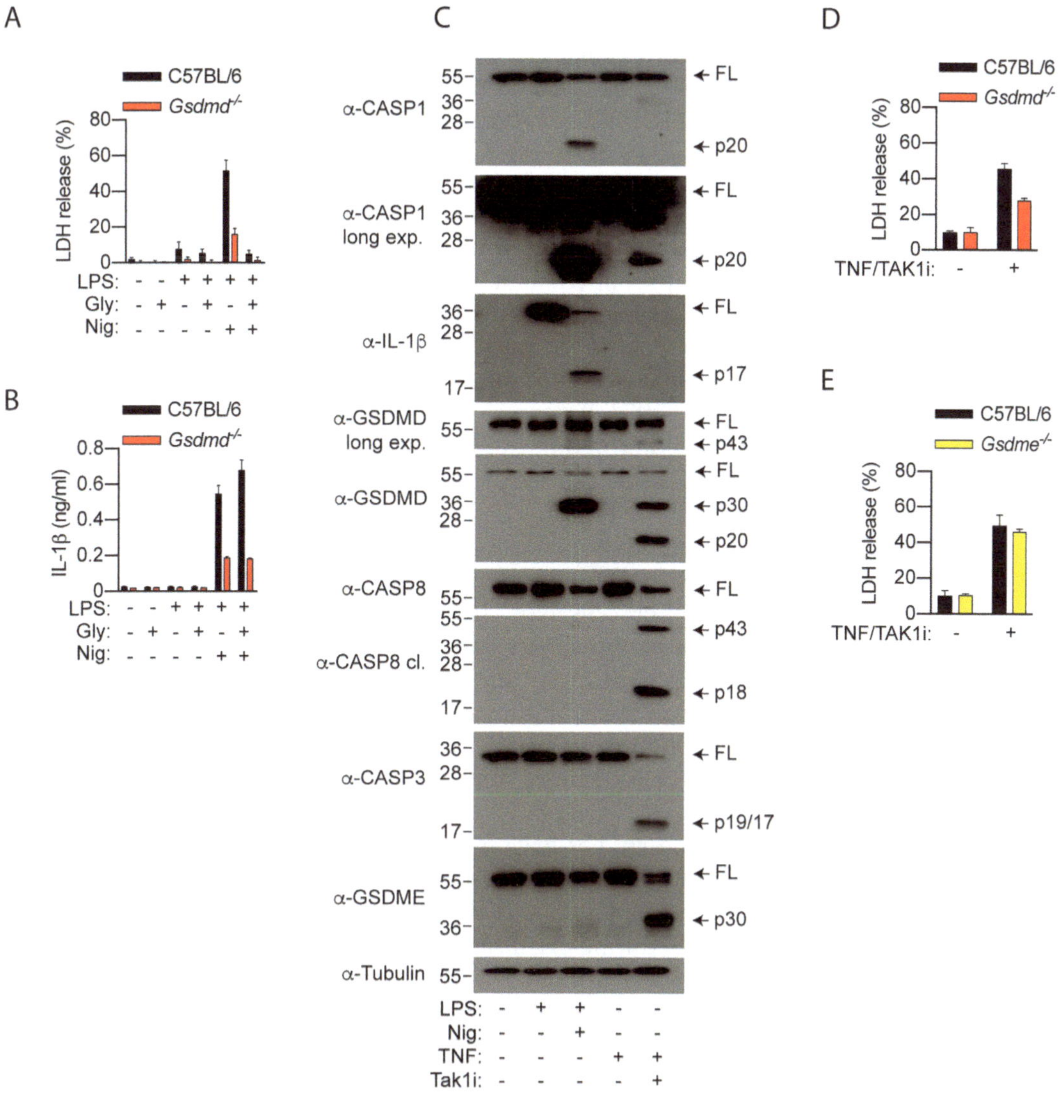

Fig. 2 (**a**, **b**) Bone marrow-derived macrophages (BMDMs) were primed with LPS (100 ng/mL) for 4 h and stimulated with nigericin (5 μM) for 1 h. Where indicated, glycine (10 mM) was added in the last 30 min of priming. (**a**) LDH and (**b**) IL-1β release in the supernatant were measured. (**c**) Mixed supernatant and cell lysate samples from wild-type BMDMs primed with LPS (4 h, 100 ng/mL) and treated with nigericin (5 μM, 1 h) or co-stimulated with TNF (100 ng/mL) and TAK1i (125 nM) for 4 h were examined by immunoblotting. (**d**, **e**) BMDMs were co-stimulated with TNF (100 ng/mL) and TAK1i (125 nM) for 4 h, and LDH release in the supernatant was measured. (**a**, **b**, **d**, and **e**) Data are mean + standard deviation of triplicate technical repeats and are representative of three to four independent experiments. (**c**) Immunoblot is representative of two independent experiments

demonstrating that cleaved IL-1β is released through GSDMD pores without the requirement of cell lysis. This can be explained by the fact that only small proteins (<50 kDa) are released through GSDMD pores, while larger proteins such as LDH (147 kDa) can only be detected in the supernatant after cell lysis [39]. A recent study validated these results by proteomics analysis of proteins that are released in a GSDMD-dependent manner but without cell lysis [84]. Importantly, pyroptotic cell death, LDH, and IL-1β detection are dependent on GSDMD (Fig. 2a, b). Interestingly, it has been recently proposed that nerve injury-induced protein 1 (NINJ1) is required for cell lysis [40]. Similar to glycine-pretreated cells, LDH is not released in *Ninj1*$^{-/-}$ cells, but IL-1β levels in the supernatant are equivalent to those in WT cells. Similarly, neutrophils, dendritic cells, and human monocytes have been also reported to secrete IL-1β without commitment of cell death under certain circumstances, a cellular state termed as "hyperactivation" [85–93]. Pyroptotic cell death (LPS/nigericin treatment) can also be monitored by immunoblotting against caspase-1 (caspase-1 p20 fragment) and the processing of its substrates such as GSDMD (GSDMD^{p30}) and IL-1β (p17) (Fig. 2c). Overall, it is important to emphasize that quantification of LDH release does not correlate with GSDMD pore formation, and it is only detected in the supernatant after cells lyse. In other words, when LDH cannot be detected in the supernatant, that does not mean that inflammasomes, GSDMD activation, and GSDMD pore formation did not occur. The study of intracellular content that is exclusively released via GSDMD pores can be studied using the addition of exogenous glycine, cells lacking NINJ1 or in hyperactivated cells.

1.4.2 Diverse GSDMD Fragments Detected by Immunoblotting After Processing by Inflammatory Caspase-1 or Apoptotic Caspases-3/-7/-8

Emerging studies have demonstrated that not only inflammatory caspases but also apoptotic caspases can cleave GSDMD. For instance, caspase-8 directly activates GSDMD to promote lytic cell death during extrinsic apoptosis triggered by TNF in combination with chemotherapeutical drugs such as TAK1 inhibitor or by *Yersinia* infection [67, 71–73] (Fig. 1). Caspase-8-mediated GSDMD activation contributes to cell lysis upon extrinsic apoptosis induction, where cells lacking GSDMD release significantly less LDH compared to WT cells (Fig. 2d). In parallel to GSDMD activation, caspase-8 also proteolytically activates executor caspases-3/-7 that, paradoxically, suppress pyroptosis by cleaving murine GSDMD at D88 (D87 in human) [66, 67]. This cleavage generates two unconventional fragments detected by immunoblotting (GSDMD^{p20} and GSDMD^{p43}) in parallel to the cytotoxic GSDMD^{p30} fragment, which is generated by caspase-1 and caspase-8 (Figs. 2c and 3). Opposite to apoptosis, GSDMD is exclusively detected as full length (FL) or GSDMD^{p30} in pyroptotic LPS-primed cells treated with nigericin (Fig. 2c). This is explained by the fact that active caspase-3 is undetectable in pyroptosis, presumably because pyroptosis-proficient cells rapidly activate caspase-1 and undergo cell death without having the time to activate apoptotic caspase.

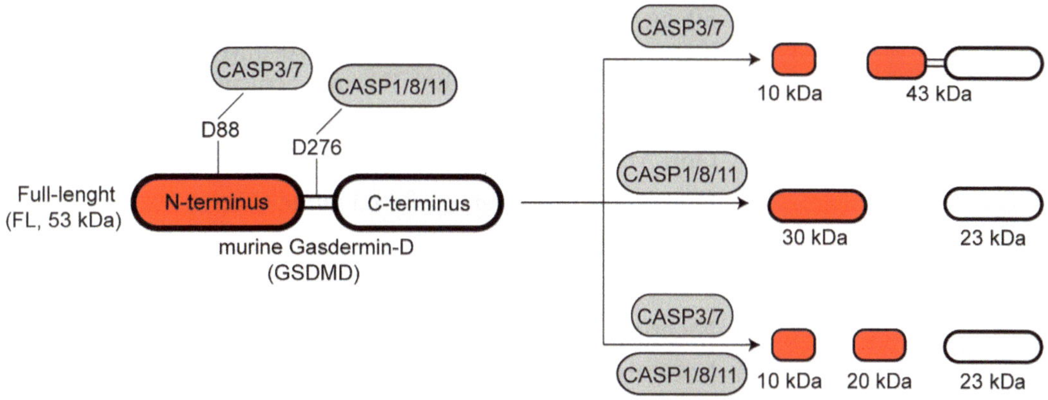

Fig. 3 Schematical overview of full-length (FL) murine GSDMD and its generated fragments after processing by inflammatory caspases-1/-11 and apoptotic caspases-8/-3/-7. Inflammatory caspase-1, murine caspase-11 (caspase-4/-5 in human), and apoptotic caspase-8 cleave GSDMD at aspartic acid (D) 276 (D277 in human) that generates GSDMD fragments with a molecular weight of 30 kDa (N-terminal domain) and 23 kDa (C-terminal domain). Apoptotic executor caspases-3/-7 cleave GSDMD at D88 (D87 in human) that generates GSDMD fragments with a molecular weight of 10 and 43 kDa. When GSDMD is cleaved in both sites (D276 and D88), three GSDMD fragments are detected with a molecular weight of 10, 20, and 23 kDa (C-terminal domain)

1.4.3 GSDME Processing Does Not Translate Into Lytic Cell Death Suring Extrinsic Apoptosis in Primary Murine Macrophages

Recent studies have demonstrated that GSDME, cleaved and activated by caspases-3/-7, drives secondary necrosis in apoptotic cells, especially in certain cancer cell lines, where GSDME is highly expressed [77–79]. The results in myeloid cells such as primary macrophages have been controversial. We and other groups have not observed a GSDME-mediated lytic cell death contribution despite GSDME processing to its cytotoxic fragment [53, 67, 72, 80–82]. *Gsdme*$^{-/-}$ BMDMs displayed similar LDH levels compared to WT counterparts when extrinsic apoptosis was induced by TNF in combination with TAK1 inhibition despite its processing to its cytotoxic p30 fragment (Fig. 2c, d). It is tempting to speculate that GSDME-induced cell lysis strongly depends on the endogenous cellular GSDME concentration and the magnitude of caspases-3/-7 activation. These results demonstrate that GSDME processing to its cytotoxic fragment does not exclusively translate into lytic cell death in WT cells. It is then crucial to validate in experiments whether the lytic cell death is influenced by the deletion of GSDME and to not assume that detection of the GSDME p30 fragment automatically correlates with engagement of lytic cell death.

2 Materials

2.1 Pyroptosis and Apoptosis Induction

1. Isolated bone marrow-derived macrophages (isolation protocol described in [94]).
2. Macrophage medium: Dulbecco's Modified Eagle's Medium (DMEM) supplemented with 20% 3T3 supernatant (or with

50 ng/mL recombinant M-CSF, ImmunoTools), 10% heat-inactivated fetal calf serum (FCS), 10-mM Hepes, 1% penicillin/streptomycin (10,000 IU/mL penicillin/10,000 μg/mL streptomycin), and 1% nonessential amino acids.
3. Opti-MEM reduced serum medium (Gibco).
4. Dulbecco's phosphate-buffered saline (PBS).
5. Tissue-cultured treated flat bottom 96-well plates.
6. Ultrapure *E. coli* 055:B5 LPS (InvivoGen). Stored at −20 °C.
7. Nigericin (Sigma-Aldrich). Stored at −20 °C.
8. Recombinant murine TNF-α (PeproTech). Stored at −20 °C.
9. 5Z7-Oxozeaenol (TAK1 inhibitor, Sigma-Aldrich). Stored at −20 °C.
10. Glycine.
11. Cell culture incubator (37 °C, 5% CO_2).
12. Cell culture hood.

2.2 Quantification of LDH Release

1. 96-well plate.
2. Multichannel pipette.
3. LDH cytotoxicity detection kit (Takara).
4. Lysis solution: 10% Triton X-100 in distilled water (dH_2O).
5. Microplate reader.
6. Stop solution: 2-M acetic acid in dH_2O. Stored at 4 °C.

2.3 Quantification of IL-1β Release by ELISA

1. IL-1β ELISA Kit (R&D Systems).
2. Nunc MaxiSorp ELISA plates (Thermo Fisher Scientific).
3. Blocking buffer: 1% bovine serum albumin (BSA) in PBS.
4. Washing buffer: 0.05% Tween20 (Sigma-Aldrich) in PBS.
5. Stop solution: 2-M sulfuric acid (H_2SO_4) in dH_2O.
6. TMB substrate chemiluminescence set (BD Biosciences).
7. Microplate reader (BioTek).

2.4 SDS-PAGE

1. 4× Sample buffer (4× NuPAGE LDS buffer, Thermo Fisher): 106-mM Tris-Cl, 141-mM Tris base, 2% lithium dodecyl sulfate (LDS), 10% glycerol, 0.51-mM EDTA, 0.22-mM SERVA Blue G250, 0.175-mM phenol red, pH 8.5.
2. Lysis buffer: 66-mM Tris-Cl (pH 7.4), 2% SDS in dH_2O.
3. Dithiothreitol DTT (1 M). Stored at −20 °C.
4. Methanol.
5. Chloroform.

6. Resolving buffer (4×): 1.5-M Tris (pH 8.8), 0.4% SDS in dH_2O.
7. Stacking buffer (4×): 0.5-M Tris (pH 6.8), 0.4% SDS (AppliChem) in dH_2O.
8. Acrylamide 4K solution (30%)—mix 37.5:1 (AppliChem).
9. Ammonium persulfate (APS, 10%) in dH_2O. Stored at −20 °C.
10. N,N,N′,N′-tetramethylethylenediamine (TEMED). Stored at 4 °C.
11. 2-propanol.
12. Casting stand.
13. Casting frame.
14. Glass short plate.
15. Glass spacer plate (1 mm thickness).
16. Gel combs (15-well 1 mm thickness).
17. Gel loading pipette tips.
18. Prestained protein ladder.
19. Polyacrylamide gels (14%, *see* Subheading 3.5).
20. Electrophoresis cell: buffer tank, electrode assembly, lid with power cables, and mini cell buffer dam.
21. Electrophoresis power supply.
22. Running buffer (1×): 0.024-M Tris, 0.2-M glycine, 0.1% SDS, pH 8.3 in dH_2O.

2.5 Gel Transfer

1. Trans-Blot turbo transfer system kit (Bio-Rad).
2. Nitrocellulose membrane 0.45 μm (Bio-Rad, provided in kit).
3. Transfer Stacks (Bio-Rad, provided in kit).
4. Gel trays (Bio-Rad, provided in kit).
5. Transfer buffer (1×): 20% 5× Trans-Blot Turbo (Bio-Rad, provided in kit) and 20% ethanol in dH_2O. Stored at 4 °C.
6. Blot roller.
7. Tweezers.
8. Western blot box.
9. Washing buffer: 0.1% Tween20 in 1× TBS (10-mM Tris, 150-mM NaCl, pH 7.4).
10. Orbital shaker.

2.6 Immunoblotting

1. Blocking buffer: 5% skim milk powder (local grocery store) in washing buffer. Stored at 4 °C.
2. Clarity Western ECL Substrate (Bio-Rad).
3. SuperSignal West Femto Chemiluminescent Substrate (Thermo Fisher Scientific). Stored at 4 °C.

4. Medical X-ray film 18 × 24 cm.
5. X-ray cassette.
6. Dark room and Western blot development equipment.
7. Primary antibodies: GSDMD (1:1000; EPR19828, Abcam), caspase-1 (1:1000; casper-1, AdipoGen), full-length caspase-8 (1:1000; #4927, Cell Signaling), cleaved caspase-8 (1:1000; #9429, Cell Signaling), caspase-3 (1:1000; #9662, Cell Signaling), pro-IL-1β (1:1000; AF-401-NA, R&D), and alpha-tubulin (1:5000; DM1A, Abcam). Store at −20 °C.
8. Secondary antibodies: Horseradish peroxidase-conjugated (HRP)-anti-Goat IgG (1:3000, SouthernBiotech), HRP-anti-Mouse IgG (1:3000, SouthernBiotech), and HRP-anti-Rabbit IgG (1:3000, SouthernBiotech). Store at 4 °C.
9. Membrane stripping solution (Restore™ PLUS Western Blot Stripping Buffer (Thermo Fisher Scientific)).
10. 0.2% sodium azide in dH_2O. Stored at 4 °C.
11. 30% hydrogen peroxide.
12. 5% BSA in washing buffer. Stored at 4 °C.

3 Methods

3.1 Pyroptosis (LPS/Nigericin) and Apoptosis (TNF/TAK1i) Induction in Bone Marrow-Derived Macrophages

1. Plate bone marrow-derived macrophages at a density of 5×10^4 per well (96-well plate) in 200 μL of macrophage medium (*see* Subheading 2.1, **item 2**) in a cell culture hood (all following steps are performed in a cell culture hood to avoid any contamination to the samples) (*see* **Note 1**).
2. Incubate the plate overnight in a cell culture incubator at 37 °C in a 5% CO_2 atmosphere.
3. Remove media and replace with 200-μL Opti-MEM. For LPS-primed cells, add LPS to a final concentration of 100 ng/mL and incubate plates for 4 h in a cell culture incubator.
4. For the glycine treatment, add glycine in the last 30 min of LPS priming to a final concentration of 10 mM.
5. For nigericin treatment, add nigericin to a final concentration of 5 μM and incubate plate in a cell culture incubator for 1 h (*see* **Note 2**).
6. For TNF/TAK1i-treated cells, add TNF to a final concentration of 100 ng/mL and TAKi to a final concentration of 125 nM and incubate plate in a cell culture incubator for 4 h (*see* **Note 3**).
7. After incubation with nigericin and TNF/TAK1i, proceed with quantification of LDH and IL-1β release (*see* Subheadings 3.2

and 3.3) and protein preparation for SDS gel electrophoresis (*see* Subheading 3.4).

3.2 Quantification of LDH Release in the Supernatant

1. Remove the plate with the cells from the incubator.
2. Add lysis buffer solution to a final concentration of 0.1% Triton X-100 into the 100% lysis control well (*see* **Note 4**).
3. Mix gently by pipetting up and down.
4. Transfer 25 μL of supernatant from each well to a new 96-well plate with a multichannel pipette (*see* **Note 5**).
5. Measure LDH release immediately on fresh supernatant samples according to the manufacturer's protocol (*see* **Note 6**).
6. Subtract the value obtained from medium alone (blank) to the other values. Quantification of LDH release is expressed as a percentage of total cellular LDH (100% lysis).

3.3 Quantification of IL-1β Release in the Supernatant by ELISA

1. Remove the plate from the incubator.
2. Transfer 25 μL of supernatant from each well to a new 96-well plate with a multichannel pipette (*see* **Note 7**).
3. Perform the corresponding dilution using blocking buffer (*see* Subheading 2.3, **item 3**) (*see* **Note 8**).
4. Add 50 μL of each sample to the MaxiSorp ELISA plate that has already been coated and blocked following the manufacturer's recommendations.
5. Continue with the protocol following the manufacturer's recommendations.
6. After stopping the reaction with 50 μL of stop solution, measure the absorbance in the plate reader at the wavelengths indicated by the manufacturer.
7. To calculate the detected IL-1β concentration in the supernatant and compare the values between different cell lines, normalize the values according to the number of cells (100% lysis obtained in quantification of LDH release, Subheading 3.2) (*see* **Note 9**).

3.4 Protein Preparation for SDS Gel Electrophoresis (Cell Lysate and Supernatant)

1. Prepare 1× sample lysis buffer: Dilute the sample buffer (4×, Subheading 2.4, **item 1**) to 1× sample buffer using the lysis buffer (*see* Subheading 2.4, **item 2**). Add DTT to the 1× sample lysis buffer to a final concentration of 10 mM (*see* **Note 10**).
2. Carefully, transfer the supernatant to another 96-well plate using a multichannel pipette (*see* **Note 11**).
3. Add 30 μL of 1× sample lysis buffer into the cells per each 96-well plate (*see* **Note 12**).
4. Transfer the cell lysate to a 1.5-mL tube.

5. Store cell lysates at −80 °C until next use.
6. When SDS gel electrophoresis will be performed, thaw the samples at room temperature.
7. Heat the samples at 95 °C for 5 min.
8. Spin samples at 14.000 rpm, 5 min, RT (*see* **Note 13**).
9. To precipitate the supernatant, collect the supernatant into a 1.5-mL tube (volume 1).
10. Add methanol (same volume as your total volume 1) and chloroform (0.3 volume from your total volume 1) and vortex each sample for 10–15 s (*see* **Note 14**).
11. Centrifuge samples for at 14.000 rpm, 10 min, 4 °C.
12. After spinning, three phases (aqueous, interphase, organic) will be observed. Discard the upper part (aqueous) of the solution without touching the interphase (proteins).
13. Add methanol (1.3 volume from your total volume 1) and vortex each sample for 10–15 s.
14. Centrifuge samples at 14.000 rpm, 10 min, 4 °C.
15. Carefully aspirate the supernatant without touching the precipitated pellet.
16. Dry the pellets for 20–30 min at RT.
17. Mix the cell lysates (**step 8** in Subheading 3.7) with the dried pellets (*see* **Note 15**).
18. Load the samples into the acrylamide gels (*see* Subheading 3.5).

3.5 SDS Polyacrylamide Gel Preparation

1. Before making the gels, clean the glass plates with ethanol first and then dH_2O. Dry the plates with a paper tissue that does not generate paper particles.
2. Place the casting frame on the bench, open the clamps, and introduce the glass spacer and short plates (glass spacer plate in the back) into the casting frame. The bottom of the plates has to be aligned and in contact with the bench.
3. Close the casting frame clamps and adjust the casting frame into the casting stand (*see* **Note 16**).
4. Prepare the resolving gel. The following amounts are calculated to prepare six gels: 9.2-mL resolving buffer (4×), 16.8-mL acrylamide (30%), 9.2-mL dH_2O, 200-μL APS (10%), and 100-μL TEMED (*see* **Notes 17** and **19**).
5. Mix the solution gently and pour immediately into the space between the plates leaving 1 cm without filling for the stacking solution after on (*see* **Note 20**).
6. Add 1 mL of 2-propanol on top of the resolving solution in the space between the plates (*see* **Note 21**).

7. Wait until the resolving solution polymerizes, about 20–30 min (*see* **Note 22**).
8. Remove the 2-propanol, wash with dH_2O, and tilt the casting frame to the side to absorb the remaining water using a tissue.
9. Prepare the stacking gel. The following the amounts are calculated to prepare six gels: 3.2-mL stacking buffer (4×), 1.6-mL acrylamide (30%), 7.2-mL dH_2O, 80-μL APS (10%), 32-μL TEMED (*see* **Notes 17** and **20**).
10. Mix the solution gently and pour immediately into the space between the plates on the top of the polymerized resolving gel until the stacking solution mix reaches the top of the short plate.
11. Immediately place the gel comb on top (*see* **Note 23**).
12. Wait until the stacking solution polymerizes (*see* **Note 22**).
13. Remove the casting frame from the casting stand.
14. Remove the glass plates with the polymerized gel from the casting frame.
15. Use the gels for the electrophoresis. Alternatively, store gels at 4 °C (*see* **Note 18**).

3.6 SDS Polyacrylamide Gel Electrophoresis

1. Set up the electrophoresis cell and place the acrylamide gel into one side of the electrode assembly. If only one gel will be used, place the plastic surface provided by the company on the other side of the electrode and close the electrode assembly clamps.
2. Fill the space between the two gels and the electrophoresis cell with running buffer. Fill to the mark printed on the electrophoresis cell.
3. Remove the gel comb and load the prestained protein ladder in one well of the gel (we normally use 5 μL of prestained protein ladder). For samples, load up to 20 μL into each well of the gel (*see* **Notes 24** and **25**).
4. Put the lid on the tank and insert the electrical leads into the electrophoresis power supply, connecting to the corresponding color.
5. Turn on the power supply. Run the blot at a constant voltage until the blue dye front is close to the bottom of the gel (*see* **Note 26**).
6. Turn off the electrophoresis power supply, remove the lid, take the electrode assembly out of the electrophoresis cell, open its clamps, and take out the plates that contain the gel.
7. Clean the electrophoresis cell with dH_2O to avoid limescale formation that can damage the equipment.
8. Continue with the gel transfer protocol (*see* Subheading 3.7).

3.7 Gel Transfer

1. Remove the gel from the plates, insert a small spatula into the space between the plates, and exert a little pressure to separate one of the plates from the gel (*see* **Note 27**).
2. Use the spatula to carefully remove the stacking gel part.
3. Place the resolving gel into the gel tray containing dH_2O.
4. Place a transfer stack into a gel tray containing transfer buffer (*see* **Note 28**).
5. Using tweezers, place the nitrocellulose membrane on to the stack and add transfer buffer until the membrane is wet (*see* **Note 29**).
6. Place the gel on the nitrocellulose membrane and properly align the gel to the membrane. Use the blot roller to remove air bubbles.
7. Place a second transfer stack on top of the gel and soak it with transfer buffer to finish the Western blot "sandwich."
8. Transfer the Western blot sandwich to a cassette from the trans-blot turbo transfer system.
9. Use the roller to remove any air bubbles that could interfere with the transfer in the assembled sandwich and carefully roll pressing from the center to the sides, avoiding moving the gel.
10. Close the cassette lid and place the cassette into the trans-blot turbo transfer system.
11. Select the transfer protocol in the trans-blot turbo transfer system and start the program according to the protein size (*see* **Note 30**).
12. Once the program is finished, remove the cassette from the instrument and open the lid.
13. Gently, disassemble the blotting "sandwich" and plate the membrane in a Western blot box with washing buffer using tweezers (*see* **Notes 31** and **32**).
14. Place the Western blot box in an orbital shaker and perform two washes of 5 min each to clean up the membrane from the transfer buffer before continuing with the immunoblotting procedure (*see* Subheading 3.8).
15. Clean the cassettes and the used materials with dH_2O to avoid limescale formation that can damage the equipment.

3.8 Immunoblotting

1. Add blocking buffer to the membrane-containing Western blot box. Fill until covering the membrane surface, and incubate in an orbital shaker (45 rpm, all next steps are performed at 45 rpm) for 1 h at RT.

2. Prepare the primary antibody solution at an appropriate concentration: dilute the primary antibodies in blocking solution supplemented with 0.2% of sodium azide (*see* **Notes 33** and **34**).
3. Remove the blocking solution.
4. Add the primary antibody solution and incubate in an orbital shaker overnight at 4 °C.
5. Remove the primary antibody solution from the Western blot and store it at 4 °C for future experiments.
6. Wash the membrane with washing buffer for 2 h on an orbital shaker at RT. Change the washing buffer four to six times during this period.
7. Prepare the secondary antibody solution at an appropriate concentration: dilute the secondary antibodies in blocking solution.
8. Remove the washing buffer.
9. Add the secondary antibody solution and incubate in an orbital shaker overnight at 4 °C.
10. Remove the secondary antibody solution.
11. Wash the membrane with washing buffer for 2 h on an orbital shaker at room temperature. Change the washing buffer four to six times during this period.
12. On a clean piece of parafilm, add 400 μL of each substrate of the Clarity Western ECL Substrate or SuperSignal West Femto Chemiluminescent Substrate, and mix (*see* **Note 35**).
13. Dry the membrane with a paper tissue.
14. Place the membrane facing side with the transferred proteins on top of the substrate solution and incubate for 1 min at RT.
15. Remove the excess substrate solution on the membrane with a paper tissue.
16. Place the membrane into an X-ray cassette.
17. In a dark room, place an X-ray film (see Subheading 2.6.4) on top of on the membrane and close the cassette. Perform multiple exposures from 1 s to 30 min in a dark room to determine the optimum exposure time (*see* **Note 36**).
18. After developing and before removing the membranes from the X-ray cassette, label your films properly.
19. Put back the membrane into the Western blot box containing washing buffer.
20. It is possible to re-probe the same membrane with additional antibodies. To remove the secondary antibody, pour 30% hydrogen peroxide solution (*see* Subheading 2.6.11), covering

the membrane surface, and incubate for 30 min at 37 °C, wash the membrane for 10 min with washing buffer in an orbital shaker, and repeat steps starting from Subheading 3.8, **step 1**. To remove the primary antibody, pour stripping buffer solution, covering the membrane surface, and incubate with agitation for 5 min at room temperature. Wash the membrane for 10 min with washing buffer in an orbital shaker, and repeat steps starting from Subheading 3.8, **step 1** (*see* **Note 37**).

4 Notes

1. After plating out the macrophages, leave the plate for 20–30 min under the hood at RT. That allows the cells to settle and uniformly attach to the surface before transferring the plate in the cell culture incubator at 37 °C. We have observed that if we introduce the plate immediately in the cell culture incubator after plating, cells tend to migrate to the exterior parts of the wells before attaching, and cell distribution is not homogenous next day.
2. We normally prepare 1000× stock nigericin aliquots (10 μL each aliquot, 10 mM) that we store at −20 °C. For treatment, we add 90 μL Opti-MEM media to the aliquot to obtain our 100× stock nigericin solution. From that, we add 2 μL to the 200 μL medium located in each well to obtain our final concentration of 5 μM.
3. We prepare TNF and TAK1 to a 1× stock solution in Opti-MEM in a Falcon tube. Then, we replace the macrophage medium used for plating the cells, for 200 μL Opti-MEM containing TNF and TAK1 (co-stimulation) with a final concentration of 100 ng/mL and 125 nM, correspondingly.
4. We normally have at least three controls in our experiments: background control (media alone), samples from untreated cells, and cells lysed with triton X-100 to obtain the total amount of LDH in our wells. When we perform different time points, we have each of these controls for every time point.
5. It is also possible to store the samples at 4 °C (not −20 °C, LDH cannot be quantified if samples were frozen) during the day. That means, when we have time points, we harvest the 25 μL of supernatant for each time point in a 96-well plate, and we perform the LDH assay at the end of the experiment. However, if the incubation time is longer than 8 h, we perform the LDH detection assay separately to avoid the comparison of fresh samples with older samples, where the LDH might not be stable anymore.

6. Before starting to measure the absorbance of the samples, we check that air bubbles are not present in the wells as they might interfere with the absorbance reading. If necessary, we use a syringe to burst the air bubbles.
7. We normally block the ELISA plate overnight starting one day before the experiment. By doing so, we can then add the fresh supernatants directly onto the ELISA plate on the experiment day.
8. When we trigger classical pyroptosis in BMDMs, we usually perform a 1:10 dilution of our samples using the blocking buffer. If not, then our values are over the standard curve. However, this strongly depends on the stimulation magnitude and incubation time. If we test experiments with new cell death inducers, we first perform a range of dilutions (1:2, 1:5, and 1: 10).
9. When we perform an ELISA assay, we also perform an LDH release quantification assay to obtain the 100% lysis values. We then use these values to normalize our ELISA values throughout the different cell lines. For that, we calculate a factor by using the 100% lysis of the highest cell line divided by the other 100% lysis from the other cell lines, and we include this factor in our calculation for the IL-1β release quantification.
10. We normally prepare aliquots of the 1× sample lysis buffer with 10 mM DTT and store them at −20 °C. Since DTT is highly sensitive to temperature, we keep DTT aliquots at −20 °C, and we avoid freeze-thaw cycles.
11. To avoid cells in the supernatant, we slowly aspirate the supernatant without touching the bottom of the plate. We also ensure that all the supernatant is removed, before adding the 1× sample lysis buffer.
12. We normally use 30 μL of 1× sample lysis buffer for 50.000 BMDMs/well (96-well plate), plated one day before of the experiment. By doing so, we obtain a good signal/background in our immunoblotting. We then scale up our 1× sample lysis buffer according to the cellular growth surface for the other well plates.
13. After adding the 1× sample lysis buffer, samples become more viscous. Normally, we do not have problems loading the gel after the samples were frozen and heated to 95 °C for 5 min. In case that the samples are still too viscous to be loaded in the wells for the SDS gel electrophoresis, we use a syringe to pipette the samples up and down three to four times, and then we centrifuge samples at 14.000 rpm, 5 min, RT. This considerably reduces the sample viscosity.

14. We normally have 200 μL in total for each 96-plate well. We then take 25 μL for LDH release quantification (*see* Subheading 3.2) and 25 μL for ELISA (*see* Subheading 3.3). We then pool the remaining 150 μL into a 1.5-mL tube (450 μL in total, technical triplicates). For Subheading 3.4, **step 10**, we add 450 μL of methanol and 150 μL of chloroform. Then, for Subheading 3.4, **step 13**, we add 600 μL of methanol.
15. At this point, we can also load cell lysate and supernatant samples separately. For that, we add 15 μL of 1× sample lysis buffer for each precipitated 96-well plate (45 μL 1× sample lysis buffer for the supernatant purification of technical triplicates). We analyze cell lysate and supernatant samples separately when we investigate which proteins are released. When we investigate the protein processing, we mix cell lysate and supernatant samples to detect the total protein in the samples.
16. To confirm that the glass plate has been well assembled into the casting frame, it is recommended to fill the space between the plates with dH_2O and wait some minutes to confirm that the liquid is not leaking. After this step, remove the dH_2O, inverting the casting frame, and use a paper tissue to remove the residual dH_2O.
17. We normally prepare APS aliquots and store them at −20 °C for 6 months. Avoid freeze-thaw cycles. We avoid keeping TEMED/APS too long (>1 h) at RT.
18. We normally make six gels at a time. Then, we keep the gels at 4 °C for 1 week maximum. To store them properly, we wrap gels with paper moistened with dH_2O and then with cling film. We recommend preparing gels one day before performing electrophoresis to ensure complete gel polymerization.
19. We normally use 14% acrylamide gels to detect the proteins (with a molecular weight between 15 and 55 kDa) described in this chapter. As a general rule, large molecular weight proteins (>70 kDa) are best detected in low percentage acrylamide gels such as 8% or 10% acrylamide gels, while small molecular weight proteins are best analyzed on a higher concentration such as 14% acrylamide gels.
20. We normally use a 50-mL tube to prepare our SDP-PAGE solutions. First, we add the resolving buffer (4×), acrylamide (30%), and dH_2O. We then mix the solution gently by inverting the tube. Then, we add APS and gently mix to homogenize the reagent in the solution. Finally, we add TEMED and mix gently again before pouring the solution into the plates. Importantly, avoid the generation of bubbles during mixing, since the oxygen interferes with gel polymerization.
21. We normally use glass spacer plates with 1 mm thickness. Then, we pour 4.5 mL of resolving solution into the space between

the plates. This amount leaves enough space for the stacking gel later on. The use of 2-propanol is highly recommended as it flattens the top of the resolving solution, creating a completely straight end.

22. After pouring the resolving/stacking solution mix, we keep the remaining resolving mix in a 50-mL tube. This helps us to monitor the polymerization process. Normally, our resolving gel is polymerized after 20–30 min.
23. Before placing the comb, remove all the air bubbles that might be generated on top of the stalking solution.
24. We recommend cleaning the wells before loading the samples. To do so, use a syringe and carefully load running buffer into the wells after filling the tank with running buffer. It is important to load the same volume of sample per well. If there are empty wells, it is recommended to fill them up with sample buffer. By doing so, samples will run at the same speed, and the sample front will not be tilted.
25. When we have different cell lines, we use the 100% lysis values from the LDH release quantification assay (*see* Subheading 3.2) to normalize our total cell number throughout the cell lines. For that, we calculate a normalization factor using the highest values from the 100% lysis divided by the values from the other cell lines. By doing so, we are able to load the specific volume of each cell line to have the same amount of protein for each cell line in each well.
26. We run the gel 10 min at a voltage (V) of 80, until the samples leave the stacking gel. This allows the samples to line up before increasing the voltage. After, we normally increase the voltage to 120 V. The total electrophoresis takes approximately 90 min for a 14% acrylamide gel. While running the samples, if the sample front shows a smile shape, it is recommended to reduce the voltage, as this phenome is an indication of running buffer overheating. Another reason why a smiling front can form is due to an excessive loading concentration of samples.
27. Carefully manipulate the gel touching the corners wearing clean gloves, where no proteins are present to avoid any damage that might negatively affect the final result.
28. Prepare the transfer buffer in advance and store it at 4 °C before using it. If the buffer is not cold, it might overheat during the transfer which can affect the transfer efficiency.
29. Before the membrane is wet, we label the membrane using a ball point pen on the bottom where proteins will not be transferred. This helps us to keep track of each membrane, especially when multiple experiments are performed at the same time.

30. We normally perform the transfer for 7 min, 25 V. For smaller proteins (<15 kDa), we decrease the transfer time to 5 min to avoid any possible loss of proteins during the transfer by going through the membrane. For larger proteins (>70 kDa), we increase the transfer time to 10 min to ensure efficient transfer to the membrane.
31. To verify that the transfer worked successfully, we check whether the prestained protein ladder is observed on the nitrocellulose membrane.
32. Make sure that the washing buffer is properly mixed before using it. We use a 5-L plastic container to store our washing buffer, and before preparing a new washing buffer, we make sure to fully clean the residues of the previous buffer to avoid any tween sedimentation or limescale formation.
33. Each primary antibody has an optimal working dilution which depends on different factors such as antibody quality and concentration, cell line, or amount of loaded protein. To start with, we usually use a 1:1000 dilution of our primary antibodies, and later we modify this concentration according to our experience during the immunoblotting (signal/background ratio).
34. It is possible to reuse the primary antibody solutions. For that, we store the antibodies in blocking buffer containing 0.02% sodium azide, and we store them at 4 °C. The period and frequency of usage depend on each antibody. We then prepare a new antibody solution when we observe that the signal decreases after developing the membranes.
35. We recommend to initially develop the Western using Clarity Western ECL Substrate (*see* Subheading 2.6.2). In our hands, it gives a good signal/background ratio. In cases where the signal detection is weak, we change to SuperSignal West Femto Chemiluminescent substrate (*see* Subheading 2.6.3) that has a higher sensitivity, but it also increases the background in some cases.
36. Some antibodies give more background than others. When we have these problems, we increased the time of the washing steps or we use 5% BSA instead of skim milk as blocking buffer (e.g., for cleaved caspase-8 (1:1000, #9429, Cell Signaling)) or we reduce the concentration of primary/secondary antibodies.
37. We normally re-blot with antibodies raised in different species using 30% hydrogen peroxide (*see* Subheading 2.6.11) which eliminates the signal from the secondary antibody exclusively. Only when we want to re-blot with an antibody raised in the same species as one used before, we incubate the membrane with stripping solution (*see* Subheading 2.6.9). We avoid stripping the membrane multiple times, since we have observed that

the detection quality of the membrane decreases after treatment. Thus, we run two blots from the same samples when we aim to blot for multiple targets, and we prefer to use 30% hydrogen peroxide instead of stripping buffer.

Acknowledgments

This work was supported by the European Research Council grant (ERC2017-CoG-770988-InflamCellDeath) and a Swiss National Science Foundation grant (SNSF 175576).

References

1. Cookson BT, Brennan MA (2001) Pro-inflammatory programmed cell death. Trends Microbiol 9(3):113–114
2. Martinon F, Burns K, Tschopp J (2002) The inflammasome: a molecular platform triggering activation of inflammatory caspases and processing of proIL-beta. Mol Cell 10(2):417–426. https://doi.org/10.1016/s1097-2765(02)00599-3
3. Fernandes-Alnemri T, Yu JW, Juliana C, Solorzano L, Kang S, Wu J, Datta P, McCormick M, Huang L, McDermott E, Eisenlohr L, Landel CP, Alnemri ES (2010) The AIM2 inflammasome is critical for innate immunity to Francisella tularensis. Nat Immunol 11(5):385–393. https://doi.org/10.1038/ni.1859
4. Jones JW, Kayagaki N, Broz P, Henry T, Newton K, O'Rourke K, Chan S, Dong J, Qu Y, Roose-Girma M, Dixit VM, Monack DM (2010) Absent in melanoma 2 is required for innate immune recognition of Francisella tularensis. Proc Natl Acad Sci U S A 107(21): 9771–9776. https://doi.org/10.1073/pnas.1003738107
5. Sauer JD, Witte CE, Zemansky J, Hanson B, Lauer P, Portnoy DA (2010) Listeria monocytogenes triggers AIM2-mediated pyroptosis upon infrequent bacteriolysis in the macrophage cytosol. Cell Host Microbe 7(5): 412–419. https://doi.org/10.1016/j.chom.2010.04.004
6. Mariathasan S, Newton K, Monack DM, Vucic D, French DM, Lee WP, Roose-Girma-M, Erickson S, Dixit VM (2004) Differential activation of the inflammasome by caspase-1 adaptors ASC and Ipaf. Nature 430(6996): 213–218. https://doi.org/10.1038/nature02664
7. Franchi L, Amer A, Body-Malapel M, Kanneganti TD, Ozoren N, Jagirdar R, Inohara N, Vandenabeele P, Bertin J, Coyle A, Grant EP, Nunez G (2006) Cytosolic flagellin requires Ipaf for activation of caspase-1 and interleukin 1beta in salmonella-infected macrophages. Nat Immunol 7(6):576–582. https://doi.org/10.1038/ni1346
8. Miao EA, Alpuche-Aranda CM, Dors M, Clark AE, Bader MW, Miller SI, Aderem A (2006) Cytoplasmic flagellin activates caspase-1 and secretion of interleukin 1beta via Ipaf. Nat Immunol 7(6):569–575. https://doi.org/10.1038/ni1344
9. Miao EA, Mao DP, Yudkovsky N, Bonneau R, Lorang CG, Warren SE, Leaf IA, Aderem A (2010) Innate immune detection of the type III secretion apparatus through the NLRC4 inflammasome. Proc Natl Acad Sci U S A 107(7):3076–3080. https://doi.org/10.1073/pnas.0913087107
10. Sun YH, Rolan HG, Tsolis RM (2007) Injection of flagellin into the host cell cytosol by Salmonella enterica serotype Typhimurium. J Biol Chem 282(47):33897–33901. https://doi.org/10.1074/jbc.C700181200
11. Kofoed EM, Vance RE (2011) Innate immune recognition of bacterial ligands by NAIPs determines inflammasome specificity. Nature 477(7366):592–595. https://doi.org/10.1038/nature10394
12. Rauch I, Tenthorey JL, Nichols RD, Al Moussawi K, Kang JJ, Kang C, Kazmierczak BI, Vance RE (2016) NAIP proteins are required for cytosolic detection of specific bacterial ligands in vivo. J Exp Med 213(5): 657–665. https://doi.org/10.1084/jem.20151809
13. Rayamajhi M, Zak DE, Chavarria-Smith J, Vance RE, Miao EA (2013) Cutting edge: mouse NAIP1 detects the type III secretion system needle protein. J Immunol 191(8):

3986–3989. https://doi.org/10.4049/jimmunol.1301549
14. Yang J, Zhao Y, Shi J, Shao F (2013) Human NAIP and mouse NAIP1 recognize bacterial type III secretion needle protein for inflammasome activation. Proc Natl Acad Sci U S A 110(35):14408–14413. https://doi.org/10.1073/pnas.1306376110
15. Zhao Y, Yang J, Shi J, Gong YN, Lu Q, Xu H, Liu L, Shao F (2011) The NLRC4 inflammasome receptors for bacterial flagellin and type III secretion apparatus. Nature 477(7366): 596–600. https://doi.org/10.1038/nature10510
16. Yu JW, Wu J, Zhang Z, Datta P, Ibrahimi I, Taniguchi S, Sagara J, Fernandes-Alnemri T, Alnemri ES (2006) Cryopyrin and pyrin activate caspase-1, but not NF-kappaB, via ASC oligomerization. Cell Death Differ 13(2): 236–249. https://doi.org/10.1038/sj.cdd.4401734
17. Seshadri S, Duncan MD, Hart JM, Gavrilin MA, Wewers MD (2007) Pyrin levels in human monocytes and monocyte-derived macrophages regulate IL-1beta processing and release. J Immunol 179(2):1274–1281. https://doi.org/10.4049/jimmunol.179.2.1274
18. McKee CM, Coll RC (2020) NLRP3 inflammasome priming: a riddle wrapped in a mystery inside an enigma. J Leukoc Biol 108:937–952. https://doi.org/10.1002/JLB.3MR0720-513R
19. Kayagaki N, Stowe IB, Lee BL, O'Rourke K, Anderson K, Warming S, Cuellar T, Haley B, Roose-Girma M, Phung QT, Liu PS, Lill JR, Li H, Wu J, Kummerfeld S, Zhang J, Lee WP, Snipas SJ, Salvesen GS, Morris LX, Fitzgerald L, Zhang Y, Bertram EM, Goodnow CC, Dixit VM (2015) Caspase-11 cleaves gasdermin D for non-canonical inflammasome signalling. Nature 526(7575):666–671. https://doi.org/10.1038/nature15541
20. Shi J, Zhao Y, Wang K, Shi X, Wang Y, Huang H, Zhuang Y, Cai T, Wang F, Shao F (2015) Cleavage of GSDMD by inflammatory caspases determines pyroptotic cell death. Nature 526(7575):660–665. https://doi.org/10.1038/nature15514
21. Aglietti RA, Estevez A, Gupta A, Ramirez MG, Liu PS, Kayagaki N, Ciferri C, Dixit VM, Dueber EC (2016) GsdmD p30 elicited by caspase-11 during pyroptosis forms pores in membranes. Proc Natl Acad Sci U S A 113(28): 7858–7863. https://doi.org/10.1073/pnas.1607769113
22. Ding J, Wang K, Liu W, She Y, Sun Q, Shi J, Sun H, Wang DC, Shao F (2016) Pore-forming activity and structural autoinhibition of the gasdermin family. Nature 535(7610): 111–116. https://doi.org/10.1038/nature18590
23. Liu X, Zhang Z, Ruan J, Pan Y, Magupalli VG, Wu H, Lieberman J (2016) Inflammasome-activated gasdermin D causes pyroptosis by forming membrane pores. Nature 535(7610): 153–158. https://doi.org/10.1038/nature18629
24. Sborgi L, Ruhl S, Mulvihill E, Pipercevic J, Heilig R, Stahlberg H, Farady CJ, Muller DJ, Broz P, Hiller S (2016) GSDMD membrane pore formation constitutes the mechanism of pyroptotic cell death. EMBO J 35(16): 1766–1778. https://doi.org/10.15252/embj.201694696
25. Kayagaki N, Warming S, Lamkanfi M, Vande Walle L, Louie S, Dong J, Newton K, Qu Y, Liu J, Heldens S, Zhang J, Lee WP, Roose-Girma M, Dixit VM (2011) Non-canonical inflammasome activation targets caspase-11. Nature 479(7371):117–121. https://doi.org/10.1038/nature10558
26. Broz P, Ruby T, Belhocine K, Bouley DM, Kayagaki N, Dixit VM, Monack DM (2012) Caspase-11 increases susceptibility to Salmonella infection in the absence of caspase-1. Nature 490(7419):288–291. https://doi.org/10.1038/nature11419
27. Rathinam VA, Vanaja SK, Waggoner L, Sokolovska A, Becker C, Stuart LM, Leong JM, Fitzgerald KA (2012) TRIF licenses caspase-11-dependent NLRP3 inflammasome activation by gram-negative bacteria. Cell 150(3):606–619. https://doi.org/10.1016/j.cell.2012.07.007
28. Meunier E, Dick MS, Dreier RF, Schurmann N, Kenzelmann Broz D, Warming S, Roose-Girma M, Bumann D, Kayagaki N, Takeda K, Yamamoto M, Broz P (2014) Caspase-11 activation requires lysis of pathogen-containing vacuoles by IFN-induced GTPases. Nature 509(7500):366–370. https://doi.org/10.1038/nature13157
29. Pilla DM, Hagar JA, Haldar AK, Mason AK, Degrandi D, Pfeffer K, Ernst RK, Yamamoto M, Miao EA, Coers J (2014) Guanylate binding proteins promote caspase-11-dependent pyroptosis in response to cytoplasmic LPS. Proc Natl Acad Sci U S A 111(16): 6046–6051. https://doi.org/10.1073/pnas.1321700111
30. Man SM, Karki R, Sasai M, Place DE, Kesavardhana S, Temirov J, Frase S, Zhu Q, Malireddi RKS, Kuriakose T, Peters JL, Neale G, Brown SA, Yamamoto M, Kanneganti TD (2016) IRGB10 liberates bacterial ligands

for sensing by the AIM2 and caspase-11-NLRP3 inflammasomes. Cell 167(2): 382–396.e317. https://doi.org/10.1016/j.cell.2016.09.012

31. Santos JC, Dick MS, Lagrange B, Degrandi D, Pfeffer K, Yamamoto M, Meunier E, Pelczar P, Henry T, Broz P (2018) LPS targets host guanylate-binding proteins to the bacterial outer membrane for non-canonical inflammasome activation. EMBO J 37(6):e98089. https://doi.org/10.15252/embj.201798089
32. Finethy R, Luoma S, Orench-Rivera N, Feeley EM, Haldar AK, Yamamoto M, Kanneganti TD, Kuehn MJ, Coers J (2017) Inflammasome activation by bacterial outer membrane vesicles requires guanylate binding proteins. mBio 8 (5):e01188-17. https://doi.org/10.1128/mBio.01188-17
33. Santos JC, Boucher D, Schneider LK, Demarco B, Dilucca M, Shkarina K, Heilig R, Chen KW, Lim RYH, Broz P (2020) Human GBP1 binds LPS to initiate assembly of a caspase-4 activating platform on cytosolic bacteria. Nat Commun 11(1):3276. https://doi.org/10.1038/s41467-020-16889-z
34. Wandel MP, Kim BH, Park ES, Boyle KB, Nayak K, Lagrange B, Herod A, Henry T, Zilbauer M, Rohde J, MacMicking JD, Randow F (2020) Guanylate-binding proteins convert cytosolic bacteria into caspase-4 signaling platforms. Nat Immunol 21(8):880–891. https://doi.org/10.1038/s41590-020-0697-2
35. Eren E, Planes R, Bagayoko S, Bordignon PJ, Chaoui K, Hessel A, Santoni K, Pinilla M, Lagrange B, Burlet-Schiltz O, Howard JC, Henry T, Yamamoto M, Meunier E (2020) Irgm2 and Gate-16 cooperatively dampen Gram-negative bacteria-induced caspase-11 response. EMBO Rep 21(11):e50829. https://doi.org/10.15252/embr.202050829
36. Finethy R, Dockterman J, Kutsch M, Orench-Rivera N, Wallace GD, Piro AS, Luoma S, Haldar AK, Hwang S, Martinez J, Kuehn MJ, Taylor GA, Coers J (2020) Dynamin-related Irgm proteins modulate LPS-induced caspase-11 activation and septic shock. EMBO Rep 21(11):e50830. https://doi.org/10.15252/embr.202050830
37. Sakaguchi N, Sasai M, Bando H, Lee Y, Pradipta A, Ma JS, Yamamoto M (2020) Role of Gate-16 and Gabarap in prevention of caspase-11-dependent excess inflammation and lethal endotoxic shock. Front Immunol 11: 561948. https://doi.org/10.3389/fimmu.2020.561948
38. Ruhl S, Broz P (2015) Caspase-11 activates a canonical NLRP3 inflammasome by promoting K(+) efflux. Eur J Immunol 45(10):2927–2936. https://doi.org/10.1002/eji.201545772
39. Heilig R, Dick MS, Sborgi L, Meunier E, Hiller S, Broz P (2018) The Gasdermin-D pore acts as a conduit for IL-1beta secretion in mice. Eur J Immunol 48(4):584–592. https://doi.org/10.1002/eji.201747404
40. Kayagaki N, Kornfeld OS, Lee BL, Stowe IB, O'Rourke K, Li Q, Sandoval W, Yan D, Kang J, Xu M, Zhang J, Lee WP, McKenzie BS, Ulas G, Payandeh J, Roose-Girma M, Modrusan Z, Reja R, Sagolla M, Webster JD, Cho V, Andrews TD, Morris LX, Miosge LA, Goodnow CC, Bertram EM, Dixit VM (2021) NINJ1 mediates plasma membrane rupture during lytic cell death. Nature 591(7848): 131–136. https://doi.org/10.1038/s41586-021-03218-7
41. Gong YN, Guy C, Olauson H, Becker JU, Yang M, Fitzgerald P, Linkermann A, Green DR (2017) ESCRT-III acts downstream of MLKL to regulate necroptotic cell death and its consequences. Cell 169(2):286–300.e216. https://doi.org/10.1016/j.cell.2017.03.020
42. Ruhl S, Shkarina K, Demarco B, Heilig R, Santos JC, Broz P (2018) ESCRT-dependent membrane repair negatively regulates pyroptosis downstream of GSDMD activation. Science 362(6417):956–960. https://doi.org/10.1126/science.aar7607
43. Kerr JF, Wyllie AH, Currie AR (1972) Apoptosis: a basic biological phenomenon with wide-ranging implications in tissue kinetics. Br J Cancer 26(4):239–257. https://doi.org/10.1038/bjc.1972.33
44. Galluzzi L, Vitale I, Aaronson SA, Abrams JM, Adam D, Agostinis P, Alnemri ES, Altucci L, Amelio I, Andrews DW, Annicchiarico-Petruzzelli M, Antonov AV, Arama E, Baehrecke EH, Barlev NA, Bazan NG, Bernassola F, Bertrand MJM, Bianchi K, Blagosklonny MV, Blomgren K, Borner C, Boya P, Brenner C, Campanella M, Candi E, Carmona-Gutierrez D, Cecconi F, Chan FK, Chandel NS, Cheng EH, Chipuk JE, Cidlowski JA, Ciechanover A, Cohen GM, Conrad M, Cubillos-Ruiz JR, Czabotar PE, D'Angiolella V, Dawson TM, Dawson VL, De Laurenzi V, De Maria R, Debatin KM, DeBerardinis RJ, Deshmukh M, Di Daniele N, Di Virgilio F, Dixit VM, Dixon SJ, Duckett CS, Dynlacht BD, El-Deiry WS, Elrod JW, Fimia GM, Fulda S, Garcia-Saez AJ, Garg AD, Garrido C, Gavathiotis E, Golstein P, Gottlieb E, Green DR, Greene LA, Gronemeyer H, Gross A, Hajnoczky G, Hardwick JM, Harris IS, Hengartner MO, Hetz C,

Ichijo H, Jaattela M, Joseph B, Jost PJ, Juin PP, Kaiser WJ, Karin M, Kaufmann T, Kepp O, Kimchi A, Kitsis RN, Klionsky DJ, Knight RA, Kumar S, Lee SW, Lemasters JJ, Levine B, Linkermann A, Lipton SA, Lockshin RA, Lopez-Otin C, Lowe SW, Luedde T, Lugli E, MacFarlane M, Madeo F, Malewicz M, Malorni W, Manic G, Marine JC, Martin SJ, Martinou JC, Medema JP, Mehlen P, Meier P, Melino S, Miao EA, Molkentin JD, Moll UM, Munoz-Pinedo C, Nagata S, Nunez G, Oberst A, Oren M, Overholtzer M, Pagano M, Panaretakis T, Pasparakis M, Penninger JM, Pereira DM, Pervaiz S, Peter ME, Piacentini M, Pinton P, Prehn JHM, Puthalakath H, Rabinovich GA, Rehm M, Rizzuto R, Rodrigues CMP, Rubinsztein DC, Rudel T, Ryan KM, Sayan E, Scorrano L, Shao F, Shi Y, Silke J, Simon HU, Sistigu A, Stockwell BR, Strasser A, Szabadkai G, Tait SWG, Tang D, Tavernarakis N, Thorburn A, Tsujimoto Y, Turk B, Vanden Berghe T, Vandenabeele P, Vander Heiden MG, Villunger A, Virgin HW, Vousden KH, Vucic D, Wagner EF, Walczak H, Wallach D, Wang Y, Wells JA, Wood W, Yuan J, Zakeri Z, Zhivotovsky B, Zitvogel L, Melino G, Kroemer G (2018) Molecular mechanisms of cell death: recommendations of the Nomenclature Committee on Cell Death 2018. Cell Death Differ 25(3): 486–541. https://doi.org/10.1038/s41418-017-0012-4

45. Fadok VA, Voelker DR, Campbell PA, Cohen JJ, Bratton DL, Henson PM (1992) Exposure of phosphatidylserine on the surface of apoptotic lymphocytes triggers specific recognition and removal by macrophages. J Immunol 148(7):2207–2216

46. Elliott MR, Chekeni FB, Trampont PC, Lazarowski ER, Kadl A, Walk SF, Park D, Woodson RI, Ostankovich M, Sharma P, Lysiak JJ, Harden TK, Leitinger N, Ravichandran KS (2009) Nucleotides released by apoptotic cells act as a find-me signal to promote phagocytic clearance. Nature 461(7261):282–286. https://doi.org/10.1038/nature08296

47. Chekeni FB, Elliott MR, Sandilos JK, Walk SF, Kinchen JM, Lazarowski ER, Armstrong AJ, Penuela S, Laird DW, Salvesen GS, Isakson BE, Bayliss DA, Ravichandran KS (2010) Pannexin 1 channels mediate "find-me" signal release and membrane permeability during apoptosis. Nature 467(7317):863–867. https://doi.org/10.1038/nature09413

48. Sandilos JK, Chiu YH, Chekeni FB, Armstrong AJ, Walk SF, Ravichandran KS, Bayliss DA (2012) Pannexin 1, an ATP release channel, is activated by caspase cleavage of its pore-associated C-terminal autoinhibitory region. J Biol Chem 287(14):11303–11311. https://doi.org/10.1074/jbc.M111.323378

49. Poon IK, Lucas CD, Rossi AG, Ravichandran KS (2014) Apoptotic cell clearance: basic biology and therapeutic potential. Nat Rev Immunol 14(3):166–180. https://doi.org/10.1038/nri3607

50. Green DR, Ferguson T, Zitvogel L, Kroemer G (2009) Immunogenic and tolerogenic cell death. Nat Rev Immunol 9(5):353–363. https://doi.org/10.1038/nri2545

51. Yatim N, Cullen S, Albert ML (2017) Dying cells actively regulate adaptive immune responses. Nat Rev Immunol 17(4):262–275. https://doi.org/10.1038/nri.2017.9

52. Vince JE, Wong WW, Gentle I, Lawlor KE, Allam R, O'Reilly L, Mason K, Gross O, Ma S, Guarda G, Anderton H, Castillo R, Hacker G, Silke J, Tschopp J (2012) Inhibitor of apoptosis proteins limit RIP3 kinase-dependent interleukin-1 activation. Immunity 36(2):215–227. https://doi.org/10.1016/j.immuni.2012.01.012

53. Vince JE, De Nardo D, Gao W, Vince AJ, Hall C, McArthur K, Simpson D, Vijayaraj S, Lindqvist LM, Bouillet P, Rizzacasa MA, Man SM, Silke J, Masters SL, Lessene G, Huang DCS, Gray DHD, Kile BT, Shao F, Lawlor KE (2018) The mitochondrial apoptotic effectors BAX/BAK activate caspase-3 and -7 to trigger NLRP3 inflammasome and caspase-8-driven IL-1beta activation. Cell Rep 25(9): 2339–2353.e2334. https://doi.org/10.1016/j.celrep.2018.10.103

54. Lawlor KE, Khan N, Mildenhall A, Gerlic M, Croker BA, D'Cruz AA, Hall C, Kaur Spall S, Anderton H, Masters SL, Rashidi M, Wicks IP, Alexander WS, Mitsuuchi Y, Benetatos CA, Condon SM, Wong WW, Silke J, Vaux DL, Vince JE (2015) RIPK3 promotes cell death and NLRP3 inflammasome activation in the absence of MLKL. Nat Commun 6:6282. https://doi.org/10.1038/ncomms7282

55. Wicki S, Gurzeler U, Wei-Lynn Wong W, Jost PJ, Bachmann D, Kaufmann T (2016) Loss of XIAP facilitates switch to TNF alpha-induced necroptosis in mouse neutrophils. Cell Death Dis 7(10):e2422. https://doi.org/10.1038/cddis.2016.311

56. Chauhan D, Bartok E, Gaidt MM, Bock FJ, Herrmann J, Seeger JM, Broz P, Beckmann R, Kashkar H, Tait SWG, Muller R, Hornung V (2018) BAX/BAK-induced apoptosis results in caspase-8-dependent IL-1beta maturation in macrophages. Cell Rep 25(9):2354–2368.e2355. https://doi.org/10.1016/j.celrep.2018.10.087

57. Chen KW, Lawlor KE, von Pein JB, Boucher D, Gerlic M, Croker BA, Bezbradica JS, Vince JE, Schroder K (2018) Cutting edge: blockade of inhibitor of apoptosis proteins sensitizes neutrophils to TNF- but not lipopolysaccharide-mediated cell death and IL-1beta secretion. J Immunol 200(10): 3341–3346. https://doi.org/10.4049/jimmunol.1701620
58. Malireddi RKS, Gurung P, Mavuluri J, Dasari TK, Klco JM, Chi H, Kanneganti TD (2018) TAK1 restricts spontaneous NLRP3 activation and cell death to control myeloid proliferation. J Exp Med 215(4):1023–1034. https://doi.org/10.1084/jem.20171922
59. Yabal M, Muller N, Adler H, Knies N, Gross CJ, Damgaard RB, Kanegane H, Ringelhan M, Kaufmann T, Heikenwalder M, Strasser A, Gross O, Ruland J, Peschel C, Gyrd-Hansen M, Jost PJ (2014) XIAP restricts TNF- and RIP3-dependent cell death and inflammasome activation. Cell Rep 7(6):1796–1808. https://doi.org/10.1016/j.celrep.2014.05.008
60. Dondelinger Y, Jouan-Lanhouet S, Divert T, Theatre E, Bertin J, Gough PJ, Giansanti P, Heck AJ, Dejardin E, Vandenabeele P, Bertrand MJ (2015) NF-kappaB-independent role of IKKalpha/IKKbeta in preventing RIPK1 kinase-dependent apoptotic and necroptotic cell death during TNF signaling. Mol Cell 60(1):63–76. https://doi.org/10.1016/j.molcel.2015.07.032
61. Kuriakose T, Man SM, Malireddi RK, Karki R, Kesavardhana S, Place DE, Neale G, Vogel P, Kanneganti TD (2016) ZBP1/DAI is an innate sensor of influenza virus triggering the NLRP3 inflammasome and programmed cell death pathways. Sci Immunol 1(2):aag2045. https://doi.org/10.1126/sciimmunol.aag2045
62. Miwa K, Asano M, Horai R, Iwakura Y, Nagata S, Suda T (1998) Caspase 1-independent IL-1beta release and inflammation induced by the apoptosis inducer Fas ligand. Nat Med 4(11):1287–1292. https://doi.org/10.1038/3276
63. Joosten LA, Netea MG, Fantuzzi G, Koenders MI, Helsen MM, Sparrer H, Pham CT, van der Meer JW, Dinarello CA, van den Berg WB (2009) Inflammatory arthritis in caspase 1 gene-deficient mice: contribution of proteinase 3 to caspase 1-independent production of bioactive interleukin-1beta. Arthritis Rheum 60(12):3651–3662. https://doi.org/10.1002/art.25006
64. Guma M, Ronacher L, Liu-Bryan R, Takai S, Karin M, Corr M (2009) Caspase 1-independent activation of interleukin-1beta in neutrophil-predominant inflammation. Arthritis Rheum 60(12):3642–3650. https://doi.org/10.1002/art.24959
65. Mandal P, Feng Y, Lyons JD, Berger SB, Otani S, DeLaney A, Tharp GK, Maner-Smith-K, Burd EM, Schaeffer M, Hoffman S, Capriotti C, Roback L, Young CB, Liang Z, Ortlund EA, DiPaolo NC, Bosinger S, Bertin J, Gough PJ, Brodsky IE, Coopersmith CM, Shayakhmetov DM, Mocarski ES (2018) Caspase-8 collaborates with caspase-11 to drive tissue damage and execution of endotoxic shock. Immunity 49(1):42–55.e46. https://doi.org/10.1016/j.immuni.2018.06.011
66. Taabazuing CY, Okondo MC, Bachovchin DA (2017) Pyroptosis and apoptosis pathways engage in bidirectional crosstalk in monocytes and macrophages. Cell Chem Biol 24(4): 507–514.e504. https://doi.org/10.1016/j.chembiol.2017.03.009
67. Chen KW, Demarco B, Heilig R, Shkarina K, Boettcher A, Farady CJ, Pelczar P, Broz P (2019) Extrinsic and intrinsic apoptosis activate pannexin-1 to drive NLRP3 inflammasome assembly. EMBO J 38(10):e101638. https://doi.org/10.15252/embj.2019101638
68. Qu Y, Misaghi S, Newton K, Gilmour LL, Louie S, Cupp JE, Dubyak GR, Hackos D, Dixit VM (2011) Pannexin-1 is required for ATP release during apoptosis but not for inflammasome activation. J Immunol 186(11):6553–6561. https://doi.org/10.4049/jimmunol.1100478
69. Chen KW, Demarco B, Broz P (2020) Pannexin-1 promotes NLRP3 activation during apoptosis but is dispensable for canonical or noncanonical inflammasome activation. Eur J Immunol 50(2):170–177. https://doi.org/10.1002/eji.201948254
70. Demarco B, Grayczyk JP, Bjanes E, Le Roy D, Tonnus W, Assenmacher CA, Radaelli E, Fettrelet T, Mack V, Linkermann A, Roger T, Brodsky IE, Chen KW, Broz P (2020) Caspase-8-dependent gasdermin D cleavage promotes antimicrobial defense but confers susceptibility to TNF-induced lethality. Sci Adv 6(47): eabc3465. https://doi.org/10.1126/sciadv.abc3465
71. Orning P, Weng D, Starheim K, Ratner D, Best Z, Lee B, Brooks A, Xia S, Wu H, Kelliher MA, Berger SB, Gough PJ, Bertin J, Proulx MM, Goguen JD, Kayagaki N, Fitzgerald KA, Lien E (2018) Pathogen blockade of TAK1 triggers caspase-8-dependent cleavage of gasdermin D and cell death. Science 362(6418): 1064–1069. https://doi.org/10.1126/science.aau2818

72. Sarhan J, Liu BC, Muendlein HI, Li P, Nilson R, Tang AY, Rongvaux A, Bunnell SC, Shao F, Green DR, Poltorak A (2018) Caspase-8 induces cleavage of gasdermin D to elicit pyroptosis during Yersinia infection. Proc Natl Acad Sci U S A 115(46):E10888–E10897. https://doi.org/10.1073/pnas.1809548115

73. Sanjo H, Nakayama J, Yoshizawa T, Fehling HJ, Akira S, Taki S (2019) Cutting edge: TAK1 safeguards macrophages against proinflammatory cell death. J Immunol 203(4): 783–788. https://doi.org/10.4049/jimmunol.1900202

74. Duprez L, Takahashi N, Van Hauwermeiren F, Vandendriessche B, Goossens V, Vanden Berghe T, Declercq W, Libert C, Cauwels A, Vandenabeele P (2011) RIP kinase-dependent necrosis drives lethal systemic inflammatory response syndrome. Immunity 35(6): 908–918. https://doi.org/10.1016/j.immuni.2011.09.020

75. Polykratis A, Hermance N, Zelic M, Roderick J, Kim C, Van TM, Lee TH, Chan FKM, Pasparakis M, Kelliher MA (2014) Cutting edge: RIPK1 Kinase inactive mice are viable and protected from TNF-induced necroptosis in vivo. J Immunol 193(4): 1539–1543. https://doi.org/10.4049/jimmunol.1400590

76. Kaufmann T, Jost PJ, Pellegrini M, Puthalakath H, Gugasyan R, Gerondakis S, Cretney E, Smyth MJ, Silke J, Hakem R, Bouillet P, Mak TW, Dixit VM, Strasser A (2009) Fatal hepatitis mediated by tumor necrosis factor TNFalpha requires caspase-8 and involves the BH3-only proteins Bid and Bim. Immunity 30(1):56–66. https://doi.org/10.1016/j.immuni.2008.10.017

77. Rogers C, Fernandes-Alnemri T, Mayes L, Alnemri D, Cingolani G, Alnemri ES (2017) Cleavage of DFNA5 by caspase-3 during apoptosis mediates progression to secondary necrotic/pyroptotic cell death. Nat Commun 8:14128. https://doi.org/10.1038/ncomms14128

78. Wang Y, Gao W, Shi X, Ding J, Liu W, He H, Wang K, Shao F (2017) Chemotherapy drugs induce pyroptosis through caspase-3 cleavage of a gasdermin. Nature 547(7661):99–103. https://doi.org/10.1038/nature22393

79. Zhou B, Zhang JY, Liu XS, Chen HZ, Ai YL, Cheng K, Sun RY, Zhou D, Han J, Wu Q (2018) Tom20 senses iron-activated ROS signaling to promote melanoma cell pyroptosis. Cell Res 28(12):1171–1185. https://doi.org/10.1038/s41422-018-0090-y

80. Lee BL, Mirrashidi KM, Stowe IB, Kummerfeld SK, Watanabe C, Haley B, Cuellar TL, Reichelt M, Kayagaki N (2018) ASC- and caspase-8-dependent apoptotic pathway diverges from the NLRC4 inflammasome in macrophages. Sci Rep 8(1):3788. https://doi.org/10.1038/s41598-018-21998-3

81. Zheng M, Williams EP, Malireddi RKS, Karki R, Banoth B, Burton A, Webby R, Channappanavar R, Jonsson CB, Kanneganti TD (2020) Impaired NLRP3 inflammasome activation/pyroptosis leads to robust inflammatory cell death via caspase-8/RIPK3 during coronavirus infection. J Biol Chem 295(41): 14040–14052. https://doi.org/10.1074/jbc.RA120.015036

82. Tixeira R, Shi B, Parkes MAF, Hodge AL, Caruso S, Hulett MD, Baxter AA, Phan TK, Poon IKH (2018) Gasdermin E does not limit apoptotic cell disassembly by promoting early onset of secondary necrosis in jurkat T cells and THP-1 monocytes. Front Immunol 9:2842. https://doi.org/10.3389/fimmu.2018.02842

83. Dickson RC, Bronk SF, Gores GJ (1992) Glycine cytoprotection during lethal hepatocellular injury from adenosine triphosphate depletion. Gastroenterology 102(6): 2098–2107. https://doi.org/10.1016/0016-5085(92)90338-y

84. Phulphagar K, Kuhn LI, Ebner S, Frauenstein A, Swietlik JJ, Rieckmann J, Meissner F (2021) Proteomics reveals distinct mechanisms regulating the release of cytokines and alarmins during pyroptosis. Cell Rep 34(10):108826. https://doi.org/10.1016/j.celrep.2021.108826

85. Zanoni I, Tan Y, Di Gioia M, Broggi A, Ruan J, Shi J, Donado CA, Shao F, Wu H, Springstead JR, Kagan JC (2016) An endogenous caspase-11 ligand elicits interleukin-1 release from living dendritic cells. Science 352(6290): 1232–1236. https://doi.org/10.1126/science.aaf3036

86. Wolf AJ, Reyes CN, Liang W, Becker C, Shimada K, Wheeler ML, Cho HC, Popescu NI, Coggeshall KM, Arditi M, Underhill DM (2016) Hexokinase is an innate immune receptor for the detection of bacterial peptidoglycan. Cell 166(3):624–636. https://doi.org/10.1016/j.cell.2016.05.076

87. Gaidt MM, Ebert TS, Chauhan D, Schmidt T, Schmid-Burgk JL, Rapino F, Robertson AA, Cooper MA, Graf T, Hornung V (2016) Human monocytes engage an alternative inflammasome pathway. Immunity 44(4): 833–846. https://doi.org/10.1016/j.immuni.2016.01.012

88. Vigano E, Diamond CE, Spreafico R, Balachander A, Sobota RM, Mortellaro A

(2015) Human caspase-4 and caspase-5 regulate the one-step non-canonical inflammasome activation in monocytes. Nat Commun 6: 8761. https://doi.org/10.1038/ncomms9761

89. Chen KW, Gross CJ, Sotomayor FV, Stacey KJ, Tschopp J, Sweet MJ, Schroder K (2014) The neutrophil NLRC4 inflammasome selectively promotes IL-1beta maturation without pyroptosis during acute Salmonella challenge. Cell Rep 8(2):570–582. https://doi.org/10.1016/j.celrep.2014.06.028

90. Karmakar M, Katsnelson MA, Dubyak GR, Pearlman E (2016) Neutrophil P2X7 receptors mediate NLRP3 inflammasome-dependent IL-1beta secretion in response to ATP. Nat Commun 7:10555. https://doi.org/10.1038/ncomms10555

91. Evavold CL, Ruan J, Tan Y, Xia S, Wu H, Kagan JC (2018) The pore-forming protein gasdermin D regulates interleukin-1 secretion from living macrophages. Immunity 48(1):35–44.e36. https://doi.org/10.1016/j.immuni.2017.11.013

92. Kang TB, Yang SH, Toth B, Kovalenko A, Wallach D (2013) Caspase-8 blocks kinase RIPK3-mediated activation of the NLRP3 inflammasome. Immunity 38(1):27–40. https://doi.org/10.1016/j.immuni.2012.09.015

93. Monteleone M, Stanley AC, Chen KW, Brown DL, Bezbradica JS, von Pein JB, Holley CL, Boucher D, Shakespear MR, Kapetanovic R, Rolfes V, Sweet MJ, Stow JL, Schroder K (2018) Interleukin-1beta maturation triggers its relocation to the plasma membrane for gasdermin-D-dependent and -independent secretion. Cell Rep 24(6):1425–1433. https://doi.org/10.1016/j.celrep.2018.07.027

94. Broz P, Monack DM (2013) Measuring inflammasome activation in response to bacterial infection. Methods Mol Biol 1040:65–84. https://doi.org/10.1007/978-1-62703-523-1_6

Chapter 15

Visualizing Effector Triggered Immunity in Response to Pore-Forming Toxins by Live-Cell Imaging

Seong H. Chow and Thomas Naderer

Abstract

The NLRP3 inflammasome senses the activity of pore-forming toxins secreted by *Staphylococcus aureus*. The bacterial toxins compromise plasma membrane integrity which activates the NLRP3 inflammasome to induce host pore-forming proteins and cellular suicide, termed pyroptosis. Host cell death rates are routinely determined at pre-defined time points and on whole cell populations. To capture the dynamic interactions between bacterial pore-forming toxins and host cell death factors, we have applied live-cell imaging techniques capable of analyzing single cell events in real time. Here, we describe methods using live-cell imaging to determine the host responses, such as plasma membrane integrity, mitochondrial health, and apoptotic caspases, towards pore-forming toxins.

Key words *S. aureus*, Widefield fluorescent microscope, Macrophages, NLRP3, Inflammasome, Cell death, Pyroptosis

1 Introduction

Staphylococcus aureus, and in particular methicillin-resistant *S. aureus* (MRSA), secrete toxins, termed leukocidins, that kill host innate immune cells. The Panton-Valentine leukocidin (PVL) and leukocidin AB (LukAB) assemble into pore-forming complexes at the plasma membrane, causing lytic cell death of macrophages and neutrophils [1–5]. Recent studies have shown that the loss of the plasma membrane integrity due to PVL and LukAB activity is sensed by the cytosolic NOD-like receptor protein 3 (NLRP3), referred to as effector triggered immunity. Active NLRP3 triggers the formation of the inflammasome resulting in the proteolytical release of the N-terminal region of gasdermin D (GSDMD), which forms membrane pores. GSDMD pores sufficiently disrupt plasma membrane integrity resulting in lytic host cell death, termed pyroptosis, and the secretion of inflammatory cytokines, such as interleukin-1β (IL-1β) after PVL and LukAB

Thomas A. Kufer and Maria Kaparakis-Liaskos (eds.), *Effector-Triggered Immunity: Methods and Protocols*, Methods in Molecular Biology, vol. 2523, https://doi.org/10.1007/978-1-0716-2449-4_15,

exposure. Typically, host cell death rates are determined at pre-defined time points and on whole cell populations, based on the release of cytosolic enzymes via ELISA or flow cytometry of DNA-stained cells. These methods, however, fail to capture the full extent of the dynamic interactions between bacterial toxins and host cell death factors.

Live-cell imaging is a versatile time-lapse imaging approach that enables the detection of dynamic host–pathogen interactions. In particular, we have developed methods to visualize host cell responses at the single cell level and at high temporal resolution (seconds or minutes) over extended periods of times (3 days). The use of multiwell plates (96 wells) facilitates semi-high-throughput analysis which allows to control for treatments and intrinsic assay variations. Here, we describe that live-cell imaging can be used to visualize the activation of host cell death factors after the treatment with *S. aureus* pore-forming toxins. To minimize imaging-induced toxicity, cells are visualized with a widefield fluorescent microscope system containing an incubation chamber set at 37 °C, 5% CO_2, >90% relative humidity (RH), and an inverted, fully-motorized stage driven by application software. The use of commercially available fluorescent probes, such as tetramethylrhodamine methyl ester (TMRM), CellEvent Caspase-3/7 Green, or DRAQ7, enables visualization of the sequential changes in mitochondrial membrane potential (ψ_m), the activity of apoptotic caspases, and loss of plasma membrane integrity, respectively, in single cells (Fig. 1). We further describe approaches to quantify single cell events that occur over extended periods of time. The use of live-cell imaging at high temporal resolution uncovers host cell responses to toxins and infections [5–8].

2 Materials

Prepare all solutions using sterilized ultrapure water (prepared by purifying deionized water, to attain a sensitivity of 18 MΩ cm at 25 °C) and tissue culture grade reagents. Diligently follow all waste disposal regulations when disposing waste materials.

2.1 Macrophages Preparation

1. Heat-inactivated fetal bovine serum (FBS) (sterile-filtered): Thaw fetal bovine serum at 4 °C overnight or at 37 °C for an hour. To heat inactivate the serum, place the bottle at 55 °C for 30 min, and mix the bottle thoroughly every 10 min. Aliquot 50 mL in each tube and store at −20 °C.
2. 1M HEPES, pH 7.0: Add 238.3 g HEPES powder to 800-mL water. Stir on magnetic stirrer and adjust pH with 10-M NaOH. Make up to 1 L with water. Autoclave or sterile filter. Store at room temperature.

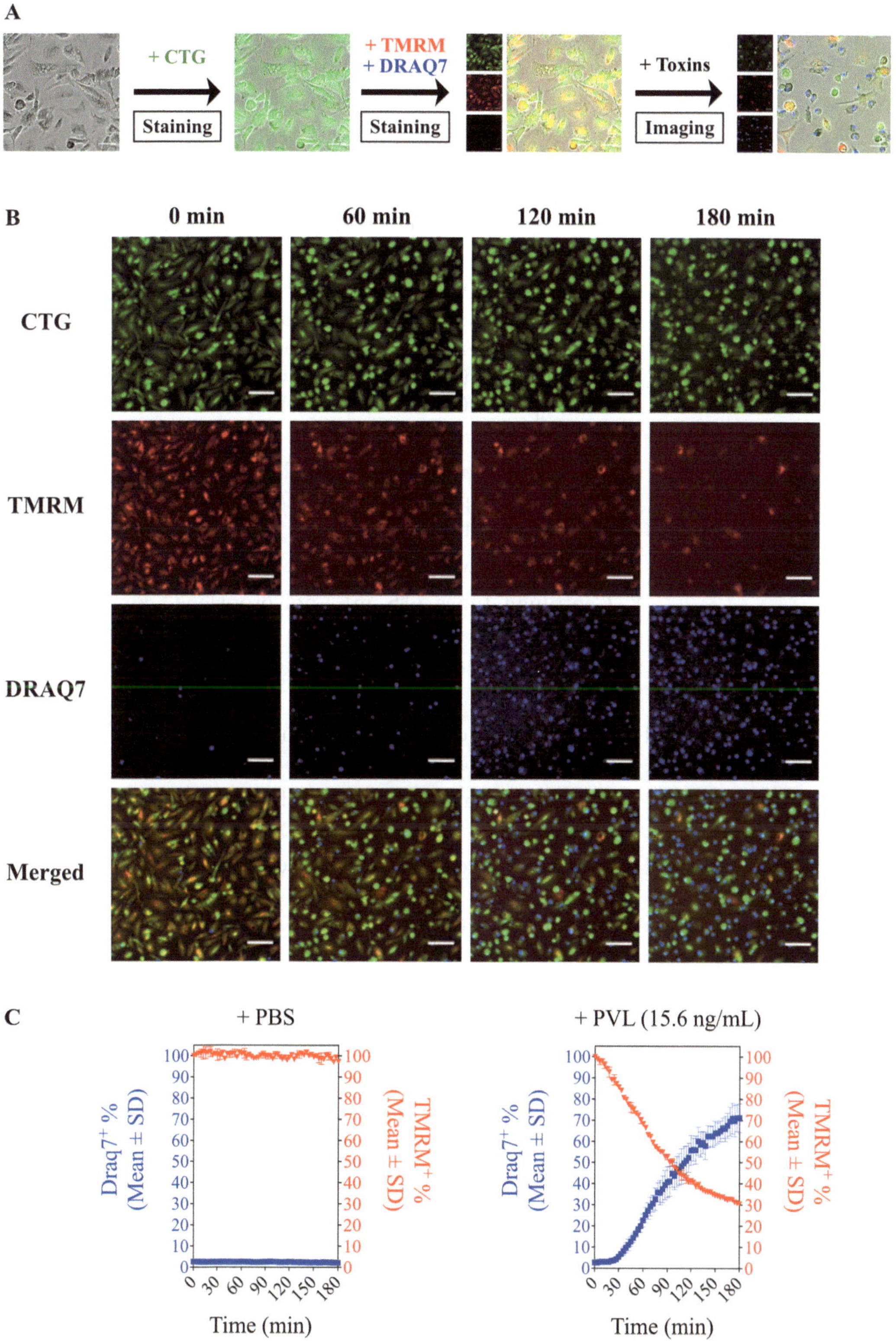

Fig. 1 Live-cell imaging analysis of human-induced pluripotent stem cell-derived macrophages (hiPSC-dMs) treated with PVL toxins. (**a**) Stepwise protocol of live-cell imaging. The cells were first stained with CTG for

3. Macrophages culture media: Roswell Park Memorial Institute (RPMI) 1640 media supplemented with 10% (v/v) FBS, 25-mM HEPES, pH 7.0. Add 50-mL heat-inactivated FBS and 12.5-mL 1M HEPES to 437.5-mL RPMI 1640 media (*see* **Notes 1** and **2**). Store at 4 °C.
4. Phosphate-buffered saline, pH 7.4 (PBS, 10×): Add 80-g NaCl, 2-g KCl, 17.8-g Na_2HPO_4, 2.4-g KH_2PO_4 to 800-mL water. Stir on magnetic stirrer and make up to 1 L with water. To make 1 L of PBS, add 100-mL 10× PBS to 900-mL water (*see* **Note 3**). Autoclave. Store at room temperature.
5. 0.5-M ethylenediaminetetraacetic acid (EDTA), pH 8.0: Add 93.06-g EDTA disodium salt to 350-mL water. Stir on magnetic stirrer and adjust pH with 10-M NaOH (*see* **Note 4**). Make up to 500 mL with water. Autoclave or sterile filter. Store at room temperature.
6. Cell dissociation buffer: 2 mM EDTA in 1× PBS. Add 2-mL 0.5 M EDTA, pH 8.0 to 498-mL 1× PBS. Sterile filter. Store at room temperature.
7. Cell scrapper with 25-cm handle and 1.8-cm blade, sterile.
8. Centrifuge tubes (15 mL, polypropylene, conical bottom, sterile).
9. Petri dish (100 mm, non-tissue culture-treated polystyrene, sterile).
10. 96-well tissue culture plate (tissue culture-treated polystyrene, clear flat bottom, sterile).
11. Centrifuge that has a suitable adapter for 15-mL conical centrifuge tubes.
12. Hemacytometer or automated cell counter.

2.2 Live-Cell Imaging

1. Dimethyl sulfoxide (DMSO): Keep in dark and store at room temperature.
2. 1-mM CTG dye (1000×): Add 108-μL DMSO to 50-μg lyophilized powder. Aliquot 12 μL in each tube, keep in dark, and store at −20 °C (*see* **Note 5**).
3. 50-μM TMRM dye (1000×): Add 998-μL DMSO to 25-μg lyophilized powder. Aliquot 50 μL in each tube, keep in dark, and store at −20 °C (*see* **Note 6**).

Fig. 1 (continued) 30 min, followed by TMRM and DRAQ7 co-staining for 30 min. Next, cells were treated with control (PBS) or PVL toxins and immediately imaged with bright-field, GFP, TX2, and Y5 filters every 3 min for up to 180 min. Scale bar is 20 μm. (**b**) Representative of time-lapse images of hiPSC-dMs treated with PVL (15.6 ng/mL) at 0, 60, 120, and 180 min. Scale bar is 100 μm. (**c**) Percentage of cell death (DRAQ7-staining) and mitochondrial health (TMRM-staining) of hiPSC-dMs treated with PBS (control, left) or PVL (15.6 ng/mL, right) overtime

4. 300-μM DRAQ7 (500×): Keep in dark and store at 4 °C (*see* **Note 7**).
5. 50-μg/mL lipopolysaccharide (LPS), ultrapure: To make 1-mg/mL LPS, add 5-mL water to 5-mg lyophilized powder. Aliquot 50 μL in each tube and store at −20 °C. To make 50-μg/mL LPS, add 50-μL 1-mg/mL LPS to 950-μL water. Aliquot 20 μL in each tube and store at −20 °C (*see* **Note 8**).
6. 50-μg/mL purified recombinant staphylococcal pore-forming toxins, endotoxin-free: Add 20-mL 1× PBS to 1-mg lyophilized powder. Aliquot 50 μL in each tube and store at −20 °C (*see* **Note 9**).
7. Serum-free media: RPMI 1640 media supplemented with 25-mM HEPES, pH 7.0. Add 12.5-mL 1M HEPES to 487.5-mL RPMI 1640 media. Store at 4 °C.
8. Macrophages culture media: RPMI 1640 media supplemented with 10% (v/v) FBS, 25-mM HEPES, pH 7.0 (*see* Subheading 2.1).
9. Centrifuge tubes (15 mL, polypropylene, conical bottom, sterile).
10. Leica DMi8 widefield fluorescent microscope system, inverted: Equipped with an incubation chamber set at 37 °C, 5% CO_2, >90% relative humidity (RH), a Leica DFC9000 GTC sCMOS camera, a 10×/0.32-NA objective lens, GFP/FITC, RFP/TRITC/TX2 and Cy5/Y5 filters, and a fully motorized stage driven by Leica LAS X software (or equivalent).
11. MetaMorph basic offline image analysis software (Molecular Devices, USA).

2.3 Monitoring Apoptosis Using Live-Cell Imaging

1. Dimethyl sulfoxide (DMSO): Keep in dark and store at room temperature.
2. CellEvent Caspase-3/7 Green ReadyProbes Reagent (Casp-3/7): Store at room temperature (*see* **Note 10**).
3. 50-μM TMRM dye (1000×) (*see* Subheading 2.2).
4. 300-μM DRAQ7 (500×) (*see* Subheading 2.2).
5. 50-μg/mL LPS, ultrapure (*see* Subheading 2.2).
6. Purified recombinant staphylococcal pore-forming toxins, endotoxin-free (*see* Subheading 2.2).
7. Macrophages culture media: RPMI 1640 media supplemented with 10% (v/v) FBS, 25-mM HEPES, pH 7.0 (*see* Subheading 2.1).
8. Centrifuge tubes (15 mL, polypropylene, conical bottom, sterile).

9. Leica DMi8 widefield fluorescent microscope system, inverted (*see* Subheading 2.2).
10. MetaMorph basic offline image analysis software (Molecular Devices, USA).

3 Methods

3.1 Macrophages Culture Preparation

1. Culture macrophages, such as THP1 or peripheral blood mononuclear cells (PBMC)-derived macrophages, in non-tissue culture-treated 100-mm petri dish with macrophage culture media, and incubate at 37 °C, 5% CO_2 incubator (*see* **Note 11**).
2. A day prior to live-cell imaging, remove the culture medium and wash the cells three times with 10-mL 1× PBS.
3. Remove 1× PBS and add 5-mL cell dissociation buffer to the petri dish (*see* **Note 12**). Incubate for 30 min at 37 °C, 5% CO_2 with gentle tapping every 5 min to detach the cells from the dish.
4. Gently scrap the cells from the dish with cell scrapper. Transfer the detached cells in a 15-mL centrifuge tube, and centrifuge at 150 × g for 5 min.
5. Count the cells using hemacytometer or automated cell counter, and seed the cells in tissue culture-treated 96-well plate (*see* **Note 13**) with macrophages culture media at a density of 5×10^5 cells/mL. Incubate the plate at 37 °C, 5% CO_2 in an incubator.

3.2 Imaging Macrophage Cell Death and Mitochondrial Membrane Potential

1. Add 12-μL 1000× CTG to 12-mL pre-warmed serum-free medium in a 15-mL centrifuge tube.
2. Replace the medium with 100-μL serum-free medium containing CTG (1 μM) to each well. Incubate the plate for 30 min at 37 °C, 5% CO_2 in the dark.
3. In the meantime, add 12-μL 50-μg/mL LPS to 12-mL pre-warmed macrophage culture medium in a 15-mL centrifuge tubes.
4. Replace the medium with 100-μL macrophage culture medium containing 50–100-ng/mL LPS per well. Incubate the plate for 3 h at 37 °C, 5% CO_2 to prime cells. Keep the plate in dark.
5. Mix 12-μL 1000× TMRM and 24-μL 500× DRAQ7 to 12-mL pre-warmed macrophage culture medium in a 15-mL centrifuge tubes. If required, include endotoxin-free toxins or specific inhibitor to macrophage culture medium containing TMRM (50 nM) and DRAQ7 (600 nM). Keep the medium in dark.

6. Replace the medium containing LPS with 100-μL macrophage culture medium containing TMRM (50 nM), DRAQ7 (600 nM), and toxins or control vehicle (PBS). Keep the plate in dark (*see* **Note 14**).
7. Immediately place the plate on the microscope stage (*see* **Note 15**) and cover the plate with lid (*see* **Note 16**).
8. Use a 10×/0.32-NA objective lens (*see* **Note 17**) for image acquisition. Check the condenser and diaphragm position.
9. Set image digitization to 12-bit depth and resolution 2048 × 2048 pixels without binning (*see* **Note 18**).
10. Setup a bright-field channel and adjust the brightness, time exposure, and aperture (*see* **Note 19**). Find the cells by adjusting the appropriate *x*-, *y*-, and *z*-position using the stage movement and focus control knob. Avoid saturation (*see* **Note 20**).
11. Setup a green channel for acquiring CTG signals using GFP/FITC filter and adjust the excitation intensity, gain, and exposure time (*see* **Note 21**). Avoid saturation.
12. Setup a red channel for acquiring TMRM signals using RFP/TRITC/TX2 filter and adjust the excitation intensity, gain, and exposure time (*see* **Note 22**). Avoid saturation.
13. Setup a far-red channel for acquiring DRAQ7 signals using Cy5/Y5 filter and adjust the excitation intensity, gain, and exposure time (*see* **Note 23**). Avoid saturation.
14. Setup the navigator panel for automated well positioning in an ordered manner, i.e., A1, A2, A3. . . . B1, B2, B3 (*see* **Note 24**).
15. Setup the autofocus function using bright-field channel to ensure a consistent *z*-position of each well over imaging session (*see* **Note 25**).
16. Setup the time interval and duration of the imaging. Start the imaging.
17. Once the imaging has ended, save raw data as TIF images (*see* **Note 26**).
18. Analyze the images in MetaMorph or ImageJ software using a custom-made journal suite incorporating the count nuclei function to segment and count the number of CTG-, TMRM-, and DRAQ7-positive cells (*see* **Note 27**).
19. Define the min width, max width, and intensity of pixels of representative image using the "Count nuclei" function in the MetaMorph (*see* **Note 28**).
20. Open the "Automated MWS v3.JNL" in the MetaMorph and then open the sub-journal "Count and log data.JNL" in the journal sub-window. Set the "Count nuclei" parameter of each fluorescent channel based on the value obtained from the representative images. Click the "Save" button below. Return

to the main journal window and click "Run Journal" to start the analysis. CTG images can be used to calculate the total cell number, typically in the first frame (*see* **Note 29**). Analyze the percentage of cell death and healthy mitochondria by dividing the number of TMRM- and DRAQ7-positive cells to the total cell number (Fig. 1c).

3.3 Imaging the Activity of Apoptotic Caspases

1. Add 12-μL 50-μg/mL LPS to 12-mL pre-warmed macrophage culture medium in a 15-mL centrifuge tube.
2. Replace the medium with 100-μL macrophage culture medium containing 50–100-ng/mL LPS per well. Incubate the plate for 3 h at 37 °C, 5% CO_2 to prime cells. Keep the plate in dark.
3. Mix 12-μL 1000× TMRM, 24-μL 500× DRAQ7, and 24 drops of Casp-3/7 Reagent to 12-mL pre-warmed macrophage culture medium in a 15-mL centrifuge tube. If required, include endotoxin-free toxins or specific inhibitor to macrophage culture medium containing Casp-3/7, TMRM (50 nM), and DRAQ7 (600 nM). Keep the media in dark.
4. Replace the media containing LPS with 100-μL macrophage culture media containing Casp-3/7, TMRM (50 nM), DRAQ7 (600 nM), and toxins. Keep the plate in dark.
5. Immediately place the plate on the microscope stage and cover the plate with lid.
6. Use a 10×/0.32-NA objective lens for image acquisition. Check the condenser and diaphragm position.
7. Set image digitization to 12-bit depth and resolution 2048 × 2048 pixels without binning.
8. Setup a bright-field channel and adjust the brightness, time exposure, and aperture. Find the cells by adjusting the appropriate *x*-, *y*-, and *z*-position using the stage movement and focus control knob. Avoid saturation.
9. Setup a green channel for acquiring Casp-3/7 signals using GFP/FITC filter and adjust the excitation intensity, gain, and exposure time (*see* **Note 30**). Avoid saturation.
10. Setup a red channel for acquiring TMRM signals using RFP/TRITC/TX2 filter and adjust the excitation intensity, gain, and exposure time. Avoid saturation.
11. Setup a far-red channel for acquiring DRAQ7 signals using Cy5/Y5 filter and adjust the excitation intensity, gain, and exposure time. Avoid saturation.
12. Setup the navigator panel for automated well positioning in an ordered manner, i.e., A1, A2, A3.... B1, B2, B3.

13. Setup the autofocus function using bright-field channel to ensure a consistent *z*-position of each well over imaging session.
14. Setup the time interval and duration of the imaging. Start the imaging.
15. Once the imaging has ended, save raw data as TIF images.
16. Analyze the images in MetaMorph or ImageJ software using a custom-made journal suite incorporating the count nuclei function to segment and count the number of Casp3/7-, TMRM-, and DRAQ7-positive cells.
17. Define the min width, max width, and intensity of pixels of representative image using the "Count nuclei" function in MetaMorph (*see* **Note 31**).
18. Open the "Automated MWS v3.JNL" in MetaMorph and then open the sub-journal "Count and log data.JNL" in the journal sub-window. Set the "Count nuclei" parameter of each fluorescent channel based on the value obtained from the representative images. Click the "Save" button below. Return to the main journal window and click "Run Journal" to start the analysis. Use only the first frame of the TMRM image to calculate the total cell number (*see* **Note 32**). Analyze the percentage of cell death, apoptotic cells, and healthy mitochondria by dividing the number of Casp3/7-, TMRM-, and DRAQ7-positive cells to the total cell number.

4 Notes

1. Phenol red-free media can be used if phenol red increases the background noise or interferes with the fluorescence signal during imaging.
2. HEPES provides supplemental buffering to culture media that is CO_2 independent, which could avoid potential fluctuating CO_2 level over imaging session.
3. The pH of the 10× PBS will be approximately 6.8, and the pH should change to 7.4 when diluted to 1× PBS. If necessary, the pH can be adjusted using HCl or NaOH.
4. EDTA is insoluble in water when pH is lower than 8.0. Add 10-M NaOH solution gradually to dissolve the salt.
5. CTG is a cell-permeant, green fluorescent dye that accumulates in the cytosol of viable cells and can be observed using a standard GFP/FITC filter. Optimal concertation of the dye varies depending upon cell types. Generally, the optimal concentration for CTG dye is 500 nM–5 μM, and it is best to use a lower dye concentration to maintain normal cellular physiology

and reduce potential artifacts. However, rapidly dividing cells or long-term experiments (more than 3 days) require a higher concentration.

6. TMRM is a cell-permeant, cationic, red-orange fluorescent dye that accumulates at healthy mitochondria and can be observed using a standard RFP/TRITC/TX2 filter. When mitochondria depolarize, TMRM signal diminished. Optimal concertation of the dye varies depending upon cell types. Generally, the optimal concentration for TMRM is 50 nM–100 nM as using a higher TMRM concentration will result in quenching.
7. DRAQ7 is a cell impermeable, far-red fluorescent DNA-binding dye that only stains the nuclei in dead and permeabilized cells and can be observed using a Cy5/Y5 filter.
8. Once thawed, discard the remaining aliquot of LPS as it is temperature sensitive.
9. Staphylococcal pore-forming toxin with an affinity tag (i.e., 6xHis, GST) can be expressed in *Escherichia coli* and purified from the soluble fraction of lysed culture. Macrophages are very sensitive to bacterial endotoxin, which is a pro-inflammatory mediator. Hence, endotoxin in the purified recombinant toxins should be removed using a high capacity endotoxin removal resin or column (Thermo Fisher Scientific) and subsequently quantified using LAL chromogenic endotoxin quantification kit. The acceptable endotoxin level in the purified toxins should be less than 0.1 EU/mL.
10. CellEvent Caspase-3/7 Green ReadyProbes reagent allows rapid detection of activated caspase-3/7 in cells, which is an indicator of apoptosis. Upon caspase-3/7 activation, the dye binds to cell nucleus, generating a green fluorescence (~530 nm) and can be observed using a standard GFP/FITC filter.
11. It is important to maintain an optimal cell condition prior to live-cell imaging, as unfavorable conditions can cause artifacts or cell death.
12. Trypsin-EDTA is not recommended for harvesting macrophages, as the proteases will degrade the surface receptors of cells, preventing the toxins binding to the cells.
13. Other tissue culture multiwell plates, such as 6-well or 24-well, can be used depending on the application.
14. For inhibitor treatment, replace the media with macrophage culture media containing TMRM (50 nM), DRAQ7 (600 nM), and specific inhibitor or control vehicle. Incubate the plate for 30 min at 37 °C, 5% CO_2 prior to toxin treatment. Also, perform at least two to three independent wells per treatment as technical replicates.

15. Setup the microscope system by turning on the fluorescence lamp, microscope power control, camera, and computer, and warm up the incubator chamber set at 37 °C, 5% CO_2, and >90% humidity level at least 30 min prior to live-cell imaging.
16. When cells are cultured in a plate with loose-fitting lid, the medium will evaporate over time, especially at the outer wells of plate. This increases the salts concentration in media that are harmful to the cells. Hence, avoid using the outer wells of plate if there are less than 60 wells for imaging. Otherwise, add 50-μL mineral oil to each well to prevent media evaporation.
17. In a widefield fluorescent microscope, a 10× objective lens is sufficient to acquire at least 1000 cells per image. For characterizing mitochondrial morphology, use a 20× objective lens.
18. Most microscope cameras are able to capture 12-bit images, which contain significantly more grayscale depth than 8-bit images (4096 vs 256 possible shades of grey). Capturing 16-bit images is also acceptable, but the image file size is significantly larger than 12-bit images.
19. We can characterize the cell morphology and cell death from the bright-field images. For example, an apoptotic cell initially appears to be shrinking, and subsequently membrane blebbing is observed, whereas a necrotic cell appears to have excessive water inflow and bursts.
20. Some important detail and color information may be lost in saturated image. Saturated pixels can be avoided by choosing appropriate light intensity and exposure duration.
21. The cytosol of viable cells is stained with CTG. Minimize the intensity and time exposure (<200 ms) to prevent photobleaching and phototoxicity. However, while imaging rapid growing cells, a higher intensity or time exposure may be required as a dividing cell will split the CTG dye to its daughter cells.
22. Cells with active mitochondria are readily stained with TMRM. Minimize the intensity and time exposure (<200 ms) to prevent photobleaching and phototoxicity.
23. Viable cells with intact plasma membrane are not stained by DRAQ7, but it is normal to have <5% DRAQ7-positive cells in the population. If no DRAQ7-positive cell is initially present, include wells that contain dead cells to enable setup for DRAQ7 imaging (i.e., 0.1% Triton X-100 treated cells). As a longer wavelength is used to excite DRAQ7, the dye is relatively stable and less phototoxic at relatively higher intensity and longer time exposure (<500 ms).
24. The automated well positioning setup varies across different software. In Leica LAS X Navigator, there is an automated well

positioning feature for a multiwell plate. Otherwise, use "Mark and Find" for well positioning.

25. As the plate bottom may not be perfectly flat, it is necessary to use the autofocus feature, but it will increase the image acquisition duration. If the plate bottom is relatively flat, run autofocus with a preference of speed (less focus steps) versus accuracy (more focus steps) to reduce time. If the plate bottom is slightly concave in shape, bright-field channel may not be ideal for autofocus. Instead, use a DRAQ7 channel for autofocus. Alternatively, use a different brand of multiwell plate or a glass bottom cell imaging plate.
26. Save 12-bit image data in a 16-bit TIF file format. Other image format, such as JPEG, is not recommended for use with scientific data.
27. We use MetaMorph software journals to perform a series of automated tasks to operate the count nuclei function on different channels and time points of image, and export the results to Microsoft Excel. The MetaMorph journals are available at https://cloudstor.aarnet.edu.au/plus/s/8XGL750dIgwLYDn, and only 16-bit TIF images can be analyzed using these journals. If the TIF images are saved as 8-bit file format, these can be converted to 16-bit file format by running a FIJI/ImageJ macro, which is available at https://cloudstor.aarnet.edu.au/plus/s/ibafMqPMwzzMxXk.
28. A representative image can be chosen from the control well (cells treated with PBS). If no DRAQ7-positive cell is present in the control well, use the well that contains dead cells to set the parameter for DRAQ7. Measure the min and max width of the pixels using the line tool and add slightly higher values when running the journals.
29. CTG signal will dissipate during cell lysis. While this can be used as an additional marker for cell death, calculating the number of cells based on CTG signal can become difficult over time. As primary macrophages rarely multiply, the total cell number will be consistent over the imaging period. Hence, it is applicable to use the first frame of CTG image to calculate the total cell number. Also, we avoid using membrane-permeant DNA-binding fluorophores, such as Hoechst 33342 and DRAQ5, for obtaining total cell number as these dyes may induce distinct DNA damage responses in cells [9]. Furthermore, repeated exposure to UV light is damaging to cells.
30. The nuclei of apoptotic cells are stained by CellEvent Caspase-3/7 Green ReadyProbes Reagent. If no Casp-3/7-positive cell is initially present, include wells that contain apoptotic cells (i.e., staurosporine-treated cells) to enable setup for Casp-3/

7 imaging. Minimize the intensity and time exposure (<200 ms) to prevent photobleaching and phototoxicity.

31. A representative image can be chosen from the control well (cells treated with PBS). To set the parameter of Casp-3/7, use the well that contains apoptotic cells.

32. A healthy cell population should have at least 90% TMRM-positive cells; thus it is appropriate to use the first frame of TMRM image to calculate the total cell number. Alternatively, CellTracker Orange CMRA can be used to substitute TMRM to calculate the total cell number.

Acknowledgment

We acknowledge the expert help from Keith E. Schulze and members of Monash Micro Imaging facilities for writing the MetaMorph journal scripts and ImageJ macro. The work in the Naderer Lab is supported by the National Health and Medical Research Council (grant number: 1163556). TN is an Australian Research Council Future Fellow (FT170100313).

References

1. Holzinger D, Gieldon L, Mysore V, Nippe N, Taxman DJ, Duncan JA, Broglie PM, Marketon K, Austermann J, Vogl T, Foell D, Niemann S, Peters G, Roth J, Loffler B (2012) *Staphylococcus aureus* Panton-Valentine leukocidin induces an inflammatory response in human phagocytes via the NLRP3 inflammasome. J Leukoc Biol 92(5):1069–1081. https://doi.org/10.1189/jlb.0112014
2. Perret M, Badiou C, Lina G, Burbaud S, Benito Y, Bes M, Cottin V, Couzon F, Juruj C, Dauwalder O, Goutagny N, Diep BA, Vandenesch F, Henry T (2012) Cross-talk between *Staphylococcus aureus* leukocidins-intoxicated macrophages and lung epithelial cells triggers chemokine secretion in an inflammasome-dependent manner. Cell Microbiol 14(7):1019–1036. https://doi.org/10.1111/j.1462-5822.2012.01772.x
3. Labrousse D, Perret M, Hayez D, Da Silva S, Badiou C, Couzon F, Bes M, Chavanet P, Lina G, Vandenesch F, Croisier-Bertin D, Henry T (2014) Kineret(R)/IL-1ra blocks the IL-1/IL-8 inflammatory cascade during recombinant Panton Valentine Leukocidin-triggered pneumonia but not during *S. aureus* infection. PLoS One 9(6):e97546. https://doi.org/10.1371/journal.pone.0097546
4. Diep BA, Chan L, Tattevin P, Kajikawa O, Martin TR, Basuino L, Mai TT, Marbach H, Braughton KR, Whitney AR, Gardner DJ, Fan X, Tseng CW, Liu GY, Badiou C, Etienne J, Lina G, Matthay MA, DeLeo FR, Chambers HF (2010) Polymorphonuclear leukocytes mediate *Staphylococcus aureus* Panton-Valentine leukocidin-induced lung inflammation and injury. Proc Natl Acad Sci U S A 107(12):5587–5592. https://doi.org/10.1073/pnas.0912403107
5. Chow SH, Deo P, Yeung ATY, Kostoulias XP, Jeon Y, Gao M-L, Seidi A, Olivier FAB, Sridhar S, Nethercott C, Cameron D, Robertson AAB, Robert R, Mackay CR, Traven A, Jin Z-B, Hale C, Dougan G, Peleg AY, Naderer T (2020) Targeting NLRP3 and Staphylococcal pore-forming toxin receptors in human-induced pluripotent stem cell-derived macrophages. J Leukoc Biol 108(3):967–981. https://doi.org/10.1002/JLB.4MA0420-497R
6. Deo P, Chow SH, Han M-L, Speir M, Huang C, Schittenhelm RB, Dhital S, Emery J, Li J, Kile BT, Vince JE, Lawlor KE, Naderer T (2020) Mitochondrial dysfunction caused by outer membrane vesicles from Gram-negative bacteria activates intrinsic apoptosis and inflammation. Nat Microbiol 5(11):1418–1427. https://doi.org/10.1038/s41564-020-0773-2

7. Deo P, Chow SH, Hay ID, Kleifeld O, Costin A, Elgass KD, Jiang J-H, Ramm G, Gabriel K, Dougan G, Lithgow T, Heinz E, Naderer T (2018) Outer membrane vesicles from *Neisseria gonorrhoeae* target PorB to mitochondria and induce apoptosis. PLoS Pathog 14(3): e1006945. https://doi.org/10.1371/journal.ppat.1006945
8. Speir M, Lawlor KE, Glaser SP, Abraham G, Chow S, Vogrin A, Schulze KE, Schuelein R, O'Reilly LA, Mason K, Hartland EL, Lithgow T, Strasser A, Lessene G, Huang DCS, Vince JE, Naderer T (2016) Eliminating *Legionella* by inhibiting BCL-XL to induce macrophage apoptosis. Nat Microbiol 1:15034. https://doi.org/10.1038/nmicrobiol.2015.34
9. Zhao H, Traganos F, Dobrucki J, Wlodkowic D, Darzynkiewicz Z (2009) Induction of DNA damage response by the supravital probes of nucleic acids. Cytometry A 75A(6):510–519. https://doi.org/10.1002/cyto.a.20727

Chapter 16

Non-apoptotic Cell Death Control of Neutrophil Extracellular Trap Formation

Akshay A. D'Cruz, Maria Ericsson, and Ben A. Croker

Abstract

Neutrophil extracellular traps (NETs) are networks of chromatin and microbicidal proteins released by neutrophils in response to infection and tissue damage. Although classically viewed as a discrete biochemical and cellular process in neutrophils, the effector pathways integrating diverse upstream activating signals to control the formation of NETs (NETosis) are poorly defined. Cell death is one such common unifying endpoint of neutrophils, with several bona fide non-apoptotic cell death agonists now described to initiate NETosis. Integrating these new genetic findings into our existing knowledge of NETosis will likely reveal varied cellular and biochemical processes controlling NET release and specific anti-microbial and inflammatory effector functions of NETs triggered by specific non-apoptotic cell death. To facilitate investigation of regulated cell death pathways in NETosis, we offer a detailed protocol for neutrophil purification from mouse bone marrow and human blood, analysis of NETs by flow cytometry, and validation by immunogold electron microscopy. Future studies may better define cell death-specific forms of NETosis and their influence on inflammation and autoimmunity.

Key words Neutrophil, Neutrophil extracellular traps, NETosis, Necroptosis, Pyroptosis, Cell death, MLKL, RIPK1, RIPK3, PAD4, GSDMD

1 Introduction

Neutrophil extracellular traps (NETs) are networks of extracellular DNA, histones, and microbicidal proteins released from neutrophils [1, 2]. NETs can restrict the replication of *Staphylococcus aureus*, *Salmonella typhimurium*, *Streptococcus pneumoniae*, and *Shigella flexneri* [2–5], and their release and/or prolonged presence has been implicated in the pathogenesis of systemic lupus erythematosus [6], rheumatoid arthritis [7], diabetes [8], acute lung injury [9], and thrombosis [10–12]. However, the initiating stimuli and physiological regulators of NETosis in disease settings remain unclear.

The capacity of some neutrophils to retain viability and movement following NETosis, at least in the short term, signifies that an

Thomas A. Kufer and Maria Kaparakis-Liaskos (eds.), *Effector-Triggered Immunity: Methods and Protocols*,
Methods in Molecular Biology, vol. 2523, https://doi.org/10.1007/978-1-0716-2449-4_16,

active biochemical process controls this effector function of neutrophils, rather than being a passive outcome of lytic or necrotic death. This is exemplified in *Staphylococcus aureus* infection in vivo, where neutrophils can continue to move following NET release [13]. Anuclear neutrophils were also described as a form of "vital" NETosis that is associated with the presence of membrane-bound vesicles of decondensed chromatin [14].

Pyroptosis- and necroptosis-associated proteins can stimulate NET formation. These inflammatory non-apoptotic cell death regulators are activated by distinct stimuli and appear to generate subtly different NET structures (Fig. 1). The phorbol ester PMA and intracellular LPS engage neutrophil elastase and caspase-11, both of which are capable of cleaving gasdermin D to initiate NETosis [15, 16]. TNF-dependent necroptosis engages RIPK1 kinase activity and phosphorylation of RIPK3 and MLKL to initiate NET formation via pMLKL-mediated membrane disruption (Fig. 2) [17]. The role of non-apoptotic cell death pathways should not be viewed as binary, as membrane repair mechanisms are known to counteract pore-forming proteins in pyroptotic and necroptotic pathways, providing capacity to execute neutrophil effector functions including NETosis in the absence of loss of cell viability [18]. Moreover, a gradual accumulation of pore-forming proteins in the plasma membrane may also facilitate the deployment of neutrophil effector functions including NETs prior to loss of plasma membrane integrity and cell death [17]. Thus, recruitment of specialized membrane disrupting complexes for NETosis has been demonstrated to serve critical effector functions prior to complete loss of cell activity. These established roles of pyroptosis and necroptosis in NET formation do not exclude alternative cell death pathways in NET formation; rather, alternate non-apoptotic cell death pathways likely contribute to these phenomena and could generate subtly different chromatin structures in the extracellular space upon regulated lytic death providing adaptability in effector outputs of neutrophils.

Necroptosis can be experimentally induced in mouse neutrophils with combinations of z-VAD-fmk, an irreversible pan-caspase inhibitor, and birinapant, a second mitochondria-derived activator of caspases (SMAC) mimetic that predominately degrades cellular inhibitor of apoptosis 1 proteins cIAP1 and cIAP2 [19, 20]. This activity prevents caspase-8 activity and RIPK1 ubiquitination leading to activation of the RIPK1 kinase domain [21]. MLKL has emerged as a central player in the execution of RIPK3-dependent necroptotic death, but its precise role in membrane disruption is not well established [22]. Recruitment of multimeric complexes of MLKL to the membrane is associated with permeabilization of the plasma membrane and cell death [23–25]. Phosphorylation of the activation loop of MLKL induces a conformational change disrupting an autoinhibitory interaction between the pseudokinase

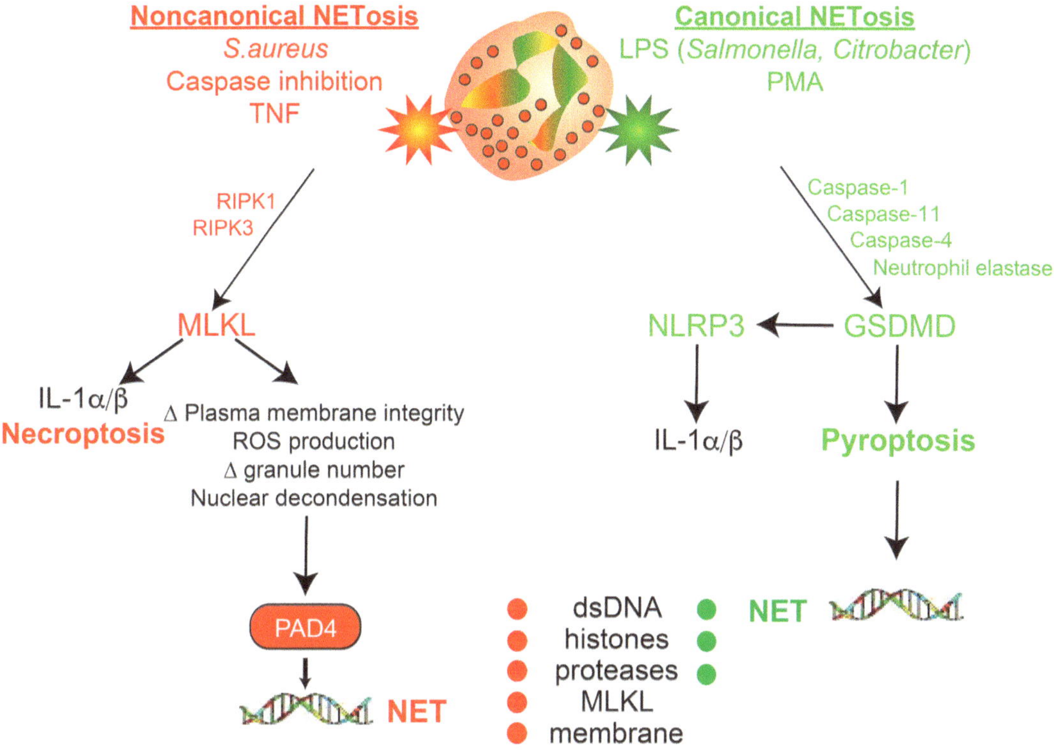

Fig. 1 Modes of NETosis involving GSDMD and MLKL in DNA release as part of regulated non-apoptotic cell death signaling. The composition of NETs appears to differ subtly following GSDMD- and MLKL-dependent NETosis. In both cases, NETosis may be accompanied by the processing and release of IL-1alpha and/or IL-1beta

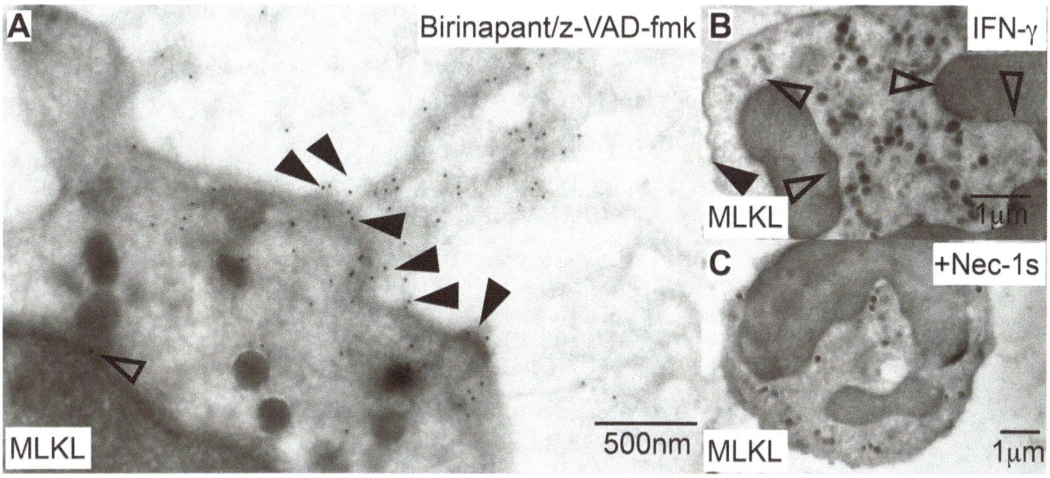

Fig. 2 Necroptotic NETosis is mediated by clusters of MLKL in the plasma membrane. (**a–c**) Immunogold electron microscopy of murine bone marrow neutrophils treated with IFN-gamma/birinapant/z-VAD-fmk in the presence or absence of Necrostatin-1s. Arrows indicate immunogold particles: filled arrowheads highlight NETs, and open arrowheads highlight nuclear or cytoplasmic staining of MLKL. (Data originally presented in D'Cruz et al., *Science Signaling* 11, eaao1716 (2018) [17])

domain and the four-helix bundle, which in turn promotes an interaction with phosphatidylinositol phosphates at the plasma membrane [23, 25]. A key to our future understanding of pyroptotic and necroptotic NETosis lies with an understanding of the upstream regulators of the RIPK1–RIPK3–MLKL pathway including ZBP1 and the physiological role of ESCRT-dependent membrane repair to finesse the extent and duration of MLKL membrane disruption and GSDMD pore formation at the neutrophil cell surface [18, 26–29]. ESCRT-dependent membrane repair mechanisms may modulate NETosis to control the function and survival of neutrophils and neutrophil cytoplasts following partial or complete extrusion of nuclear or mitochondrial DNA. We present protocols to help delineate the intricacies of the relationship between cell death pathways and NET formation in mouse and human neutrophils.

2 Materials

All solutions are prepared using Milli-Q water and ACS-grade reagents. Solutions for neutrophil purification are stored at 4 °C unless otherwise noted. Percoll gradients are prepared using 14-mL polypropylene tubes, 3-mL syringes, and 18-gauge 1.5-inch needles.

2.1 Solutions

1. *Red Cell Removal Buffer (RCRB):* 7.79-g NH_4Cl, 0.037-g disodium EDTA, 1-g $NaHCO_3$, in 1-L H_2O. Filter sterilize and store at 4 °C.
2. *HBSS-FBS:* Hanks' Balanced Salt Solution, 1% FBS.
3. *HBSS-BSA:* Hanks' Balanced Salt Solution, 1% BSA (Fraction V).

2.2 Percoll Stock Solutions

1. *100% Percoll:* 10% v/v 10 × HBSS, 90% v/v Percoll-Plus solution.
2. *HBSS-EDTA:* Hanks' Balanced Salt Solution, 1% FBS, 15-mM EDTA (*see* **Note 1**).
3. *78% Percoll:* 39-mL 100% Percoll + 11-mL HBSS-EDTA + 117-μL 2N HCl.
4. *68% Percoll:* 34-mL 100% Percoll + 16-mL HBSS-EDTA + 102-μL 2N HCl (*see* **Note 2**).
5. *52% Percoll:* 26-mL 100% Percoll + 24-mL HBSS-EDTA + 78-μL 2N HCl.

2.3 Antibodies and Fluorescent Stains for NET Detection

1. Anti-ds DNA antibody (Abcam, ab27156).
2. Quant-iT PicoGreen.

3. Anti-Histone H3 (citrulline R2 + R8 + R17) antibody (Abcam, ab5103).

2.4 Reagents for EM Embedding

1. Fixation buffer: 2.5% formaldehyde, 2.5% glutaraldehyde in 0.1-M sodium cacodylate buffer, pH 7.4.
2. 2.3-M sucrose solution in water.
3. 16% paraformaldehyde solution.
4. 25% glutaraldehyde solution.
5. Picric acid.
6. 2% osmium tetroxide solution.
7. Maleic acid.
8. Uranyl acetate.
9. Propylene oxide.
10. Dodecenylsuccinic anhydride (DDSA).
11. Methyl-5-norbornene-2,3-dicarboxylic anhydride (NMA).
12. 2,4,6-Tris(dimethylaminomethyl)phenol (DMP-30).
13. Taab 812 Resin.

3 Methods

Multiple methods of neutrophil purification have been used successfully in the literature. This description of Percoll purification is an adaptation of a method originally described by Boxio et al. [30]. This Percoll-based method is rapid, cheap, and robust for purification of granulocytes from mouse bone marrow, mouse peripheral blood, and human peripheral blood samples. This protocol removes cell debris and dead or damaged cells introduced by RBC lysis and maintains neutrophil function for downstream superoxide, chemotaxis, NET, cytokine production, cell death, and degranulation assays. It does not reliably remove eosinophils, but additional bead-based purification steps can be added to further enrich neutrophils. Cytocentrifugation or flow cytometry should be used to confirm cell purity and maturation status of cell populations used for study. All procedures involving human blood are performed with universal precautions and waste treated according to institutional biosafety and waste regulations.

3.1 Mouse Neutrophil Purification

1. Isolate mouse bone marrow cells from the femurs, tibias, and iliac crests in 3-mL HBSS-EDTA. Use 15-mL conical polypropylene tubes for bone marrow collection. Handle cells on ice at all stages of purification prior to incubation.
2. Build layers in a 14-mL polypropylene tube using a 3-mL syringe and 18-gauge 1.5-inch needle to underlay 2 mL of

each density layer per tube/gradient. Start with 2 mL of the 52% layer and then underlay 2 mL of the 68% layer and then 2 mL of the 78% layer (*see* **Note 3**).

3. Overlay 2 mL of bone marrow cells on to the three-layer Percoll gradient (*see* **Note 4**).
4. Centrifuge at 400 × *g*, 30 min, 4 °C without braking (or set to the lowest acceleration and deceleration setting) (*see* **Note 5**).
5. Carefully aspirate off the top layer, the 52% layer, and the interface of the 52% and 68% layer (*see* **Note 6**).
6. Collect 2–2.5 mL of neutrophils from the 68%/78% interface into 11 mL of HBSS-FBS.
7. Spin at 400 × *g*, 5 min, 4 °C (with braking).
8. Resuspend pellet immediately in 1-mL HBSS-FBS. Add 2 mL of RCRB for 30 s precisely (*see* **Note 7**).
9. Add 11 mL of HBSS-FBS to dilute the RCRB. Spin 400 × *g*, 5 min, 4 °C.
10. Repeat wash in **Step 12** to remove residual EDTA, or proceed to label neutrophils with a fluorescent cell-tracing stain if appropriate.
11. Resuspend neutrophils in desired cell culture media for assays at desired volume for cell counting.
12. Check purity of neutrophil preparations by cytocentrifugation and May-Grünwald Giemsa staining, or flow cytometry (*see* **Note 8**).

3.2 Human Neutrophil Purification

Human neutrophil purification can proceed using the protocol above, except that RBC are removed prior to Percoll density purification. The latter RBC lysis step after Percoll density centrifugation is then avoided.

1. Mix 3 mL of heparinized human blood with 12 mL of cold RCRB. Mix gently.
2. Immediately spin 400 × *g*, 5 min, 4 °C.
3. Discard supernatant.
4. Resuspend pellet in 15-mL RCRB, mix gently, and repeat **Steps 2–3** once.
5. Resuspend white pellet in 3-mL HBSS-EDTA, and overlay on a three-layer Percoll gradient as above.
6. Proceed to **Step 4**, Subheading 3.1, to complete neutrophil purification (*see* **Note 9**).

3.3 Flow Cytometry

Flow cytometric analysis of neutrophil extracellular traps (NETs) may be performed on live or fixed samples using published methods and a stringent double-positive gating strategy [17, 31]. Methods to validate and assess the morphology and function of bona fide NETs may be warranted and include imaging flow cytometry, electron microscopy, immunogold electron microscopy, immunofluorescence microscopy, and microbial killing assays in the presence or absence of DNase and PAD4 inhibitors.

1. Processing of samples can be done in 96-well plates by first pelleting cells for 3 min at 400 × *g*, 4 °C.
2. Remove the supernatant with a single flick.
3. Wash with HBSS-BSA and repeat **Steps 1–3** once prior to staining.
4. Stain neutrophils for 30 min with primary antibodies specific to double-stranded DNA and hypercitrullinated histone H3 (1: 200). Detection of (extracellular) dsDNA may also be achieved using 1/2000 of Quant-iT PicoGreen dsDNA Reagent.
5. Wash twice in HBSS-BSA.
6. For primary antibody detection, use an appropriate secondary antibody for 15 min on ice.
7. Wash twice in HBSS-BSA as above.
8. Resuspend in an appropriate volume of HBSS-BSA for analysis by flow cytometry.

3.3.1 Gating Strategy

1. Use FSC and SSC to identify neutrophils.
2. Gate for neutrophils that are double-positive for dsDNA/PicoGreen and also for hypercitrullinated histone H3 (Fig. 3). Additional markers of NETs can also be used including antibodies specific to neutrophil elastase (*see* **Note 10**).

3.4 Electron Microscopy

An important tool for validating NETosis and investigating the ultrastructural changes and biochemical players in canonical and non-canonical NET formation is electron microscopy. Immunogold electron microscopy offers the resolution necessary to discern key intracellular changes accompanying the regulated formation of NETs and the effects of chemical inhibitors, activating stimuli, and host genetic changes. The resolution of EM enables a number of important features to be discerned including chromatin decondensation, changes in granule content and subcellular localization, movement of dsDNA and H3Cit across the plasma membrane, and translocation of pore-forming (or membrane-disrupting) proteins including MLKL (Fig. 2). The regulated and sequential changes now defined for MLKL-dependent NETosis will help define other forms of bona fide NETosis from passive release of DNA by dead neutrophils. In this section, we outline the methods required for sample preparation and some markers appropriate for characterizing the early ultrastructural changes to the cells and the NETs emerging during this period (Fig. 2).

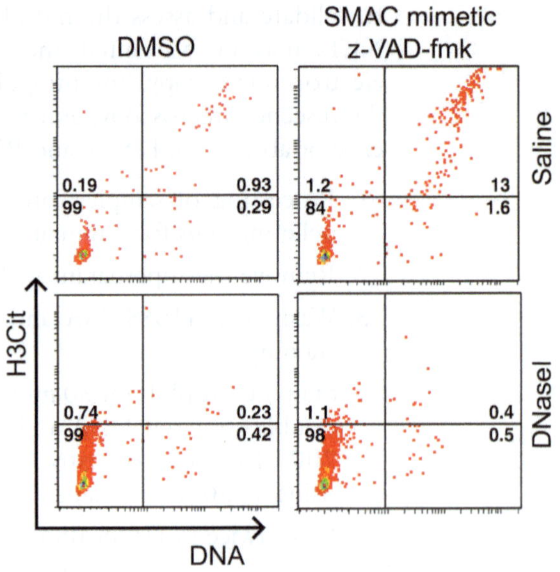

Fig. 3 NET formation assessed by flow cytometry. Neutrophils are stained with a DNA-specific dye such as PicoGreen and a H3Cit antibody. Deoxyribonuclease I (DNase I) is used to degrade extracellular dsDNA structures and validate the presence of NETs. (Data originally presented in D'Cruz et al., *Science Signaling* 11, eaao1716 (2018) [17])

3.4.1 Transmission Electron Microscopy

1. Fix the neutrophil pellet for at least 2 h at room temperature in fixation buffer.
2. Wash the cell pellets in 0.1-M cacodylate buffer.
3. Post-fix with 1% osmium tetroxide (OsO_4)/1.5% potassium ferrocyanide ($KFeCN_6$) for 1 h.
4. Wash in water three times and incubate in 1% aqueous uranyl acetate for 1 h followed by two washes in water.
5. Conduct a stepwise dehydration in grades of ethanol (10 min each; 50%, 70%, 90%; 2 × 10 min in 100%).
6. Treat the samples with propylene oxide for 1 h.
7. Infiltrate overnight in a 1:1 mixture of propylene oxide and TAAB Epon (TAAB Laboratories Equipment).
8. The next day, embed samples in TAAB Epon and polymerize at 60 °C for 48 h.
9. Cut ultrathin sections (about 60 nm) on a microtome, and pick up on copper grids. Before imaging, stain sections with 0.2% lead citrate.
10. Examine using an electron microscope and record images using an CCD camera.

3.4.2 Immunogold Electron Microscopy (Tokuyasu Method)

Gold labeling can be carried out at room temperature on a piece of parafilm. All antibodies and protein A-gold are diluted in PBS + 1% BSA.

1. For preparation of cryosections, fix the cell pellets in 4% paraformaldehyde for 2 h at room temperature.
2. Replace the fixative with PBS.
3. Infiltrate with 2.3 M sucrose in PBS for 15 min.
4. Transfer the cryoprotected cell pellet to a Cryo-specimen pin, remove excess sucrose, and freeze in liquid nitrogen.
5. Section the frozen samples (~80 nm) at −120 °C, pick up on a drop of 2.3-M sucrose in a loop, and transfer to formvar carbon-coated copper grids.
6. Float the grids on PBS until the immunogold labeling is carried out.
7. Centrifuge diluted antibody solution for 1 min at 14,000 × *g* prior to labeling to remove possible aggregates.
8. Float the grids on drops of 1% BSA for 10 min to block nonspecific labeling.
9. Transfer to 5-μL drops of primary antibody and incubate for 30 min.
10. Wash the grids in four drops of PBS for a total of 15 min.
11. Transfer to 5-μL drops of protein A-gold and incubate for 20 min.
12. Wash in four drops of PBS and six drops of double distilled water for a total of 20 min.
13. Contrast and embed the labeled grids on ice in 0.3% uranyl acetate in 2% methyl cellulose for 10 min.
14. Pick up the grids and remove excess liquid by streaking on filter paper, leaving a thin coat of methyl cellulose.
15. Examine the grid on an electron microscope and record images using an CCD camera.

4 Notes

1. EDTA should be dissolved from powder and not 0.5-M EDTA solution to avoid changes in tonicity of Hanks' Balanced Salt Solution.
2. Minor changes to the density of the 68% layer can have significant effects on the yield and purity of neutrophil fractions, so attention should be given to this layer in particular.
3. Be careful when the needle enters and exits the Percoll layers. Add Percoll in a smooth slow motion so as not to mix the

layers. Defined boundaries should be evident between each layer within the density gradient.

4. Gentle. Do not jet the cells into the gradient but rather add them dropwise to the top of the gradient.
5. A shorter spin time is also suitable for effective cell separation.
6. Do not disturb the boundary between the 68% and 78% layers.
7. This is a gentle lysis step suitable for contaminating erythroid cells present in this fraction from mouse bone marrow cells.
8. Granulocyte purity should exceed 95%, with contaminating cells typically comprising lymphocytes or erythroid cells.
9. Expect 5–15 $\times$ 10^6 neutrophils per mouse and 2–5 $\times$ 10^6 neutrophils per mL of human peripheral blood.
10. Size alone is an unreliable indicator of NETs due to the variability in the size of NETs detected using this method. These assumptions can be validated using alternative methodologies including imaging flow cytometry, live cell imaging, immunofluorescence microscopy, western blot, electron microscopy, and immunogold electron microscopy.

Acknowledgments

This work was supported by NIH Grant 1R01AI155869 (BAC) and the United States-Israel Binational Science Foundation grant (BAC).

References

1. Fuchs TA, Abed U, Goosmann C et al (2007) Novel cell death program leads to neutrophil extracellular traps. J Cell Biol 176:231–241
2. Brinkmann V, Reichard U, Goosmann C et al (2004) Neutrophil extracellular traps kill bacteria. Science 303:1532–1535
3. Beiter K, Wartha F, Albiger B et al (2006) An endonuclease allows Streptococcus pneumoniae to escape from neutrophil extracellular traps. Curr Biol 16:401–407
4. Berends ET, Horswill AR, Haste NM et al (2010) Nuclease expression by Staphylococcus aureus facilitates escape from neutrophil extracellular traps. J Innate Immun 2:576–586
5. Thammavongsa V, Missiakas DM, Schneewind O (2013) Staphylococcus aureus degrades neutrophil extracellular traps to promote immune cell death. Science 342:863–866
6. Hakkim A, Furnrohr BG, Amann K et al (2010) Impairment of neutrophil extracellular trap degradation is associated with lupus nephritis. Proc Natl Acad Sci U S A 107: 9813–9818
7. Khandpur R, Carmona-Rivera C, Vivekanandan-Giri A et al (2013) NETs are a source of citrullinated autoantigens and stimulate inflammatory responses in rheumatoid arthritis. Sci Transl Med 5:178ra40
8. Wong SL, Demers M, Martinod K et al (2015) Diabetes primes neutrophils to undergo NETosis, which impairs wound healing. Nat Med 21:815–819
9. Thomas GM, Carbo C, Curtis BR et al (2012) Extracellular DNA traps are associated with the pathogenesis of TRALI in humans and mice. Blood 119:6335–6343
10. Martinod K, Demers M, Fuchs TA et al (2013) Neutrophil histone modification by peptidylarginine deiminase 4 is critical for deep vein thrombosis in mice. Proc Natl Acad Sci U S A 110:8674–8679

11. Fuchs TA, Brill A, Duerschmied D et al (2010) Extracellular DNA traps promote thrombosis. Proc Natl Acad Sci U S A 107:15880–15885
12. Demers M, Krause DS, Schatzberg D et al (2012) Cancers predispose neutrophils to release extracellular DNA traps that contribute to cancer-associated thrombosis. Proc Natl Acad Sci U S A 109:13076–13081
13. Yipp BG, Petri B, Salina D et al (2012) Infection-induced NETosis is a dynamic process involving neutrophil multitasking in vivo. Nat Med 18:1386–1393
14. Zhao W, Fogg DK, Kaplan MJ (2015) A novel image-based quantitative method for the characterization of NETosis. J Immunol Methods 423:104–110
15. Chen KW, Monteleone M, Boucher D et al (2018) Noncanonical inflammasome signaling elicits gasdermin D-dependent neutrophil extracellular traps. Sci Immunol 3:eaar6676
16. Sollberger G, Choidas A, Burn GL et al (2018) Gasdermin D plays a vital role in the generation of neutrophil extracellular traps. Sci Immunol 3:eaar6689
17. D'Cruz AA, Speir M, Bliss-Moreau M et al (2018) The pseudokinase MLKL activates PAD4-dependent NET formation in necroptotic neutrophils. Sci Signal 11:eaao1716
18. Gong YN, Guy C, Olauson H et al (2017) ESCRT-III acts downstream of MLKL to regulate necroptotic cell death and its consequences. Cell 169:286–300 e16
19. Lawlor KE, Feltham R, Yabal M et al (2017) XIAP loss triggers RIPK3- and Caspase-8-driven IL-1beta activation and cell death as a consequence of TLR-MyD88-induced cIAP1-TRAF2 degradation. Cell Rep 20:668–682
20. Moujalled DM, Cook WD, Okamoto T et al (2013) TNF can activate RIPK3 and cause programmed necrosis in the absence of RIPK1. Cell Death Dis 4:e465
21. Wong WW, Vince JE, Lalaoui N et al (2014) cIAPs and XIAP regulate myelopoiesis through cytokine production in an RIPK1- and RIPK3-dependent manner. Blood 123:2562–2572
22. Sun L, Wang H, Wang Z et al (2012) Mixed lineage kinase domain-like protein mediates necrosis signaling downstream of RIP3 kinase. Cell 148:213–227
23. Hildebrand JM, Tanzer MC, Lucet IS et al (2014) Activation of the pseudokinase MLKL unleashes the four-helix bundle domain to induce membrane localization and necroptotic cell death. Proc Natl Acad Sci U S A 111: 15072–15077
24. Cai Z, Jitkaew S, Zhao J et al (2014) Plasma membrane translocation of trimerized MLKL protein is required for TNF-induced necroptosis. Nat Cell Biol 16:55–65
25. Dondelinger Y, Declercq W, Montessuit S et al (2014) MLKL compromises plasma membrane integrity by binding to phosphatidylinositol phosphates. Cell Rep 7:971–981
26. Rühl S, Shkarina K, Demarco B et al (2018) ESCRT-dependent membrane repair negatively regulates pyroptosis downstream of GSDMD activation. Science 362:956–960
27. Lin J, Kumari S, Kim C et al (2016) RIPK1 counteracts ZBP1-mediated necroptosis to inhibit inflammation. Nature 540:124–128
28. Kuriakose T, Man SM, Malireddi RKS et al (2016) ZBP1/DAI is an innate sensor of influenza virus triggering the NLRP3 inflammasome and programmed cell death pathways. Sci Immunol 1:aag2045
29. Jiao H, Wachsmuth L, Kumari S et al (2020) Z-nucleic-acid sensing triggers ZBP1-dependent necroptosis and inflammation. Nature 580: 391–395
30. Boxio R, Bossenmeyer-Pourie C, Steinckwich N et al (2004) Mouse bone marrow contains large numbers of functionally competent neutrophils. J Leukoc Biol 75:604–611
31. Gavillet M, Martinod K, Renella R et al (2015) Flow cytometric assay for direct quantification of neutrophil extracellular traps in blood samples. Am J Hematol 90:1155–1158

Chapter 17

Analysis of Bacteria-Triggered Inflammasome: Activation in Neutrophils by Immunoblot

Rémi Planès, Karin Santoni, and Etienne Meunier

Abstract

Detection of microbes relies on the expression of germline-encoded pattern recognition receptors (PRRs). While PRRs can directly sense conserved pattern expressed by various microbes, they can also induce effector-triggered immunity (ETI) by sensing pathogenic alterations of cellular homeostasis. One consequence of ETI is the death of the infected cell through the induction of inflammasome-dependent cell death, namely, pyroptosis. Such process can be easily studied in macrophages and epithelial cells, yet neutrophils encode an arsenal of proteolytic enzymes that imped easy and reliable study of ETI-triggered inflammasome response. Here, we describe an immunoblotting methodology to study both ETI- and PRR-driven inflammasome responses in neutrophils upon bacterial infections. This method is also transposable to other microbial pathogen- and toxin-induced inflammasome response in neutrophils.

Key words Neutrophils, Pyroptosis, Inflammasome, Infections

1 Introduction

Germline-encoded receptors, namely, the pattern recognition receptors (PRRs), constitute the foundation of innate immune responses [1]. PRRs monitor for the presence of microbial or host damage-derived ligands following two main ways. They can either detect the presence of the microbe-associated molecular patterns (MAMPs) or the damage-associated molecular patterns (DAMPs), but they can also behave as guard PRRs, able to react to host perturbations induced upon infections, hence triggering an immune response called effector-triggered immunity (ETI) [2–5]. This potent mechanism of response, initially studied in plants, relies on the ability of some PRRs to detect alterations in crucial cellular pathways induced by various virulence factors released by infectious agents [6–8]. In this context, PRR-triggered ETI does

Rémi Planès and Karin Santoni contributed equally with all other contributors.

Thomas A. Kufer and Maria Kaparakis-Liaskos (eds.), *Effector-Triggered Immunity: Methods and Protocols*, Methods in Molecular Biology, vol. 2523, https://doi.org/10.1007/978-1-0716-2449-4_17,

not rely on the direct detection of a MAMP or DAMP but rather on the sensing of the inactivation of intracellular signalling processes. Hence, ETI provides cells with a supplemental layer of immune response to virulence factors expressed by pathogens rather than responding the same way to pathogenic or commensal organisms. In this context, guard proteins can trigger an innate immune response characterized by the engagement of multiple inflammatory signaling pathways such as but not restricted to mitogen-activated protein kinases (MAPKs), nuclear factor-kappa B (NF-κB), interferon, or cell death [2, 7]. Specifically cell death constitutes a very common consequences of various virulence factor-induced ETI both in plants and metazoans [2, 7]. Among others, some mammalian cytosolic receptors/sensors that belong to the inflammasome family specifically behave as guard proteins and can trigger host cell death (pyroptosis) but also interleukin (IL)-1 family cytokine maturation and release [9]. Upon activation sensor/receptor-forming inflammasomes assemble multiprotein complexes composed of the receptor/sensor, the adaptor protein ASC and the protease caspase-1 are able to cleave and mature the potent inflammatory cytokines IL-1β and IL-18 [9]. Unique to the caspase-11 non-canonical inflammasome and the CARD8 inflammasome is their independence from the ASC protein to promote respectively caspase-11- and caspase-1-driven pyroptosis [10]. Once activated caspase-1 and caspase-11 directly cleave and activate the pyroptosis inducer gasdermin D (GSDMD) [11]. The generated GSDMD N-terminal fragment will insert into membranes and form pores, hence leading to the induction of pyroptosis, a process that will be terminated by the final shrinkage of the cell involving the protein Ninjurin-1 [12].

Hence, Rho GTPase activating toxins or toxin-induced ionic perturbations are sensed by the NOD-like receptor (NLR) containing a pyrin (PYD)-3 (NLRP3) inflammasome, whereas toxin-inactivated Rho GTPases are detected by another NLR, the PYRIN (also named TRIM20) inflammasome [5, 7, 8, 13–18]. Microbial ubiquitin ligases and protease activity can also be detected by the NLRP1 and CARD8 inflammasomes following a model of proteasome-mediated ETI [4, 19–30]. Similarly, microbial inhibition of various innate and cell death pathways (RIPK1/caspases) leads to the induction of compensatory/alternative responses characterized by the induction of inflammatory cell deaths necroptosis and gasdermin-driven pyroptosis [31–33]. Complementary to this, ETI also embeds mechanisms to detect intracellular membrane disruption/alterations upon microbial infections [5]. In this context, the intracellular lectins and galectins, the ubiquitin system, and the interferon-inducible GTPases behave as guard proteins able to detect membrane damages and to assemble supramolecular complexes that will activate innate immune responses such as autophagy or inflammasome responses [34]. Among IFN-inducible GTPases, the guanylate-binding proteins (GBPs) have received major attention due to their importance at both detecting intracellular and microbial components and indirectly promoting the assembly of inflammasome complexes upon infections [34, 35].

While most of those ETI processes can occur both in professional and non-professional immune cells, studying their importance in each cell type remains challenging due to the specificity of each cell. Specifically, granulocyte neutrophils are professional microbicidal cells that are extremely hard to study due to the expression of various proteolytic enzymes, which may render difficult the interpretation of the results [36]. Indeed, the expression of numerous intracellular and secreted microbicidal molecules such as the serine proteases neutrophil elastase (NE)/cathepsin G (CSTG)/proteinase 3 (PR3), the strong levels of NADPH oxidase and myeloperoxidase (MPO), the various types of microbicidal granules, and the calprotectin protein makes the neutrophil the most biotoxic cell of the organism [36]. However, several pathogens successfully use virulence strategies to escape microbicidal processes developed by neutrophils, including *Francisella tularensis*, *Francisella tularensis* spp. *novicida*, *Anaplasma phagocytophilum*, *Salmonella enterica* subsp. *enterica* ser. Typhimurium, and *Staphylococcus aureus* [37].

In this context, endomembrane destabilization in neutrophils and also microbial targeting of microbicidal and innate immune pathways can be translated into inflammasome responses characterized by the cleavage and activation the protease caspase-1 and caspase-11, the cleavage and activation of the pore-forming protein gasdermin D, as well as the maturation and release of IL-1 cytokines [38–40]. However, study of ETI-triggered inflammasome response in neutrophils constitutes a challenge due to the very potent proteolytic activity of neutrophil-specific proteases that will degrade extensively most neutrophil proteins, a process that can barely be inhibited using classical protease inhibitor tablets (Fig. 1). This renders extremely challenging both the analysis and the interpretation of results.

Here, we describe a methodology to study both ETI- and MAMP-/DAMP-triggered inflammasome response in neutrophils, upon GBP-driven inflammasome detection of the bacteria *Francisella novicida* [41–43]. This method avoids neutrophil proteases to artificially degrade proteins of interest and can also be applied to other microbial and sterile activators of inflammasomes in neutrophils (Fig. 2).

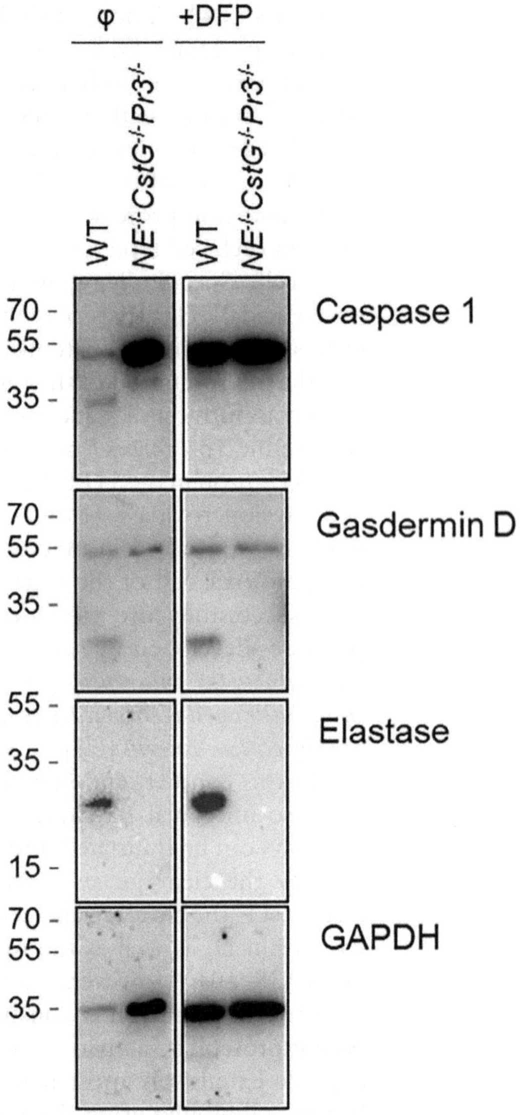

Fig. 1 Bone marrow neutrophils isolated from WT and *elane*$^{-/-}$ *ctsg*$^{-/-}$ *pr3*$^{-/-}$ (KO) mice were treated and processed as described in this method or by omitting DFP. Results show detection of the pro-form of caspase-1, gasdermin D, neutrophil elastase, and GAPDH proteins by immunoblotting in resting unstimulated BMN in presence or absence of DFP

2 Materials

2.1 Generation of Murine Bone Marrow Neutrophils (BMNs)

1. Cell culture facility and equipment including water-jacketed CO_2 incubator and laminar flow hood/chemical hood.
2. CO_2 tank including flow regulation.
3. Mice.

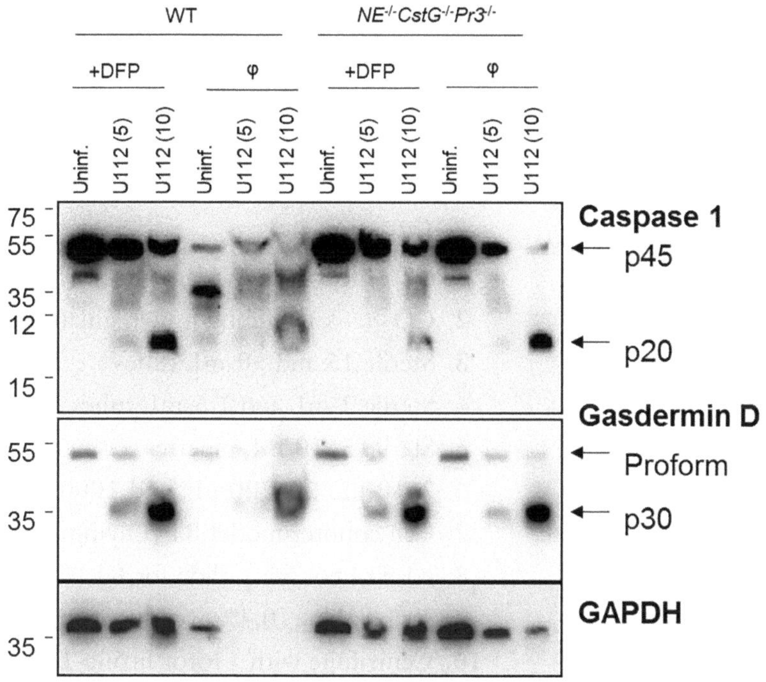

Fig. 2 Bone marrow neutrophils isolated from WT and *NE*$^{-/-}$ *Ctsg*$^{-/-}$ *Pr3*$^{-/-}$ (KO) mice were infected with *Francisella novicida* at the MOI of 5 or 25 for 3 h. After infection cells were treated and processed as described in this method in the presence or absence of DFP. Results show detection of caspase-1 and gasdermin D, in their processed or pro-form in presence or absence of DFP

4. 70% ethanol.
5. Autoclaved scissors and forceps.
6. Paper towels.
7. Sterile 50-mL tubes.
8. 2-mL and 1.5-sterile tubes.
9. Sterile 5, 10-mL pipettes.
10. Cell Culture Petri Dishes (10 cm^2).
11. Centrifuge with a rotor fitting 1.5-mL tubes.
12. Ice-cold 1× phosphate-buffered saline (PBS), tissue culture grade.
13. MACS buffer: 0.5% bovine serum albumin (BSA), 2-mM EDTA in PBS (1X), sterile 0.2 μm filtered.

2.2 Purification of Murine Bone Marrow Neutrophils (BMNs)

1. Cell culture facility and equipment including water-jacketed CO_2 incubator and laminar flow hood/chemical hood.
2. The QuadroMACS™ Separator (Miltenyi Biotec).
3. LS Columns (Miltenyi Biotec).

4. Anti-Ly-6G MicroBeads UltraPure, mouse (Miltenyi Biotec).
5. Ice-cold MACS buffer: 0.5% bovine serum albumin (BSA), 2-mM EDTA in PBS (1X), sterile 0.2 μm filtered.
6. Centrifuge with a rotor fitting 15-mL tubes.
7. Opti-MEM reduced serum medium, unsupplemented.

2.3 Seeding Murine Bone Marrow Neutrophils (BMNs)

1. Cell culture facility and equipment including water-jacketed CO_2 incubator and laminar flow hood.
2. CO_2 tank including flow regulation.
3. Sterile 15 and 50-mL tubes.
4. Sterile 2-mL and 1.5-mL tubes.
5. Sterile 5, 10-mL pipettes.
6. 2–10-μL, 20–200-μL, and 1000-μL pipettes.
7. Cell counter model R1 (Olympus).
8. R1-SLI counting slide for R1 (Olympus).
9. Trypan blue (0.4%, weight/weight).
10. Centrifuge with a rotor fitting 15-mL tubes.
11. 6-well plates.
12. Opti-MEM reduced serum medium, unsupplemented.

2.4 Infection/Stimulation of Murine Bone Marrow Neutrophils (BMNs)

1. Cell culture facility and equipment including water-jacketed CO_2 incubator and laminar flow hood.
2. CO_2 tank including flow regulation.
3. Sterile 2-mL and 1.5-mL Eppendorf tubes.
4. Sterile 10-mL tubes.
5. 2–10-μL, 20–200-μL, and 1000-μL pipettes.
6. Opti-MEM reduced serum medium, unsupplemented.
7. Brain Heart Infusion (BHI) broth.
8. L-cysteine.
9. O.D 600 nm spectrophotometer.
10. Polystyrene Spectrophotometry Cuvettes.

2.5 Processing BMN Samples for Immunoblotting

1. Cell culture facility and equipment including water-jacketed CO_2 incubator and laminar flow hood.
2. Ice bucket.
3. Cell scrapper.
4. Sterile 2-mL and 1.5-mL tubes.
5. 2–10-μL, 20–200-μL, and 1000-μL pipettes.
6. Centrifuge with a rotor fitting 1.5–2-mL tubes.
7. 5-M diisopropylfluorophosphate (DFP) solution in water.

8. RIPA buffer: 150-mM NaCl, 50-mM Tris-HCl, 1% Triton X-100, 0.5% Na-deoxycholate, 5-mM of DFP, and cOmplete™ Protease Inhibitor Cocktail (Roche).
9. 100% trichloroacetic acid (TCA) solution.
10. Acetone.
11. Laemmli buffer (5×): 300-mM Tris-HCl (pH 8.0), 10% SDS, 20-mM EDTA, 25% ß-mercaptoethanol, 0.1% bromophenol blue, 50% glycerol.
12. 1-M Tris-HCl, pH 12.
13. Thermoblock (95 °C) fitting 1.5–2-mL tubes.

2.6 Western Blot of Murine Bone Marrow Neutrophil (BMN) Samples

1. Running buffer: Tris-HCl Glycine-SDS (1X) in sterile water.
2. Transfer buffer: 25-mM Tris, 190-mM glycine, 20% methanol, SDS 0.1%, in sterile water. Stored at 4 °C.
3. 12% SDS-PAGE gel.
4. TEMED.
5. Acrylamide/bisacrylamide 37.5/1 30%.
6. Ammonium persulfate (APS).
7. Tris-HCl, pH 6.8.
8. Tris base pH 8.8.
9. PVDF membrane.
10. Methanol.
11. Wash buffer (TBS-T): Tris-HCl-buffered saline solution, 0.1% Tween 20, pH 8.8.
12. Blocking buffer: 5% milk in TBS-T.
13. Primary antibodies.
14. Secondary antibodies—HRP conjugate.
15. Prestained Protein Ladder.
16. ECL detection kit.
17. A chemiluminescent machine compatible for Western blotting revelation.

3 Methods

3.1 Isolation of Bone Marrow Neutrophils (BMNs)

1. Isolate total BMNs as described previously [44], (*see* **Notes 1** and **2**).
2. Centrifuge the BMNs at 300 × *g* for 10 min.
3. Open the tube containing BMNs in the laminar flow hood.
4. Aspirate and remove cell supernatant.

5. Resuspend cell pellet corresponding to the total bone marrow cells from one mouse in 450 μL of MACS buffer, and homogenize cell suspension by gently pipetting up and down five times with a P1000 micropipette.
6. Add 50 μL of Anti-Ly-6G MicroBeads UltraPure.
7. Mix gently by pipetting up and down five times with a P200 micropipette.
8. Incubate for 10 min in the refrigerator (2–8 °C).
9. Wash the cells by adding 5 mL of MACS Buffer in a new sterile 15-mL Falcon tube, under the laminar flow hood. Close the tube and mix by inverting five times.
10. Centrifuge at 300 × *g* for 10 min.
11. Aspirate and eliminate cell supernatant.
12. Resuspend the pellet with 500 μL of MACS buffer and homogenize cell suspension by gently pipetting up and down five times with a P1000 micropipette.
13. Proceed to magnetic separation in the laminar flow hood.
14. Place LS column in the magnetic field of a MACS separator.
15. Place a 15-mL collection tube under the column to collect flow-through.
16. Prepare column by rinsing with 3 mL of MACS buffer. Eliminate flow-through. Never let the column dry.
17. Apply cell suspension onto the column.
18. Collect flow-through containing unlabeled cells.
19. Then, wash column with 3 mL of MACS buffer three times. Collect unlabeled washed out cells.
20. To elute Ly6G+ BMNs, remove column from the separator and place it on a sterile 15-mL collection tube.
21. Add 5 mL of MACS buffer into the column.
22. Collect BMNs flushed out of the column by pushing the plunger.
23. Centrifuge at 300 × *g* for 10 min.

3.2 Seeding Murine Bone Marrow Neutrophils (BMNs)

1. Resuspend BMN pellet in 1 mL of Opti-MEM culture medium (*see* **Note 3**) and homogenize cell suspension by gently pipetting up and down five times with a P1000 micropipette.
2. In a sterile 1.5-mL Eppendorf tube, mix 10 μL of cell suspension with 10 μL of Trypan blue by pipetting up and down ten times with a P10 micropipette.
3. Pipette 10 μL of the mix with Trypan blue and deposit in one counting slide.

4. Determine living cell number with trypan blue cell death exclusion using automated cell counter. Make sure that cells are in focus while counting. For optimal results, make sure that cell viability is ≥70%.
5. Complete the 990-μL cell suspension of freshly isolated BMNs with Opti-MEM medium to obtain a 1 × 10^6 cells/mL suspension.
6. Seed 2 mL (2 × 10^6 cells/mL) per well in 6-well plate.
7. Place the plate into the incubator set at 37 °C and 5% CO_2 (*see* **Notes 4** and **5**).

3.3 Infection of Murine Bone Marrow Neutrophils (BMNs)

1. One day before your experiment, start a 2-mL culture of bacteria. Take out your frozen bacteria glycerol stock tube. Open the tube under the laminar flow hood. Scratch the frozen bacteria stock with one 100-μL tips and deposit tip in a sterile tube containing 2 mL of BHI medium supplemented with 0.2% L-cysteine (*see* **Notes 6** and 7).
2. Incubate the culture overnight at 37 °C under constant orbital agitation at 150 rpm.
3. The following day start a bacteria sub-culture by diluting 200 μL of the overnight bacterial culture into a sterile 50-mL Falcon tube containing 5 mL of BHI + 0.2% L-cysteine medium.
4. Incubate the 5.2-mL sub-culture at 37 °C under constant orbital agitation 150 rpm until OD600 nm reached 0.6–1 (this usually takes around 2 h of incubation).
5. To measure the OD600 nm bacterial culture, work under the laminar flow hood. First pipette 1 mL of sterile medium in Polystyrene Spectrophotometry Cuvettes. Read the OD600nm and set it as reference. Then pipette 1 mL of your bacterial cell culture in Polystyrene Spectrophotometry Cuvettes. Read the OD600nm.
6. Prepare a bacterial dilution in 37 °C pre-warmed Opti-MEM medium to reach a multiplicity of infection (MOI) of 25 per 50 μL. Note that an O.D. of 1 correspond to a bacteria concentration of about 10^9 bacteria per mL in the frame of *Francisella novicida* experiments, but this has to be determined for each bacterial specie.
7. Transfer the 6-well plate containing BMN to the laminar flow hood and infect the cells with 50 μL of bacteria dilution.
8. Incubate the 6-well plate at 37 °C, 5% CO_2 the desired time.

3.4 Processing BMN Samples for Immunoblotting

1. At the end of the infection, remove the plate from the incubator.
2. Check the BMNs under the microscope. Most of the BMN should be adherent.

3. Add 2 μL of a 5-M solution of diisopropylfluorophosphate (DFP) serine protease inhibitor to each well (*see* **Note 8**).
4. Collect cell supernatant into a sterile 2-mL tube with a P1000 micropipette.
5. Lyse adherent cells in the well by adding 80 μL of RIPA buffer to each well.
6. Gently swirl the plate to make sure buffer cover most of the well surface.
7. Place plate on ice for 10 min.
8. During that time, centrifuge the 2-mL tube filled with cell supernatant at 300 × *g* for 10 min.
9. At the end of centrifugation, collect supernatant in a new sterile 2-mL tube and place on ice while processing cell lysate. Keep remaining cell pellet on ice.
10. Scrape the surface of the well with cell scraper to ensure optimal recovery of cell lysate and gather the lysate on one side of the well (*see* **Note 9**).
11. Collect cell lysate with a P100 micropipette and mix with corresponding cell pellet in the 2-mL tube kept on ice containing pelleted cells from cell-free supernatant.
12. Pipette up and down ten times to homogenize cell lysate.
13. Clarify cell lysate from insoluble fraction by centrifugation for 10 min, 14.000 × *g* at 4 °C.
14. Collect soluble fraction in a new 1.5-mL tube and keep it on ice while processing the cell supernatant.

3.5 Precipitation of Cell Supernatant

1. Add 200 μL of a 100% trichloroacetic acid (TCA) solution to 1800 μL of BMN supernatant fraction (*see* **Note 10**).
2. Close the lid and vortex 30 s.
3. Incubate on ice for 60 min.
4. Spin down at 11,000 × *g* for 45 min, at 4 °C.
5. Eliminate supernatant gently without touching pellet.
6. Wash pellet with 1 mL of ice-cold acetone.
7. Spin down at 11,000 × *g* for 15 min, at 4 °C.
8. Eliminate supernatant.
9. To ensure total evaporation of acetone, leave the tube open under a chemical hood for 10–15 min.
10. Resuspend precipitated cell supernatant fraction with corresponding cell lysate.
11. Pipette up and down 30 times to ensure optimal resuspension.
12. Add 20 μL of a 5X Laemmli buffer.

13. Add 5 μL of a Tris-HCl pH 12 solution.
14. Boil the sample for 15 min at 95 °C.
15. Spin down quickly the tubes 1 min at 11,000 × *g*.

3.6 Immunoblotting

1. Prepare a 12% SDS-PAGE gel as described previously by Bio-Rad "General Protocol for Western Blotting." Gel percentage selection depends on the size of the protein of interest.
2. Load equal amounts of protein (around 15 μL of sample) (*see* **Note 11**) into the wells of the SDS-PAGE gel. Load in parallel one well with a molecular weight protein ladder.
3. Run the gel for 1–2 h at 100 V (100 mA). Adjust migration time in order to optimize separation of your samples.
4. Activate the PVDF membrane with methanol for 1 min and rinse with transfer buffer before preparing the stack.
5. Incubate transfer sponges, filter paper, PVDF membrane, and gel in cold transfer buffer for 5 min.
6. Assemble the transfer sandwich in one transfer cassette in the following order: anode side/transfer sponge/two filter papers/gel/PVDF membrane/two filter papers/transfer sponge/cathode side and make sure no air bubbles are trapped in the sandwich.
7. Transfer the blot for 60 min at 4 °C and set the generator on constant voltage (100 V).
8. After transfer take the PVDF membrane out from sandwich and mark the membrane with a pen to identify transfer side.
9. Rinse membrane with TBS-T wash buffer for 5 min.
10. Block the membrane for 1 h at room temperature by immersing in blocking buffer.
11. Incubate the membrane overnight at 4 °C with the primary antibody diluted in blocking buffer solution.
12. Wash the membrane three times in TBS-T, 10 min each.
13. Incubate the membrane with the recommended dilution of secondary antibody HRP-conjugated in TBS-T, for 1 h.
14. Wash the membrane three times in TBS-T, 10 min each.
15. Briefly dry the membrane with paper towels.
16. Spread the membrane on the revelation tray of a machine compatible with chemiluminescent applications with transfer side facing up.
17. Apply the chemiluminescent substrate to the blot according to the manufacturer's instructions.

18. Capture the chemiluminescent signals using a machine compatible with chemiluminescent applications (e.g., ChemiDoc from Bio-Rad).
19. Capture colorimetric image to appreciate molecular weight of the bands.

4 Notes

1. BMN isolation usually yields between 6×10^6 and 20×10^6 BMNs per mouse; however, this amount may vary depending on age, weight, and genotype of mice.
2. It is possible to load the total bone marrow of up to three mice in one single LS column (maximal loading capacity of up to 10^8 cells/column). If larger amounts are required, use several LS column accordingly.
3. This protocol includes a step for the precipitation of the cell-free supernatant. Therefore, it is important to perform the experiment in Opti-MEM unsupplemented or other serum-free or low-protein medium. In case of the presence of a large protein pellet after precipitation of the cell-free supernatant, it may be difficult to resuspend it, especially if it is too big or you let it overdry for more than 15 min. In that case, you may need to increase the homogenization time by pipetting up and down with a p100 pipette. You may also need to vortex, sonicate, or increase the boiling time of your sample up to 30 min.
4. Study of inflammasome-dependent response may need one priming step before infection/stimulation. This can be performed with IFN, LPS, or PAM_3CSK_4 for 1 to 6 h prior to infection or stimulation without disturbing the use of the current protocol.
5. Because neutrophils are short living cell type (starting to undergo apoptosis around 8–12 h after isolation), it is important to perform the experiments straight after cell sorting.
6. Prepare fresh L-cysteine for bacterial growth.
7. *Francisella tularensis* spp. *novicida* absolutely needs L-cysteine for optimal growth.
8. In the case you cannot use diisopropylfluorophosphate (DFP) as a neutrophil serine protease inhibitor, other protease inhibitors can be used, including phenylmethylsulfonyl fluoride (PMSF).
9. The scrapping of the cells is an important step to optimize neutrophil cell lysate collection that could be very sticky. For that reason, if you need to change plate format (upscale or downscale the experiment), you need to make sure that the

scrapping step could be performed correctly with a cell scrapper fitting your cell culture plate.

10. The use of trichloroacetic acid (TCA) must be done with cautious, under a chemical hood, as TCA is a very corrosive and toxic product.

11. This protocol describes a technique to analyze both cell lysate and cell supernatant in a combined tube, but you may notice that both fractions can also be analyzed separately. In that case, suspend the precipitated pellet with 80 μL of RIPA buffer supplemented with 5 mM of DFP in addition to the cOmplete™ Protease Inhibitor Cocktail (Roche) and 20 μL of Laemmli (5X) before boiling at 95 °C. Accordingly, add 20 μL of Laemmli (5X) to 80 μL of cell lysate before boiling at 95 °C.

Acknowledgments

Christine T N Pham (Washington University, Seattle, USA) generated and provided the *NE–/–/CstG–/–/Pr3–/–* mice.
Thomas Henry (CIRI, Lyon) provided *Francisella novicida* strains.

References

1. Janeway CA, Medzhitov R (2002) Innate immune recognition. Annu Rev Immunol 20: 197–216
2. Rajamuthiah R, Mylonakis E (2014) Effector triggered immunity: activation of innate immunity in metazoans by bacterial effectors. Virulence 5:697. https://doi.org/10.4161/VIRU.29091
3. Stuart LM, Paquette N, Boyer L (2013) Effector-triggered versus pattern-triggered immunity: how animals sense pathogens. Nat Rev Immunol 13:199–206
4. Vance RE, Isberg RR, Portnoy DA (2009) Patterns of pathogenesis: discrimination of pathogenic and nonpathogenic microbes by the innate immune system. Cell Host Microbe 6:10–21
5. Fischer NL, Naseer N, Shin S, Brodsky IE (2019) Effector-triggered immunity and pathogen sensing in metazoans. Nat Microbiol 51(5):14–26. https://doi.org/10.1038/S41564-019-0623-2
6. Dangl JL, Jones JDG (2001) Plant pathogens and integrated defence responses to infection. Nature 4116839 411:826–833. https://doi.org/10.1038/35081161
7. Kufer TA, Creagh EM, Bryant CE (2019) Guardians of the cell: effector-triggered immunity steers mammalian immune defense. Trends Immunol 40:939–951. https://doi.org/10.1016/J.IT.2019.08.001
8. Boyer L, Magoc L, Dejardin S, Cappillino M, Paquette N, Hinault C, Charriere GM, Ip WKE, Fracchia S, Hennessy E, Erturk-Hasdemir D, Reichhart JM, Silverman N, Lacy-Hulbert A, Stuart LM (2011) Pathogen-derived effectors trigger protective immunity via activation of the Rac2 enzyme and the IMD or rip kinase signaling pathway. Immunity 35:536–549. https://doi.org/10.1016/J.IMMUNI.2011.08.015
9. Martinon F, KBJT (2002) The inflammasome: a molecular platform triggering activation of inflammatory caspases and processing of proIL-β. Mol Cell 10:417–426
10. Broz P, Dixit VM (2016) Inflammasomes: mechanism of assembly, regulation and signalling. Nat Rev Immunol 16:407–420. https://doi.org/10.1038/NRI.2016.58
11. Liu X, Xia S, Zhang Z, Wu H (2021) Lieberman J (2021) Channelling inflammation: gasdermins in physiology and disease. Nat Rev Drug Discov 205 20:384–405. https://doi.org/10.1038/s41573-021-00154-z

12. Kayagaki N, Kornfeld OS, Lee BL, Stowe IB, O'Rourke K, Li Q, Sandoval W, Yan D, Kang J, Xu M, Zhang J, Lee WP, McKenzie BS, Ulas G, Payandeh J, Roose-Girma M, Modrusan Z, Reja R, Sagolla M, Webster JD, Cho V, Andrews TD, Morris LX, Miosge LA, Goodnow CC, Bertram EM, Dixit VM (2021) NINJ1 mediates plasma membrane rupture during lytic cell death. Nature 5917848 591: 131–136. https://doi.org/10.1038/s41586-021-03218-7
13. Dufies O, Boyer L (2021) RhoGTPases and inflammasomes: guardians of effector-triggered immunity. PLoS Pathog 17: e1009504. https://doi.org/10.1371/JOURNAL.PPAT.1009504
14. Dufies O, Doye A, Courjon J, Torre C, Michel G, Loubatier C, Jacquel A, Chaintreuil P, Majoor A, Guinamard RR, Gallerand A, Saavedra PHV, Verhoeyen E, Rey A, Marchetti S, Ruimy R, Czerucka D, Lamkanfi M, Py BF, Munro P, Visvikis O, Boyer L (2021) Escherichia coli Rho GTPase-activating toxin CNF1 mediates NLRP3 inflammasome activation via p21-activated kinases-1/2 during bacteraemia in mice. Nat Microbiol 6:401–412. https://doi.org/10.1038/S41564-020-00832-5
15. Xu H, Yang J, Gao W, Li L, Li P, Zhang L, Gong Y, Peng X, Xi J, Chen S, Wang F, Shao F (2014) Innate immune sensing of bacterial modifications of Rho GTPases by the Pyrin inflammasome. Nature 513:237–241. https://doi.org/10.1038/NATURE13449
16. Aubert DF, Xu H, Yang J, Shi X, Gao W, Li L, Bisaro F, Chen S, Valvano MA, Shao F (2016) A Burkholderia type VI effector Deamidates Rho GTPases to activate the Pyrin inflammasome and trigger inflammation. Cell Host Microbe 19:664–674. https://doi.org/10.1016/J.CHOM.2016.04.004
17. De Zoete MR, Flavell RA (2014) Detecting "different": Pyrin senses modified GTPases. Cell Res 24:1286–1287. https://doi.org/10.1038/CR.2014.101
18. Gao W, Yang J, Liu W, Wang Y, Shao F (2016) Site-specific phosphorylation and microtubule dynamics control Pyrin inflammasome activation. Proc Natl Acad Sci 113:E4857–E4866. https://doi.org/10.1073/PNAS.1601700113
19. Johnson DC, Okondo MC, Orth EL, Rao SD, Huang HC, Ball DP, Bachovchin DA (2020) DPP8/9 inhibitors activate the CARD8 inflammasome in resting lymphocytes. Cell Death Dis 11:1–10. https://doi.org/10.1038/s41419-020-02865-4
20. Sandstrom A (2019) Functional degradation: a mechanism of NLRP1 inflammasome activation by diverse pathogen enzymes. Science (80-) 5:eaau1330
21. BC Frew VJJM (2012) Proteolytic processing of Nlrp1b is required for inflammasome activity. PLoS Pathog 8:e1002659
22. Chavarría-Smith J, RV (2013) Direct proteolytic cleavage of NLRP1B is necessary and sufficient for inflammasome activation by anthrax lethal factor. PLoS Pathog 9:e1003452
23. Robert Hollingsworth L, David L, Li Y, Griswold AR, Ruan J, Sharif H, Fontana P, Orth-He EL, Fu TM, Bachovchin DA, Wu H (2021) Mechanism of filament formation in UPA-promoted CARD8 and NLRP1 inflammasomes. Nat Commun 12:1–13. https://doi.org/10.1038/s41467-020-20320-y
24. Linder A, Bauernfried S, Cheng Y, Albanese M, Jung C, Keppler OT, Hornung V (2020) CARD8 inflammasome activation triggers pyroptosis in human T cells. EMBO J 39: e105071. https://doi.org/10.15252/embj.2020105071
25. Wang Q, Gao H, Clark KM, Mugisha CS, Davis K, Tang JP, Harlan GH, DeSelm CJ, Presti RM, Kutluay SB, Shan L (2021) CARD8 is an inflammasome sensor for HIV-1 protease activity. Science (80-):eabe1707. https://doi.org/10.1126/science.abe1707
26. Xu Z, Zhou Y, Liu M, Ma H, Sun L, Zahid A, Chen Y, Zhou R, Cao M, Wu D, Zhao W, Li B, Jin T (2021) Homotypic CARD-CARD interaction is critical for the activation of NLRP1 inflammasome. Cell Death Dis 12:1–15. https://doi.org/10.1038/s41419-020-03342-8
27. Robinson KS, Teo DET, Sen TK, Toh GA, Ong HH, Lim CK, Lay K, Au BV, Lew TS, Chu JJH, Chow VTK, Wang DY, Zhong FL, Reversade B (2020) Enteroviral 3C protease activates the human NLRP1 inflammasome in airway epithelia. Science (80-) 370. https://doi.org/10.1126/science.aay2002
28. Tsu BV, Beierschmitt C, Ryan AP, Agarwal R, Mitchell PS, Daugherty MD (2021) Diverse viral proteases activate the nlrp1 inflammasome. elife 10:1–76. https://doi.org/10.7554/eLife.60609
29. Mitchell PS, Sandstrom A, Vance RE (2019) The NLRP1 inflammasome: new mechanistic insights and unresolved mysteries. Curr Opin Immunol 60:37–45
30. Boyden ED, WD (2006) Nalp1b controls mouse macrophage susceptibility to anthrax lethal toxin. Nat Genet 38:240–244

31. Sarhan J, Liu BC, Muendlein HI, Li P, Nilson R, Tang AY, Rongvaux A, Bunnell SC, Shao F, Green DR, Poltorak A (2018) Caspase-8 induces cleavage of gasdermin D to elicit pyroptosis during Yersinia infection. Proc Natl Acad Sci 115:E10888–E10897. https://doi.org/10.1073/PNAS.1809548115
32. Chen KW, Demarco B, Heilig R, Ramos SP, Grayczyk JP, Assenmacher C-A, Radaelli E, Joannas LD, Henao-Mejia J, Brodsky IE, Broz P (2021) RIPK1 activates distinct gasdermins in macrophages and neutrophils upon pathogen blockade of innate immune signalling. bioRxiv 2021.01.20.427379. https://doi.org/10.1101/2021.01.20.427379
33. Orning P, Weng D, Starheim K, Ratner D, Best Z, Lee B, Brooks A, Xia S, Wu H, Kelliher MA, Berger SB, Gough PJ, Bertin J, Proulx MM, Goguen JD, Kayagaki N, Fitzgerald KA, Lien E (2018) Pathogen blockade of TAK1 triggers caspase-8 dependent cleavage of Gasdermin D and cell death. Science 362:1064. https://doi.org/10.1126/SCIENCE.AAU2818
34. Huang S, Meng Q, Maminska A, MacMicking JD (2019) Cell-autonomous immunity by IFN-induced GBPs in animals and plants. Curr Opin Immunol 60:71. https://doi.org/10.1016/J.COI.2019.04.017
35. Man SM, Place DE, Kuriakose T, Kanneganti T-D (2017) Interferon-inducible guanylate-binding proteins at the interface of cell-autonomous immunity and inflammasome activation. J Leukoc Biol 101:143–150. https://doi.org/10.1189/JLB.4MR0516-223R
36. Burn GL, Foti A, Marsman G, Patel DF, Zychlinsky A (2021) The neutrophil. Immunity 54:1377–1391. https://doi.org/10.1016/J.IMMUNI.2021.06.006
37. Kobayashi SD, Malachowa N, DeLeo FR (2018) Neutrophils and bacterial immune evasion. J Innate Immun 10:432–441. https://doi.org/10.1159/000487756
38. Chen KW, Groß CJ, Sotomayor FV, Stacey KJ, Tschopp J, Sweet MJ, Schroder K (2014) The neutrophil NLRC4 Inflammasome selectively promotes IL-1β maturation without pyroptosis during acute Salmonella challenge. Cell Rep 8:570–582. https://doi.org/10.1016/J.CELREP.2014.06.028
39. Kovacs SB, Oh C, Maltez VI, McGlaughon BD, Verma A, Miao EA, Aachoui Y (2020) Neutrophil Caspase-11 is essential to defend against a cytosol-invasive bacterium. Cell Rep 32:107967. https://doi.org/10.1016/J.CELREP.2020.107967
40. Chen KW, Monteleone M, Boucher D, Sollberger G, Ramnath D, Condon N, von Pein J, Broz P, Sweet M, Schroder K (2018) Noncanonical inflammasome signaling elicits gasdermin D-dependent neutrophil extracellular traps. Sci Immunol 3. https://doi.org/10.1126/SCIIMMUNOL.AAR6676
41. Wallet P, Benaoudia S, Mosnier A, Lagrange B, Martin A, Lindgren H, Golovliov I, Michal F, Basso P, Djebali S, Provost A, Allatif O, Meunier E, Broz P, Yamamoto M, Py BF, Faudry E, Sjöstedt A, Henry T (2017) IFN-γ extends the immune functions of Guanylate binding proteins to inflammasome-independent antibacterial activities during Francisella novicida infection. PLoS Pathog 13. https://doi.org/10.1371/JOURNAL.PPAT.1006630
42. Lagrange B, Benaoudia S, Wallet P, Magnotti F, Provost A, Michal F, Martin A, Di LF, Py BF, Molinaro A, Henry T (2018) Human caspase-4 detects tetra-acylated LPS and cytosolic Francisella and functions differently from murine caspase-11. Nat Commun 91(9):1–14. https://doi.org/10.1038/s41467-017-02682-y
43. Meunier E, Wallet P, Dreier RF, Costanzo S, Anton L, Rühl S, Dussurgey S, Dick MS, Kistner A, Rigard M, Degrandi D, Pfeffer K, Yamamoto M, Henry T, Broz P (2015) Guanylate-binding proteins promote activation of the AIM2 inflammasome during infection with Francisella novicida. Nat Immunol 16:476–484. https://doi.org/10.1038/NI.3119
44. Oh H, Siano B, Diamond S (2008) Neutrophil isolation protocol. J Vis Exp 17. https://doi.org/10.3791/745

Chapter 18

Quantifying Regulated Mitochondrial Fission in Macrophages

Syeda Farhana Afroz, Nicholas D. Condon, Matthew J. Sweet, and Ronan Kapetanovic

Abstract

Mitochondria have co-evolved with eukaryotic cells for more than a billion years, becoming an important cog in their machinery. They are best known for being tasked with energy generation through the production of adenosine triphosphate, but they also have roles in several other cellular processes, for example, immune and inflammatory responses. Mitochondria have important functions in macrophages, key innate immune cells that detect pathogens and drive inflammation. Mitochondrial activity is influenced by the highly dynamic nature of the mitochondrial network, which alternates between interconnected tubular and fragmented forms. The dynamic balance between this interconnected fused network and fission-mediated mitochondrial fragmentation modulates inflammatory responses such as production of cytokines and mitochondrial reactive oxygen species. Here we describe methods to differentiate mouse bone marrow cells into macrophages and the use of light microscopy, electron microscopy, flow cytometry, and Western blotting to quantify regulated mitochondrial dynamics in these differentiated macrophages.

Key words Drp1, Fission, Fusion, Inflammation, Macrophages, Microscopy, Mitochondria, Mitochondrial dynamics

1 Introduction

Macrophages are key cellular components of the innate immune system, participating in pathogen detection, host defense, wound healing, and homeostasis. These cells use pattern recognition receptors to sense danger and environmental perturbations [1]. One family of these receptors, the toll-like receptors (TLRs), recognizes microbial products such as lipopolysaccharide (LPS) and lipoproteins, activating downstream inflammatory and antimicrobial gene expression programs [2, 3]. In addition to regulating inflammatory gene expression [4, 5] and cell metabolism [6, 7], TLR signaling also modulates mitochondrial functions, which in turn affects immune responses [8, 9].

Thomas A. Kufer and Maria Kaparakis-Liaskos (eds.), *Effector-Triggered Immunity: Methods and Protocols*, Methods in Molecular Biology, vol. 2523, https://doi.org/10.1007/978-1-0716-2449-4_18,

Mitochondria are derived from ancient bacterial ancestors, merging with eukaryotic cells more than a billion years ago [10]. Inside each cell, mitochondrial numbers are finely regulated by the processes of biogenesis and mitophagy (Fig. 1). Mitochondrial biogenesis is mediated by both nuclear and mitochondrial transcription regulators such as peroxisome proliferator-activated receptor gamma-1α and mitochondrial transcription factor A, respectively, enabling expression of genes required for both biogenesis and mitochondrial DNA replication [11]. Mitophagy involves the delivery of damaged mitochondria to autophagosomes for recycling, with the PTEN-induced kinase 1/PARKIN complex being critical in this process [12, 13]. In addition to biogenesis and mitophagy, the balance between an interconnected tubular state (fusion-skewed) and a fragmented state (fission-skewed) also regulates mitochondrial numbers. This dynamic is mediated by specific GTPases; dynamin-related protein 1 (DRP1) drives fission, while the mitofusins (MFN1, MFN2) and dominant optic atrophy (OPA1) support fusion [14]. Modulation of the mitochondrial network can be very rapid as the activation of these molecules is regulated by post-translational modifications (PTMs), for example, deacetylation of MFN1 [15] and phosphorylation/dephosphorylation of DRP1 [16].

Mitochondrial dynamics have been linked to many cellular processes, including macrophage inflammatory responses [17–20]. Bacterial components, such as LPS, modulate mitochondrial dynamics by skewing towards fission, with subsequent induction of inflammatory mediators in macrophages [21, 22]. LPS stimulation also rewires mitochondrial activity, resulting in increased mitochondrial reactive oxygen species (mitoROS) that contributes to the pro-inflammatory phenotype of LPS-activated macrophages [23]. Several reports have shown an association between mitochondrial fission and mitoROS production in an inflammatory context. For instance, pharmacological antagonism of fission abrogated mitoROS generation in osteoblasts [24]. Furthermore, a peptide-based inhibitor of fission attenuated mitoROS production in a Huntington's disease cell culture model [25]. It is worth noting that mitoROS generation and mitochondrial fission can also have a reciprocal relationship with mitoROS shown to skew mitochondrial dynamics towards fission in neurons [26]. Given this intertwined connection, there may be value in assessing mitoROS generation when characterizing regulated mitochondrial dynamics responses in macrophages.

Here we present in vitro methods to assess mitochondrial dynamics and mitoROS production in murine macrophages responding to inflammatory stimuli. We describe simple methods that can be used to visualize and quantify mitochondrial dynamics, as well as important specificity controls (Fig. 1). We focus on the use of both confocal imaging and electron microscopy to not only

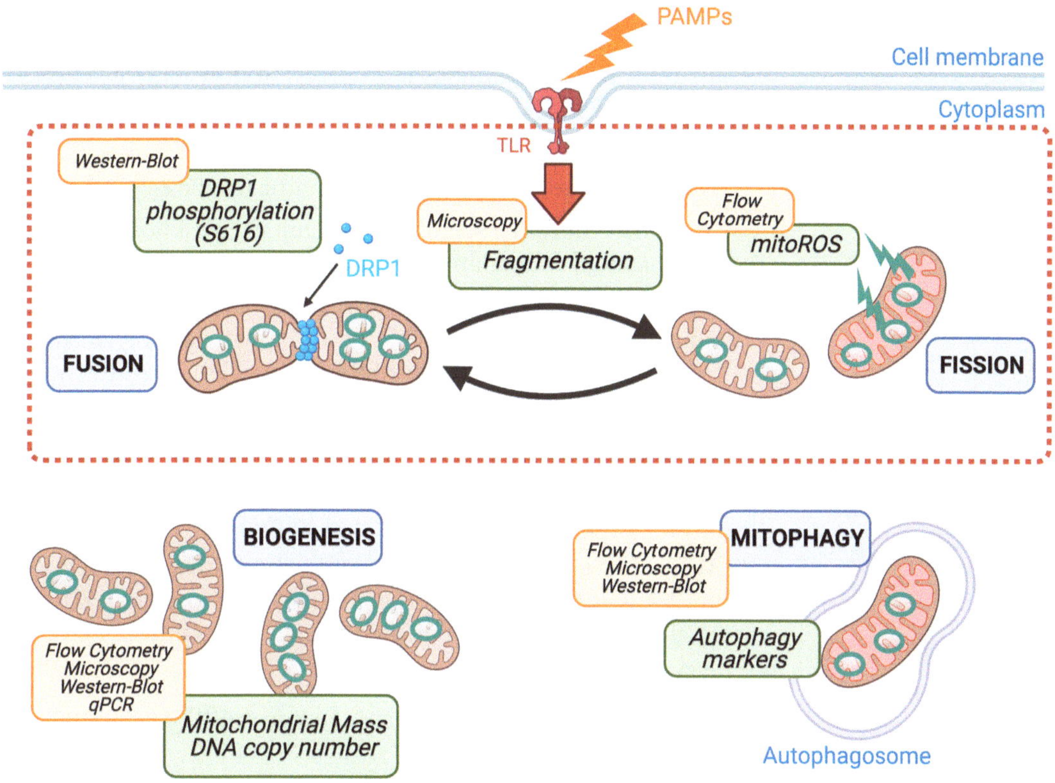

Fig. 1 TLR-inducible mitochondrial fission and methods to monitor this process. TLR stimulation alters the equilibrium between mitochondrial fusion and fission, skewing towards fragmentation of the mitochondrial network. Mitochondrial fission is accompanied by the acute phosphorylation of Drp1. Fragmented and stressed mitochondria release mitoROS and, if too damaged, are directed towards autophagy. Importantly, mitochondrial fission can be induced independently of mitochondrial biogenesis as mitochondria possess more than one copy of their DNA (small green circles). This chapter describes methods to assess inducible mitochondrial fission (in red dashed box) with specific methods (yellow boxes)

assess and enumerate mitochondrial morphology and regulated mitochondrial fission in macrophages but also outline additional approaches that can be used for further validation. These methods should be broadly applicable to other cell types and stimuli.

2 Materials

2.1 Macrophage Culture

1. 100-mm square petri dishes, 6-well/24-well tissue culture plates, 3-cm petri dishes, tweezers, scissors, 50-mL conical tubes.
2. Complete medium: RPMI 1640 with 10% heat-inactivated fetal calf serum (FCS), antibiotics (50-U/mL penicillin, 50-mg/mL streptomycin), and GlutaMAX (2 mM).

3. Recombinant human colony-stimulating factor-1 (rhCSF-1): Stock concentration is 15 μg/mL in RPMI 1640 complete medium (without antibiotics), stored at 4 °C (for 2–3 weeks, −80 °C for longer storage) and used at a final concentration of 150 ng/mL.
4. 10-mL syringe, 25G needle, 18G blunt needle.
5. Phosphate-buffered saline (PBS): Dulbecco's Phosphate Buffered Saline, [-] calcium chloride, [-] magnesium chloride.
6. Lipopolysaccharide from *Salmonella enterica* serotype minnesota (Sigma-Aldrich, L2137) or ultrapure lipopolysaccharide from *Salmonella minnesota* R595 (InvivoGen, tlrl-smlps).
7. Dimethyl sulfoxide (DMSO).
8. M1 (Sigma-Aldrich, #SML0629), reconstituted in DMSO, used at a final concentration of 20 μM, and stored at −20 °C in small aliquots to avoid repeated freeze-thaw cycles.
9. Mdivi-1 (Sigma-Aldrich, #M0199), reconstituted in DMSO, used at a final concentration of 10 μM, and stored at −20 °C in small aliquots to avoid repeated freeze-thaw cycles.
10. CO_2 incubator.
11. Centrifuge that can spin 50-mL falcon at 400 *g*.

2.2 Quantification of Mitochondrial Dynamics and mitoROS Production

1. 13-mm #1.5 coverslips, fine tweezers/forceps.
2. Mounting media (with a refractive index close to 1.4–5 when using with an oil objective): Gelvatol mounting medium (6-g polyvinyl acetate, 15-mL dH_2O, 30-mL Tris-HCl (pH 8.6), 15-g glycerol).
3. MitoSOX™ Red Mitochondrial Superoxide Indicator (Invitrogen, #M36008) abs/em ~510/580 nm.
4. MitoTracker Deep Red FM (*see* **Note 1**) (Invitrogen, #M22426) abs/em ~644/665 nm.
5. PBS-DAPI: 20 ng/mL 4′,6-diamidino-2-phenylindole (DAPI, Sigma-Aldrich) in PBS; stock concentration is 1 μg/mL, stored at −20 °C.
6. Calcein Blue AM Fluorescent Dye (Invitrogen, #C1429), stored at −20 °C in small aliquots to avoid repeated freeze-thaw cycle.
7. Primary antibodies:

   Tom20 (Rabbit, Cell Signaling Technology, #42406S), diluted 1:200 and used at a final concentration of 0.26 μg/mL.

   DRP1 (Rabbit, Cell Signaling Technology, #8570S), diluted 1: 100 and used at a final concentration of 3.66 μg/mL.

8. Secondary antibody: Goat anti-rabbit Alexa Fluor 488 (Invitrogen, #A-11008) diluted 1:400 and used at a final concentration of 5 μg/mL.
9. 0.1% Triton X-100 in PBS.
10. 0.5% bovine serum albumin (BSA) in PBS.
11. 8% paraformaldehyde (PFA) in PBS.
12. 2.5% glutaraldehyde in water.
13. Lift buffer: 1-mM EDTA, 0.1% sodium azide in PBS.
14. Confocal microscope: e.g., Carl Zeiss LSM 710 confocal scanner attached to an AxioExaminer Upright Stand using a 63 × 1.4 NA plan apochromat objective.
15. Software used for image analysis: Fiji, an open-source image-processing package based on ImageJ.
16. Flow cytometer with a blue laser (488 nm) and a detection filter set around 585 nm. We routinely use a Gallios Cytometer (Beckman Coulter), as well as its acquisition/analysis software Kaluza.

2.3 Quantification of DRP1 Phosphorylation by Western Blotting

1. Anti-DRP1 antibody (Mouse, Cell Signaling Technology, #14647S), diluted 1:1000 and used at a final concentration of 1.03 μg/mL.
2. Anti-phospho-DRP1 S616 antibody (Rabbit, Cell Signaling Technology, #3455S), diluted 1:1000 and used at a final concentration of 0.12 μg/mL.
3. Anti-rabbit IgG HRP-linked (Cell Signaling Technology, #7074S), diluted 1:2500 and used at a final concentration of 0.03 μg/mL.
4. Anti-mouse IgG HRP-linked (Cell Signaling Technology #7076S), diluted 1:2500 and used at a final concentration of 0.06 μg/mL.
5. PVDF or nitrocellulose membrane.
6. Whatman filter paper.
7. 1-mL syringes with a 25G needle.
8. Cell scrapers.
9. Phosphatase inhibitor tablets, PhosSTOP (Sigma-Aldrich, #4906837001).
10. cOmplete™ Protease Inhibitor Cocktail (Sigma-Aldrich, #11873580001).
11. NuPAGE™ LDS sample buffer (Invitrogen, #NP0008).
12. NuPAGE™ sample reducing agent (Invitrogen, #NP0009).
13. Prestained Protein Standard.

14. Western ECL substrate (clarity western luminol/enhancer solution and peroxide solution).
15. Hydrogen peroxide 30%, aqueous solution.
16. *Optional*: Benzonase, diluted 1:10 in $MgSO_4$ and used at a final concentration of 25 units/μL RIPA buffer – 50-mM Tris-HCl pH 8.0, 1-mM EDTA, 1% Triton-X100, 10% Glycerol, 0.1% sodium deoxycholate, 150-mM NaCl.
17. Wash buffer: Tris-buffered saline (TBS) with Tween 20 (TBST) (10-mM Tris-HCl pH 8.0, 150-mM NaCl, 0.05% Tween 20).
18. Running buffer: 10X stock (30.3-g/L Tris, 144-g/L glycine, 50 mL of 20% sodium dodecyl sulfate solution (SDS) in 1-L ultrapure water).
19. Transfer buffer (1 L): 200 mL of Trans-Blot Turbo 5x Transfer Buffer, 600 mL of Ethanol, 200 mL ultrapure water.
20. Blocking buffer: 5% skim milk in 1X TBST.
21. Gel Imaging System.
22. Semidry Blot System.
23. Roller (to use during blot transfer preparation).
24. Heating block to boil samples for Western blotting.
25. Centrifuge that can spin 1.5-mL reaction tubes at 17,000 g.
26. Image Lab software from Bio-Rad.

3 Methods

In this chapter, we focus on methods that do not require specific cell lines or plasmid constructs (e.g., mito-Dendra2) but rather simple mitochondrial-staining dyes that are commercially available. The first step is to collect bone marrow progenitor cells from mice and then differentiate them for 7 days with the growth factor CSF-1 to generate bone marrow-derived macrophages (BMMs). After stimulation of these cells, mitochondria are stained with MitoTracker or mitochondria markers (e.g., anti-Tom20), and the mitochondrial network is then analyzed by microscopy using Fiji software [27] to determine various parameters, for example, mitochondrial numbers per cell. Specific controls that can be incorporated to confirm that changes observed relate to changes in mitochondrial dynamics are also described. Use of electron microscopy to assess mitochondrial morphology and the level of fragmentation is also outlined, as is phosphorylation of Drp1, an indirect readout of fission. The use of flow cytometry to quantify production of mitoROS, a well-known inflammatory mediator relevant to mitochondrial fission, is also detailed. Finally, we briefly discuss the assessment of biogenesis via quantification of mitochondrial mass and mitochondrial DNA copy number, as well as mitophagy, as important controls that should be considered when quantifying mitochondrial fission (Fig. 1).

3.1 Differentiating Bone Marrow Progenitor Cells into BMMs

1. Euthanize a mouse (*see* **Note 2**), as per institutional animal ethics requirements, collect the two femurs and tibias (without breaking the bones), and place them in a small 3-cm dish filled with complete medium.
2. In a biosafety cabinet, use tweezers to remove any muscle and surrounding tissues from the bones. Soak the clean bones in 70% ethanol for 1 min and place in complete medium again (we normally place the bones in medium in a 6-well plate).
3. Carefully cut off the tips of the femurs and tibias using scissors and then flush the four bones with complete media using a 25G needle (10-mL syringe) into a 50-mL falcon tube. Continue the flushing until the bones are white in appearance. Usually, it takes 10- to 15-mL media to flush four bones.
4. Centrifuge bone marrow cells at 500 *g* for 5 min.
5. Resuspend the cell pellet in complete medium (*see* **Note 3**), split them into eight 100-mm square petri plates with 15-mL media per plate, and add recombinant human colony-stimulating factor-1 (rhCSF-1) to each plate at a final concentration of 10^4 U/mL. We use rhCSF-1 produced in a highly efficient baculovirus-insect cell expression system at the Protein Expression Facility, the University of Queensland, but commercial rhCSF-1 is also available. In both cases, we recommend optimizing the rhCSF-1 concentration needed for your own experimental conditions. Culture the cells for 6 days in a 37 °C CO_2 incubator (*see* **Note 4**).
6. On day 6, BMMs are harvested by removing media from the plates and adding 10-mL sterile PBS to each plate. Use a 10-mL syringe with an 18G blunt needle to harvest the cells from the plate, collecting the BMMs in a 50-mL falcon tube. Centrifuge the cells at 500 × *g* for 5 min, resuspend them in complete media, and count them. Be sure to add rhCSF-1 to the media when plating cells (final concentration of 10^4 U/mL), as this growth factor is essential for mouse BMM survival.

3.2 Measuring Mitochondrial Fission by Microscopy After LPS Stimulation

1. Place sterile coverslips into the wells of a 24-well tissue culture plate. Wash once with sterile PBS to clean the coverslip from any contaminants. Note that it is important to have at least duplicate coverslips for each condition.
2. Plate 2×10^5 BMMs per well in 1-mL complete media supplemented with rhCSF-1 (10^4 U/mL) and incubate overnight at 37 °C and 5% CO_2.

Table 1
list of siRNAs used to silence *Drp1* in mouse macrophages

siRNA target	Sequence (5′-3′)
mouse Drp1_1	GCCAUGCUGUCAAUUUGCUAGAUGU
mouse Drp1_2	CAGGCAACUGGAGAGGAAUGCUGAA
mouse Drp1_3	GGUGGUGCUAGGAUUUGUUAUAUUU

3. On the following day, add 100-ng/mL LPS to the designated well(s) and incubate for 6 h (or desired time points). As LPS does not solubilize very well, it is important to vortex the LPS before use for a minimum of 30 s. It is also important to note that LPS from distinct species and the methods of purification confer very different activities and may result in contamination with other bacterial products that also activate macrophages, so attention should be given to the source of LPS and the purification method used [28, 29]. Depending on the bacterial species from which it is purified, a concentration between 10 and 1000 ng/mL of LPS should be sufficient to achieve maximum activity in macrophages, whereas a concentration range of approximately 0.1 to 10 ng/mL is generally considered to be sub-maximal. For our studies, we use *Salmonella* LPS at 10 to 100 ng/mL. However, we find a similar fission response when using sub-maximal LPS concentrations (e.g., 0.5 to 1 ng/mL).
4. *Controls*: (a) BMMs can be pre-treated for 1 h with compounds that will inhibit mitochondrial fission before LPS stimulation. Mdivi-1 attenuates fission by inhibiting Drp1 [30] and M1 promotes fusion [31]. We use these compounds at a final concentration of 10 μM and 20 μM, respectively. At these concentrations, Mdivi-1 and M1 do not affect BMM viability in our hands, as assessed by lactate dehydrogenase release and 3-(4,5-dimethylthiazol-2-yl)-2,5-diphenyltetrazolium bromide reduction assays. However, we note that Mdivi-1 can also have Drp1-independent effects [32], so additional controls are strongly recommended. (b) Genetic approaches such as siRNAs that target *Drp1* (Table 1) or *Drp1*-deficient cell lines can be used as controls to confirm LPS-inducible fission [22].

3.3 Staining with MitoTracker Red FM

1. After the required time of LPS stimulation (*see* **Note 5**), replace the medium from the wells with new medium containing MitoTracker Red FM at a final concentration of 150 nM.
2. Incubate at 37 °C for 30 min. Keep the plate in the dark as much as possible.

3. Wash three times with 1-mL PBS per well (place plate for 5 min in the incubator for each wash). Be careful while washing—do not pipette PBS directly on the coverslips; rather direct it towards the wall of the wells.
4. Add 300 μL per well of 4% cold PFA diluted in PBS (*see* **Note 6**) and incubate for a minimum of 15 min in the dark (e.g., covered with foil paper) at room temperature (RT) for fixation.
5. Wash three times with 1-mL PBS (5 min/wash at RT). At this point, coverslips can be stored at 4 °C for approximately 2 days, if necessary. At this stage, BMM can also be stained for mitochondria-related proteins of interest, such as Tom20 or Drp1 (*see* Subheading 3.4).
6. Remove PBS and add 300-μL PBS-DAPI (20 ng/mL) per well.
7. Keep for 30 min at RT in the dark.
8. Wash three times with 1-mL PBS (5 min/wash).
9. Pipette 5-μL mounting media on the slides for each cover slip. Carefully grab the cover slip from the well using forceps. Remove excess PBS by carefully tapping the side of the cover-slip on a Kimwipe.
10. Place the coverslip onto the slide ensuring the side with the cells is in contact with the mounting media.
11. Keep the slides in the dark at RT overnight for drying off. On the following day, store at 4 °C in the dark until imaging is performed (*see* **Note 7**).
12. Image using a laser scanning confocal microscope with a 63 × 1.4NA plan apochromat objective.

3.4 Immunostaining for Mitochondrial Proteins (Drp1/Tom20)

1. Use PFA-fixed BMM from **Step 5** in Subheading 3.3 for the following steps.
2. Permeabilize cells by adding 500-μL 0.1% Triton X-100 in PBS per well for 5 min.
3. Wash three times with 1-mL PBS.
4. Block the cells by adding 100 μL of 0.5% bovine serum albumin (BSA) in PBS to each well. Incubate at RT for 1 h.
5. Place 30-μL droplets of diluted antibodies onto a large piece of parafilm attached inside a box, protected from light. We use anti-Drp1 and anti-Tom20 antibodies at final concentrations of 3.66 μg/mL and 0.26 μg/mL, respectively.
6. Carefully remove the coverslips from the wells using forceps, dry the edges of each coverslip on a Kimwipe, and then place coverslips cell side down on 30-μL droplets of anti-Drp1 or anti-Tom20. Be sure to place one coverslip on 30-μL droplets of 0.5% BSA in PBS without the primary antibody. This control coverslip will receive only the secondary antibody.

7. Incubate at RT for 1 h.
8. Using forceps, carefully flip the coverslips still in the box, dry the edges, and place them cell side up. Wash three times by gently adding 100-μL droplets of PBS on the coverslip (5 min per wash).
9. Stain with the secondary antibody and DAPI in a similar fashion to **Steps 5** and **6**. Prepare a master mix of secondary antibody and DAPI in 0.5% BSA in PBS.
   (a) Secondary antibody: goat anti-rabbit Alexa Fluor 488 diluted 1:400 (final concentration 5 μg/mL).
   (b) PBS-DAPI: 20 ng/mL final concentration.
10. Dab the edges of each coverslip gently on a Kimwipe and place cell side down on 30-μL droplets of the master mix.
11. Incubate in the dark at RT for 1 h.
12. Repeat **Step 8** and wash three times with 100-μL droplets of PBS for 5 min per wash (in the dark).
13. Dab the edges of each coverslip between washes to ensure excess secondary antibody and stain are removed.
14. Mount the coverslips onto slides as described previously. Nail polish can be used, but we recommend using mounting media that seals the coverslip.
15. Store at 4 °C in the dark until they are imaged (*see* **Note 7**).

3.5 Assessing Fission by Immunofluorescence: Image Acquisition

Fragmentation of the mitochondrial network can be quantified by different methods. Here we present two techniques we have routinely used in our studies (Fig. 2). Fluorescent images of fixed cells are acquired using a Carl Zeiss LSM 710 confocal scanner attached to an AxioExaminer Upright Stand using a 63 × 1.4 NA plan apochromat objective. It is extremely important that, once the acquisition settings are chosen, they are not altered. Because of this, we recommend sampling a control and an LPS-stimulated slide to ensure detected intensities don't increase above the detector range, saturating the signal. To have a full view of the mitochondrial network, acquisition of a z-stack may be useful. However, we have observed that, in macrophages in vitro, most of the mitochondrial network resides on a plane close to the basal membrane. Thus, a single high-definition scanning plane can be performed without the loss of significant data. This will allow for increased individual field of view and thus more cells at a high resolution for the duration of the imaging session. To increase the signal-to-noise ratio and the quality of the image, we recommend using a slower scanning speed and line averaging (×16).

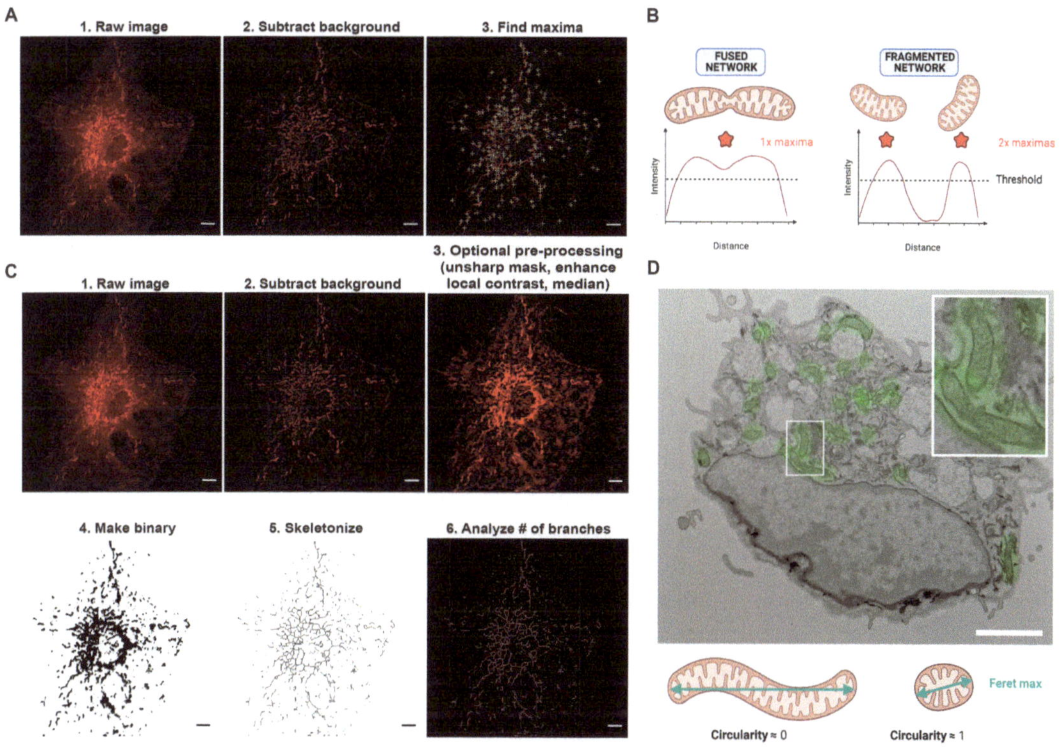

Fig. 2 Quantifying mitochondria fission by light and electron microscopy. (**a**) Maxima methods from a RAW image of a BMM, including processing using the "subtract background" and "find maxima" functions. (**b**) Schematic diagram of the maxima identification using intensity threshold. (**c**) Steps from the skeletal method to detect fragmentation of the mitochondrial network. (**d**) An electron micrograph of mitochondria in a macrophage artificially colored in green (upper). Morphology of mitochondria can be assessed using their circularity or ferret max number (lower). Scale bar 1 μm for light microscopy (**a**, **c**) and 2 μm for electron microscopy (**d**)

3.6 Assessing Fission by Microscopy: Image Quantification

We use Fiji, an open-source image-processing package based on ImageJ, for further image processing and quantification [27]. Two different toolsets in ImageJ are used to quantify regulated fission (*see* **Note 8**): "Find Maxima" and "Mitochondrial Network Analysis" (MiNA) [33], both of which are described below.

(i) *MAXIMA:* The maxima method quantifies relative numbers of single mitochondria, with this varying depending upon the threshold of fluorescence chosen (Fig. 2a, b).

1. Open an image in ImageJ.
2. Select "Image" and then "color" and "split channels." For further processing, use mitochondria-stained channel.
3. You can run the "sharpen" function to preprocess the image (optional).
4. Use "Process/Subtract Background" to reduce and normalize background staining, as per your requirements (ensuring the radius is greater than the mitochondrial cross section).

The effect of this subtraction will vary between experiments due to the staining process and the microscopy acquisition settings.

5. Using the manual selection tools, isolate one macrophage. Run "Process/Find Maxima. . . ." Set the "noise tolerance" to sampled intensities of your mitochondria (place the cursor over mitochondria and noting the intensity value displayed in the status bar of the main FIJI window). As a guide, we typically obtain around 60–100 maxima (mitochondria) per cell in BMM. However, this number can vary depending on staining, image quality, preprocessing steps, cell type, and treatments. Although variation is noted among differing experiments, it is critical that these settings should be fixed within one specific experiment and used across all samples.
6. A fully automated script has also been described [34]. If your sample size is not too large, you can manually select the cells and apply the post-acquisition steps. We usually measure maxima of 30–50 macrophages per condition per experiment to ensure that the inherent cell-to-cell variability of macrophage populations does not confound data interpretation. Each experiment is usually repeated at least three times independently.

(ii) *SKELETAL:* The skeletal method utilizes the Mitochondrial Network Analysis (MiNA) toolset, which is available in ImageJ. This approach allows a semi-automated quantification of mitochondrial networks in an in vitro cellular system (Fig. 2c).

1. Open an image in ImageJ.
2. Select "Image" and then "color" and "split channels." For further processing, use mitochondria-stained channel.
3. Preprocess the image using the following filters: Unsharp mask and enhance local contrast (CLAHE) and median. These preprocessing steps are optional; however, it allows to augment image quality before converting to binary image and producing morphological skeletons.
4. For analyzing images, you can either perform batch processing or select an individual cell. For our studies, we used individual cells for analysis. After selecting an individual cell, run the following commands: make binary and then skeletonize.
5. After skeletonizing, run the command "analyze skeleton," and then from the table of parameters, use "# branches" to calculate individual and networks of mitochondria. The sum of individual and networks gives the total number of mitochondria in one cell.
6. Repeat the same analysis for all other cells in that image, as well as for all other images from the same experiment.

3.7 Assessing Fission by Electron Microscopy (EM): Image Acquisition and Analysis

Fluorescence-based tools are a convenient way of quantifying mitochondrial numbers and assessing mitochondrial dynamics. However, EM is an important method to validate fluorescence-based imaging quantification and to provide quantitative ultrastructural information on mitochondrial morphology.

1. Seed day 6 BMM (from **Step 6**, Subheading 3.1) into two 3-cm petri dishes in complete media and leave them to rest overnight. Note that it is recommended to use a higher cell density for EM, by comparison to that used for light microscopy. We usually plate three million BMM in 3-mL complete media containing rhCSF-1 (final concentration of 10^4 U/mL).
2. On the following day, stimulate the BMMs with LPS (100 ng/mL final) or leave untreated.
3. After 4–6 h, wash cells twice with PBS and fix with PBS containing 2.5% glutaraldehyde for 10 min at RT. Remove the fixative agent and then wash the cells with PBS (*see* **Note 9**).
4. Process the samples for EM. Detailed methods for sample preparation for EM are described elsewhere [35, 36]. Collect sections on EM grids and transfer to the transmission EM.
5. Image grids at a suitable magnification (e.g., 10,000–25,000×) and capture micrographs in a random fashion across the section.
6. Image analysis: Ideally, this is performed on a set of images in a blinded fashion with the experimental condition indicated by a code. Identify mitochondria in each image (readily identified by their characteristic morphology, enclosed by a double membrane: an outer membrane and an inner membrane that is folding towards the inside of the organelle—the cristae, Fig. 2d, highlighted in green). Select the draw tool in ImageJ and highlight the circumference of each mitochondrion. Two measurements that are particularly useful for quantification, as indicated in Fig. 2d, are:
   (a) *Circularity*: ImageJ calculates a coefficient representing how close your selection is to a perfect circle (a circularity of 1 is a perfect circle). Mitochondria undergoing fission will show a higher circularity than a fused mitochondrial network.
   (b) *Feret max*: The ferret max is the longest line you can draw within your selected area. Mitochondrial fission will show a lower ferret max by comparison to the elongated mitochondria that are typically associated with fusion.
7. Manual annotation is possible, but, unfortunately, there is no easy way to automatically identify mitochondria with classic transmission EM to enable automated analyses of large data sets. However, some research groups have started to create

machine learning-based ImageJ scripts that recognize and quantify mitochondria [37, 38]. Note that, depending on your sample size, this method could take longer than conventional approaches to manually assess mitochondrial morphology.

4 Assessing Fission Response by Quantifying Drp1 Post-translational Modifications (PTM)

Drp1 activity is regulated by several PTMs, including phosphorylation, sumoylation, and nitrosylation [39]. Among these, serine phosphorylation has been intensively studied, particularly DRP1 activation by S616 phosphorylation (S635 in mouse Drp1) and S637 dephosphorylation (S656 in mouse Drp1). LPS treatment of BMM triggers rapid phosphorylation of Drp1 at S635 (within ~15–30 min). This acute response can be quantified by Western blotting using antibodies that detect phospho-S635 Drp1 in BMM (equivalent of human phospho-S616 DRP1) and total Drp1 (*see* **Note 10**).

1. Plate day 6 BMM (from **Step 6**, Subheading 3.1) at 2×10^6 cells per well of a 6-well plate in 2 mL of complete media with rhCSF-1 (10^4 U/mL) and incubate overnight at 37 °C and 5% CO2.
2. On the following day, add 100-ng/mL LPS to designated wells and incubate over a time course. Vortex the LPS before use for a minimum of 30 s.
3. After each time point, keep the plate on ice, wash the cells twice with 1-mL ice-cold PBS, and add 200 μL of RIPA buffer (including freshly added 1X phosphatase inhibitor and 1X protease inhibitor cocktail).
4. Using a cell scraper, lift all cells in RIPA buffer and transfer the lysate to a tube. Use a 1-mL syringe with a 25G needle to homogenize the lysate and to mechanically shear the DNA. Optional: If the lysate is too viscous, treat the lysates with 2-μL Benzonase per sample (diluted 1:10 in $MgSO_4$) for 20 min at RT. Then, centrifuge the samples at a high speed (~17,000 g) for 5 min. Collect the supernatant in a new tube. At this point, lysates can be frozen at −20 °C until Western blots are ready to be run.
5. If required, perform protein quantification assays of lysates (e.g., BCA or Bradford assay) to determine protein concentration and prepare your protein samples. We usually prepare 20 μL of sample for a 12-well blot (20 μL per well).
   (a) 1X NuPAGE™ LDS sample buffer (5 μL of the 4X stock solution).

(b) 1X NuPAGE™ sample reducing agent (2 μL of the 10X stock solution).

(c) 10-μg protein.

(d) Water (adjust for a final volume of 20 μL).

6. Boil the samples at 100 °C for 10 min, spin down the samples, and leave them on ice while preparing Bio-Rad gels for loading. Do not forget to remove the sticky strip from the bottom of the gel. Add 1× SDS running buffer to fill the internal space and the external space, as required. Wash and readjust the wells before loading. Load 5 μL of protein marker and the samples and then run at desired voltage and time (e.g., 200 V for approximately 30 min).
7. For transferring, we use the Trans-Blot Turbo transfer system from Bio-Rad:

(a) Cut a piece of nitrocellulose or PVDF membrane to the size of blotted gel. Nitrocellulose can be used without preparation, whereas PVDF membrane must be immersed in methanol for 1 min before use to increase hydrophilicity.

(b) Prepare Turbo transfer buffer, as described in the materials.

(c) Soak the gel, 8× filter paper, and membrane in transfer buffer and arrange onto the cassette in the following order: 4× filter paper (bottom), membrane, gel, 4× filter paper (top). Remove any bubbles and excess buffer using a roller. Soak off excess buffer from the cassette using tissue paper.

(d) Run at 25 V for 9 min.

(e) Rinse the cassette in water after use and clean the turbo lid using wet tissue paper.

8. Block the membrane for 1 h at RT or overnight at 4 °C with blocking buffer on a shaker.
9. Remove blocking buffer, wash once with 1× TBST, and then incubate the membrane with primary anti-phospho-S635 Drp1 rabbit antibody (diluted 1:1000) in 5% BSA in TBST overnight at 4 °C on a shaker.
10. On the following day, wash the membrane three times with TBST on a shaker (5 min for each wash) and incubate with the HRP-conjugated anti-rabbit secondary antibody (diluted 1: 2500) in 5% milk in TBST for 1 h at RT.
11. Wash three times with TBST (5 min at RT for each wash).
12. Set up 1.6-mL ECL reagent (50% from each solution) on parafilm. Blot the membrane on tissue to remove as much liquid as possible and place the blot protein side down on to

the ECL reagent. Leave for 1 min. Remove liquid from membrane as earlier, put on transparent plastic wrap, and image the blot using a sensitive camera system or scanner.

13. After imaging, inactivate the membrane using 30% H_2O_2 (just enough volume to cover the blot) keeping in a shaking incubator (37 °C) for approx. 30 min.
14. Wash the membrane once with TBST, add the next primary antibody (mouse anti-Drp1) diluted 1:1000 in 5% milk in TBST, and incubate overnight, as before (*see* **Note 11**).
15. Follow **Steps 10** to **13** but use HRP-conjugated anti-mouse secondary antibody (diluted 1:2500).
16. After exporting the image from the camera/scanner system, you can quantify the band volume using ImageLab (Software available from Bio-Rad). Quantify the volume of p-Drp1 S635 and total Drp1 bands for each lane to calculate relative levels of p-Drp1 S635/total Drp1.
17. You can also probe for α-tubulin and the mitochondrial outer membrane protein Tom20 as additional controls for total protein content and mitochondrial mass, respectively. Details of these antibodies are provided in materials.

5 Measuring mitoROS Production After LPS Stimulation

Reactive oxygen species play an important role in immune responses. mitoROS oxidizes and irreversibly damages intracellular microbes and promotes inflammatory responses [40]. Several studies have also linked mitochondrial fission to mitoROS production [41], so mitoROS production is often quantified in parallel with mitochondrial fission. To assess mitoROS production, we stain macrophages with MitoSOX (a dye that stains mitochondrial ROS) and quantify relative levels by flow cytometry.

1. Plate 5×10^5 of BMM (day 6) in 1 mL of complete media per well in a 24-well plate and leave them overnight. Plate enough cells to perform at least duplicates for each condition. You will also need unstained control cells (without MitoSOX) as a control.
2. On the following day, check the cells for healthy morphology using a microscope, add 100 ng/mL LPS (*see* **Note 12**) to designated wells, and incubate for 6 h (or desired time points). Vortex the LPS before use for a minimum of 30 s.
3. After 6 h of stimulation, remove the media and wash twice with PBS. Reconstitute a vial of MitoSOX in 13 μL of DMSO (5-mM stock). Dilute MitoSOX 1:1000 in lift buffer (e.g., 5 μL in 5 mL) to a final concentration of 5 μM. You will need 0.5 mL of diluted working stock per well.

4. Add 0.5 mL of MitoSOX in lift buffer (*see* Subheading 2.2, **item 13**) to the respective wells and incubate at 37 °C for 20 min. Add 0.5-mL lift buffer alone to the unstained control sample.
5. Leave plates on ice in the dark, then collect cells by pipetting up and down and transfer into FACS tubes. Keep cells on ice. Run the cells through a flow cytometer using a blue laser with the detection filter set around 585 nm.
6. Gating technique (using the acquisition software, e.g., Kaluza): First gate cells on size and granularity (SSC-A x FSC-A) and then remove any doublets (FSC-A x FSC-H). You can also exclude dead cells by using dyes such as Calcein Blue AM Fluorescent Dye when staining with MitoSOX (Calcein-AM can be used at a final concentration of 1 μM). Live cells will take up and hydrolase the Calcein-AM, leading to its intracellular accumulation. Other dyes such as 7-AAD or PI can be problematic as their emission wavelength is close to that of MitoSOX. Once the gating is configured, run your samples and measure median fluorescence intensity (MFI).

6 Controls to Assess Mitochondrial Fission Versus Mitochondrial Biogenesis and/or Mitophagy

Although this chapter focuses on mitochondrial dynamics, the quantification of biogenesis and mitophagy may also be important for appropriate interpretation of findings. Mitochondrial fission can be initiated independently of biogenesis. Individual mitochondria can contain several copies of the mitochondrial genome; thus mitochondrial DNA replication is not a prerequisite for fragmentation [22, 42]. However, any changes in mitochondrial numbers that are observed could reflect alterations in biogenesis and/or mitophagy. For these reasons, it is preferable to have controls for biogenesis and mitophagy when assessing mitochondrial fission. Several methods can be used to quantify biogenesis and total mitochondrial mass. For example, the number of copies of mitochondrial DNA can be determined by using qPCR to quantify a mtDNA-specific region versus that of a nuclear-encoded gene such as beta-2-microglobulin (B2M) [22, 43]. Alternatively, mitochondrial mass per cell can be estimated using confocal microscopy or flow cytometry through quantifying fluorescence of MitoTracker-stained cells. Levels of mitochondrial proteins such as Tom20 can also be quantified by Western blotting as an indirect read-out of mitochondrial mass [22]. Contrary to biogenesis, macrophages can discard damaged mitochondria through mitophagy, a specific autophagy pathway [44]. Mitophagy can be challenging to assess as it may be difficult to discriminate from regular autophagy. Nonetheless, traditionally mitophagy is assessed by monitoring the colocalization of LC3-positive autophagosomes and mitochondria (stained with MitoTracker or with an antibody against Tom20) [45]. Flow cytometry can also be used as a quantitative approach to measure mitophagy flux, utilizing MitoTracker Deep Red in combination with both lysosomal and mitophagy inhibitors [46].

7 Notes

1. We find that *MitoTracker Deep Red FM* works well in our studies. Because it uses the far-red channel, it also allows the simultaneous use of other dyes or antibodies. ***MitoTracker Green*** is also a mitochondrial-selective probe, but we find that its performance is affected by fixation. Tetramethylrhodamine, methyl ester (TMRM) is also widely used.
2. We usually use male or female 8–12-week-old C57Bl/6 mice. Older mice can also be used but may result in lower BMM yields because of an altered composition of their bone marrow [47].
3. At this stage, bone marrow cells can also be frozen in 90% FCS/10% DMSO for later use.
4. Depending on the activity of the rhCSF1 and the density at which bone marrow cells are initially plated, we recommend topping up cultures with 5-mL rhCSF-1-containing media per plate on the afternoon or evening of day 4 or day 5.
5. We find that LPS-inducible fission in BMM occurs rapidly and is sustained for up to 24 h; however, we routinely assess this response at 6 h post-LPS stimulation. We note that inducible fission can vary, depending on the stimuli and/or cell type. For example, MEF cells require overnight FCS starvation and a longer incubation with LPS for a maximal fission response.
6. 8% PFA can be stored at −20 °C for several months. We usually do not keep 4% PFA (diluted in PBS) at 4 °C for more than a week.
7. Don't forget to bring the slides to RT at least 30 min before imaging them, as temperature can affect image acquisition.
8. We have performed both methods on the same images to directly compare and find that they give similar data, but the maxima method is a little less variable in our hands [22].
9. At this stage, cells can be kept at 4 °C for a couple of days, if necessary, before further processing.
10. The antibody directed against phospho-DRP1 (*Cell Signaling Technology*) does not perform well for microscopy in our hands; thus, it is challenging to visualize the localization of phosphorylated Drp1 in BMM using confocal microscopy.

11. If you are using rabbit anti-Drp1, you must strip the membrane instead of using H_2O_2 for HRP inactivation. For stripping, incubate the membrane on shaker at RT with 1X antibody stripping solution in distilled water for 15 min. Then, remove the stripping solution and block the membrane for 1 h at RT with blocking buffer on a shaker before adding another antibody.
12. Bacterial infection is usually a better stimulus than LPS for mitoROS production.

Acknowledgments

RK received funding from the European Union's Horizon 2020 research and innovation program under the Marie Sklodowska-Curie grant agreement No 894690, a Rebecca L. Cooper grant (#021765). MJS is an NHMRC Leadership Fellow, supported by an NHMRC Investigator Grant (APP1194406) and an NHMRC project grant (APP1125316). SFA is supported by an Australian Government Research Training Program (RTP) Scholarship. NDC is supported as a CZI Imaging Scientist by grant number 2020-225648 from the Chan Zuckerberg Initiative DAF, an advised fund of Silicon Valley Community Foundation. Microscopy was performed at the Australian Cancer Research Foundation (ACRF)/Institute for Molecular Bioscience Cancer Biology Imaging. We thank Professor Robert Parton for insightful discussions around best practice for electron microscopy imaging and analysis.

Funded by the European Union

References

1. Kawasaki T, Kawai T (2014) Toll-like receptor signaling pathways. Front Immunol 5:461. https://doi.org/10.3389/fimmu.2014.00461
2. Stocks CJ, Schembri MA, Sweet MJ, Kapetanovic R (2018) For when bacterial infections persist: Toll-like receptor-inducible direct antimicrobial pathways in macrophages. J Leukoc Biol 103(1):35–51. https://doi.org/10.1002/JLB.4RI0917-358R
3. Satoh T, Akira S (2016) Toll-Like Receptor signaling and its inducible proteins. Microbiol Spectrum 4(6). https://doi.org/10.1128/microbiolspec.MCHD-0040-2016
4. Schroder K, Irvine KM, Taylor MS, Bokil NJ, Le Cao KA, Masterman KA et al (2012) Conservation and divergence in Toll-like receptor 4-regulated gene expression in primary human versus mouse macrophages. Proc Natl Acad Sci U S A 109(16):E944–E953. https://doi.org/10.1073/pnas.1110156109
5. Perkins DJ, Patel MC, Blanco JC, Vogel SN (2016) Epigenetic mechanisms governing innate inflammatory responses. J Interf Cytokine Res 36(7):454–461. https://doi.org/10.1089/jir.2016.0003
6. Palsson-McDermott EM, O'Neill LA (2013) The Warburg effect then and now: from cancer to inflammatory diseases. BioEssays: news and reviews in molecular, cellular and developmental biology 35(11):965–973. https://doi.org/10.1002/bies.201300084

7. O'Neill LA, Pearce EJ (2016) Immunometabolism governs dendritic cell and macrophage function. J Exp Med 213(1):15–23. https://doi.org/10.1084/jem.20151570
8. West AP, Brodsky IE, Rahner C, Woo DK, Erdjument-Bromage H, Tempst P et al (2011) TLR signalling augments macrophage bactericidal activity through mitochondrial ROS. Nature 472(7344):476–480. https://doi.org/10.1038/nature09973
9. Banoth B, Cassel SL (2018) Mitochondria in innate immune signaling. Transl Res J Lab Clin Med 202:52–68. https://doi.org/10.1016/j.trsl.2018.07.014
10. Roger AJ, Munoz-Gomez SA, Kamikawa R (2017) The origin and diversification of mitochondria. Current Biol 27(21): R1177–R1R92. https://doi.org/10.1016/j.cub.2017.09.015
11. Popov LD (2020) Mitochondrial biogenesis: an update. J Cell Mol Med 24(9):4892–4899. https://doi.org/10.1111/jcmm.15194
12. Youle RJ, Narendra DP (2011) Mechanisms of mitophagy. Nat Rev Mol Cell Biol 12(1):9–14. https://doi.org/10.1038/nrm3028
13. Eiyama A, Okamoto K (2015) PINK1/Parkin-mediated mitophagy in mammalian cells. Curr Opin Cell Biol 33:95–101. https://doi.org/10.1016/j.ceb.2015.01.002
14. Giacomello M, Pyakurel A, Glytsou C, Scorrano L (2020) The cell biology of mitochondrial membrane dynamics. Nat Rev Mol Cell Biol 21(4):204–224. https://doi.org/10.1038/s41580-020-0210-7
15. Lee JY, Kapur M, Li M, Choi MC, Choi S, Kim HJ et al (2014) MFN1 deacetylation activates adaptive mitochondrial fusion and protects metabolically challenged mitochondria. J Cell Sci 127(Pt 22):4954–4963. https://doi.org/10.1242/jcs.157321
16. Chang CR, Blackstone C (2010) Dynamic regulation of mitochondrial fission through modification of the dynamin-related protein Drp1. Ann N Y Acad Sci 1201:34–39. https://doi.org/10.1111/j.1749-6632.2010.05629.x
17. Park S, Won JH, Hwang I, Hong S, Lee HK, Yu JW (2015) Defective mitochondrial fission augments NLRP3 inflammasome activation. Sci Rep 5:15489. https://doi.org/10.1038/srep15489
18. Katoh M, Wu B, Nguyen HB, Thai TQ, Yamasaki R, Lu H et al (2017) Polymorphic regulation of mitochondrial fission and fusion modifies phenotypes of microglia in neuroinflammation. Sci Rep 7(1):4942. https://doi.org/10.1038/s41598-017-05232-0
19. Nair S, Sobotka KS, Joshi P, Gressens P, Fleiss B, Thornton C et al (2019) Lipopolysaccharide-induced alteration of mitochondrial morphology induces a metabolic shift in microglia modulating the inflammatory response in vitro and in vivo. Glia 67(6): 1047–1061. https://doi.org/10.1002/glia.23587
20. Gao F, Reynolds MB, Passalacqua KD, Sexton JZ, Abuaita BH, O'Riordan MXD (2020) The mitochondrial fission regulator DRP1 controls post-transcriptional regulation of TNF-alpha. Front Cell Infect Microbiol 10:593805. https://doi.org/10.3389/fcimb.2020.593805
21. Park J, Choi H, Min JS, Park SJ, Kim JH, Park HJ et al (2013) Mitochondrial dynamics modulate the expression of pro-inflammatory mediators in microglial cells. J Neurochem 127(2): 221–232. https://doi.org/10.1111/jnc.12361
22. Kapetanovic R, Afroz SF, Ramnath D, Lawrence GM, Okada T, Curson JE et al (2020) Lipopolysaccharide promotes Drp1-dependent mitochondrial fission and associated inflammatory responses in macrophages. Immunol Cell Biol 98(7):528–539. https://doi.org/10.1111/imcb.12363
23. Mills EL, Kelly B, Logan A, Costa ASH, Varma M, Bryant CE et al (2016) Succinate dehydrogenase supports metabolic repurposing of mitochondria to drive inflammatory macrophages. Cell 167(2):457–70 e13. https://doi.org/10.1016/j.cell.2016.08.064
24. Zhang L, Gan X, He Y, Zhu Z, Zhu J, Yu H (2017) Drp1-dependent mitochondrial fission mediates osteogenic dysfunction in inflammation through elevated production of reactive oxygen species. PLoS One 12(4):e0175262. https://doi.org/10.1371/journal.pone.0175262
25. Guo X, Disatnik MH, Monbureau M, Shamloo M, Mochly-Rosen D, Qi X (2013) Inhibition of mitochondrial fragmentation diminishes Huntington's disease-associated neurodegeneration. J Clin Invest 123(12): 5371–5388. https://doi.org/10.1172/JCI70911
26. Hung CH, Cheng SS, Cheung YT, Wuwongse S, Zhang NQ, Ho YS et al (2018) A reciprocal relationship between reactive oxygen species and mitochondrial dynamics in neurodegeneration. Redox Biol 14:7–19. https://doi.org/10.1016/j.redox.2017.08.010
27. Schindelin J, Arganda-Carreras I, Frise E, Kaynig V, Longair M, Pietzsch T et al (2012) Fiji: an open-source platform for biological-image analysis. Nat Methods 9(7): 676–682. https://doi.org/10.1038/nmeth.2019

28. Xiao X, Sankaranarayanan K, Khosla C (2017) Biosynthesis and structure-activity relationships of the lipid a family of glycolipids. Curr Opin Chem Biol 40:127–137. https://doi.org/10.1016/j.cbpa.2017.07.008
29. Hirschfeld M, Ma Y, Weis JH, Vogel SN, Weis JJ (2000) Cutting edge: repurification of lipopolysaccharide eliminates signaling through both human and murine toll-like receptor 2. J Immunol 165(2):618–622. https://doi.org/10.4049/jimmunol.165.2.618
30. Manczak M, Kandimalla R, Yin X, Reddy PH (2019) Mitochondrial division inhibitor 1 reduces dynamin-related protein 1 and mitochondrial fission activity. Hum Mol Genet 28(2):177–199. https://doi.org/10.1093/hmg/ddy335
31. Wang D, Wang J, Bonamy GM, Meeusen S, Brusch RG, Turk C et al (2012) A small molecule promotes mitochondrial fusion in mammalian cells. Angew Chem Int Ed Engl 51(37): 9302–9305. https://doi.org/10.1002/anie.201204589
32. Smith G, Gallo G (2017) To mdivi-1 or not to mdivi-1: is that the question? Dev Neurobiol 77(11):1260–1268. https://doi.org/10.1002/dneu.22519
33. Valente AJ, Maddalena LA, Robb EL, Moradi F, Stuart JA (2017) A simple ImageJ macro tool for analyzing mitochondrial network morphology in mammalian cell culture. Acta Histochem 119(3):315–326. https://doi.org/10.1016/j.acthis.2017.03.001
34. Bosch A, Calvo M (2019) Automated quantitative analysis of mitochondrial morphology. Methods Mol Biol 2040:99–115. https://doi.org/10.1007/978-1-4939-9686-5_6
35. Graham L, Orenstein JM (2007) Processing tissue and cells for transmission electron microscopy in diagnostic pathology and research. Nat Protoc 2(10):2439–2450. https://doi.org/10.1038/nprot.2007.304
36. Takasato M, Er PX, Chiu HS, Maier B, Baillie GJ, Ferguson C et al (2015) Kidney organoids from human iPS cells contain multiple lineages and model human nephrogenesis. Nature 526(7574):564–568. https://doi.org/10.1038/nature15695
37. Liu J, Li L, Yang Y, Hong B, Chen X, Xie Q et al (2020) Automatic reconstruction of mitochondria and endoplasmic reticulum in electron microscopy volumes by deep learning. Front Neurosci 14:599. https://doi.org/10.3389/fnins.2020.00599
38. Li R, Zeng X, Sigmund SE, Lin R, Zhou B, Liu C et al (2019) Automatic localization and identification of mitochondria in cellular electron cryo-tomography using faster-RCNN. BMC Bioinform 20(Suppl 3):132. https://doi.org/10.1186/s12859-019-2650-7
39. Adaniya SM, J OU, Cypress MW, Kusakari Y, Jhun BS (2019) Posttranslational modifications of mitochondrial fission and fusion proteins in cardiac physiology and pathophysiology. Am J Physiol Cell Physiol 316(5):C583–C604. https://doi.org/10.1152/ajpcell.00523.2018
40. Shekhova E (2020) Mitochondrial reactive oxygen species as major effectors of antimicrobial immunity. PLoS Pathog 16(5):e1008470. https://doi.org/10.1371/journal.ppat.1008470
41. Jezek J, Cooper KF, Strich R (2018) Reactive oxygen species and mitochondrial dynamics: the Yin and Yang of mitochondrial dysfunction and cancer progression. Antioxidants (Basel) 7(1). https://doi.org/10.3390/antiox7010013
42. Waters LR, Ahsan FM, Wolf DM, Shirihai O, Teitell MA (2018) Initial B cell activation induces metabolic reprogramming and mitochondrial remodeling. iScience 5:99–109. https://doi.org/10.1016/j.isci.2018.07.005
43. Sun X, Lee W, Vaghjiani V, St John JC (2016) Analysis of mitochondrial DNA copy number and its regulation through DNA methylation of POLGA. Methods Mol Biol 1351:131–141. https://doi.org/10.1007/978-1-4939-3040-1_10
44. Pickles S, Vigie P, Youle RJ (2018) Mitophagy and quality control mechanisms in mitochondrial maintenance. Curr Biol 28(4): R170–RR85. https://doi.org/10.1016/j.cub.2018.01.004
45. Zhu J, Dagda RK, Chu CT (2011) Monitoring mitophagy in neuronal cell cultures. Methods Mol Biol 793:325–339. https://doi.org/10.1007/978-1-61779-328-8_21
46. Mauro-Lizcano M, Esteban-Martinez L, Seco E, Serrano-Puebla A, Garcia-Ledo L, Figueiredo-Pereira C et al (2015) New method to assess mitophagy flux by flow cytometry. Autophagy 11(5):833–843. https://doi.org/10.1080/15548627.2015.1034403
47. Gibon E, Loi F, Cordova LA, Pajarinen J, Lin T, Lu L et al (2016) Aging affects bone marrow macrophage polarization: relevance to bone healing. Regen Eng Transl Med 2(2): 98–104. https://doi.org/10.1007/s40883-016-0016-5

Chapter 19

Live Cell Imaging of T Cell Pyroptosis

Andreas Linder and Veit Hornung

Abstract

Pyroptosis is a lytic, pro-inflammatory cell death program that is tightly regulated by inflammasomes in most cases. Inflammasome-dependent pyroptosis is best characterized in myeloid cells, but there is growing evidence that this cell death program also functions in human T cells. Several studies have suggested a role for inflammasome components in T-cell biology but often do not unambiguously clarify whether this means that T cells progress to pyroptosis. Pyroptosis has distinct morphological features, such as early loss of membrane integrity and ballooning, that allow it to be distinguished from apoptosis in a microscopic experiment. However, the most stringent definition of inflammasome-dependent pyroptosis is that it is genetically dependent on a pro-inflammatory caspase (caspase-1 or caspase-4) and the pore-forming protein gasdermin D (GSDMD). Therefore, using live cell imaging of T cells in combination with a genetic loss-of-function setup is the most reliable tool for us to unequivocally demonstrate that a T cell undergoes pyroptosis. Parallel live cell imaging of T cells and macrophages is limited due to the fact that T cells do not adhere while macrophages do. This can be overcome by using so-called micro-inserts that hold the cells in a limited area that can be monitored by microscopic field of view. Here we describe in detail how live cell imaging of human T cells and monocyte-derived macrophages undergoing pyroptosis can be performed.

Key words T cell, Pyroptosis, Inflammasome, CARD8, Microscopy, Live cell imaging

1 Introduction

T cells play a central role in the adaptive immune system. Programmed cell death (PCD) is an essential regulator of diverse T-cell functions ranging from thymic selection to restoration of homeostatic cell numbers [1]. The most relevant type of PCD in these settings is apoptosis. Apoptosis can be activated through extrinsic cues such as death receptor signaling or cell intrinsic mechanisms that lead to an imbalance in favor of pro-apoptotic over anti-apoptotic factors. Essential for the execution of apoptosis is a distinct set of cysteine-aspartic proteases (caspases). As such, caspase-8 is considered the initiator caspase of extrinsic apoptosis, while caspase-9 serves this function in the context of intrinsic apoptosis. Caspases-3, 7, and 6 are executioner caspases that

Thomas A. Kufer and Maria Kaparakis-Liaskos (eds.), *Effector-Triggered Immunity: Methods and Protocols*, Methods in Molecular Biology, vol. 2523, https://doi.org/10.1007/978-1-0716-2449-4_19,

become activated when cleaved by one of the initiator caspases. The activated executioner caspases in turn cleave diverse substrates that lead to the cell's demise. Apoptosis is considered immunologically silent as the potentially pro-inflammatory cellular content is shielded in so-called apoptotic bodies. A cell that commits to apoptosis acquires distinct morphological features when studied under the microscope. Among the characteristics of an apoptotic cell are cytoplasmic shrinkage, chromatin condensation, plasma membrane blebbing, and the formation of apoptotic bodies [2].

Another type of PCD is pyroptosis which is in most cases regulated through inflammasomes [3, 4]. Inflammasome-dependent pyroptosis also relies on caspases to carry out cell death, yet a distinct set of caspases is employed, namely, the pro-inflammatory initiator caspases caspase-1 and caspase-4 (or its murine orthologue caspase-11). Upstream of caspase-1, pyroptosis is regulated through inflammasomes. These are multimeric protein complexes consisting of a sensor protein and the adapter protein ASC which serves to recruit caspase-1, although some inflammasomes can activate caspase-1 in an ASC-independent fashion. The best characterized inflammasome sensor proteins are AIM2, NLRP1, NLRP3, NAIP-NLRC4, NLRP6, PYRIN, and CARD8. These sensor proteins are traditionally ascribed to the family of pattern recognition receptors (PRRs), even though some of these sensors do not directly engage with pathogen-associated molecular patterns (PAMPs). Indeed, some modes of inflammasome activation can rather be categorized as effector-triggered immunity (ETI) [5], in which cellular processes are guarded against pathogenic activity. Following its activation, caspase-1 cleaves various cytosolic substrates, including the highly pro-inflammatory cytokine pro-IL-1β, which is rendered bioactive by this cleavage step. Moreover, caspase-1 initiates pyroptosis by cleaving GSDMD, of which the N-terminal cleavage fragment assembles into a large pore complex at the plasma membrane. The GSDMD pore is big enough so that certain proteins such IL-1β can pass through the membrane. However, it also disrupts the electrochemical ion gradient across the membrane, allowing uncontrolled water influx and thus cell swelling. This leads to the loss of mechanical resilience, which renders these cells susceptible to rupture by external mechanical perturbations [6].

While inflammasome-dependent pyroptosis was long believed to be restricted to "professional" innate immune cells such as macrophages, evidence is accumulating that also T cells can commit to inflammasome-dependent pyroptosis. As such it was proposed that HIV can trigger pyroptosis of non-productively infected T cells [7] and that T cell intrinsic, inflammasome-dependent release of IL-1β plays a role in the differentiation or maintenance of Th1 or Th17 cells [8, 9]. These studies mainly rely on fluorogenic substrates to monitor pyroptotic versus apoptotic caspase activity or require measuring the release of IL-1 cytokines to infer an involvement of inflammasomes and pyroptosis. Nevertheless, FACS-based assays to monitor necrotic forms of cell death are not well suited to study cell death in which the cell is rapidly disintegrating. On the other hand, measurement of the release of IL-1β is restricted to cell types that express IL-1β.

Very recently we and others reported that T cells can commit to bona fide pyroptosis when the CARD8 inflammasome is activated [10, 11]. In this scenario the dipeptidyl-peptidase inhibitor Val-boroPro triggered a lytic type of PCD that recapitulated genetic, biochemical, and morphological features of pyroptosis. As IL-1 family cytokines remained undetectable in peripheral blood-derived T cells, we consider microscopy the most reliable read-out to study T cell pyroptosis. It combines the advantages of a single-cell resolution read-out with the possibility to monitor cells over time and the ability to assess the morphology of the dying cells. This helps on the one hand not only to identify the cell under study as a lymphocyte but also to judge on the type of cell death triggered. When combined with T cells genetically engineered to lack essential regulators of PCD, high standards to prove the involvement of pyroptosis or apoptosis are met. In this protocol, we describe how T cells and monocyte-derived macrophages (MDMs) can be studied using live cell microscopy for pyroptotic morphology. By using a special type of microscopy micro-insert, we achieve complete coverage of the experimental well using a 25× objective, which is of advantage when studying motile cells such as T cells.

2 Materials

1. Cell culture medium, RPMI 1640 supplemented with 2.5% (v/v) human serum, L-glutamine, penicillin-streptomycin, pyruvate, and HEPES.
2. Primary human T cells isolated from peripheral blood mononuclear cells (PBMCs) through MACS purification [12].
3. Primary monocytes isolated from PBMCs through $CD14^+$ MACS purification [12].
4. Recombinant human interleukin-7 (IL-7) and human interleukin-15 (IL-15).
5. Recombinant colony-stimulating factor 1 (CSF-1).
6. Accutase cell detachment solution.
7. Tweezers.
8. Ibidi micro-Inserts 4 Well FulTrac.
9. Ibidi μ-plates 24-well.

10. Propidium iodide (stock 8 μg/ml in PBS), stored at 4 °C.
11. Leica DMi8-inverted microscope equipped with an automatic stage, incubator, environmental control, water immersion pump, and the following objectives: HC PL FLUOTAR L 25×/0.80 IMM and HC PL APO 63×/1.20 W CORR CS2 or equivalent.
12. Live cell imaging chamber.
13. Val-boroPro (stock 10 mM in DMSO), stored in 5-μl aliquots at −20 °C.
14. NeedleTox (composed of recombinant protective antigen (PA), stock 1 mg/ml, and recombinant LFn-YscF, stock 5 mg/ml), each stored in 5-μl aliquots at −80 °C.
15. ABT737 (stock 10 mM in DMSO), stored in 5-μl aliquots at −20 °C.

    S63845 (stock 10 mM in DMSO), stored in 5-μl aliquots at −20 °C.

3 Methods

1. As a starting point to study pyroptosis in human T cells via microscopy, we usually rely on peripheral blood mononuclear cells (PBMCs) isolated through density centrifugation standard protocols from which T cells and monocytes are purified via magnetic-activated cell sorting (MACS) following the manufacturer's instructions (*see* **Note 1** and **Note 2**) [12].
2. The T cells are cultured in human serum-supplemented RPMI medium in the presence of IL-7 and IL-15 (5 ng/mL each) at a density between one and four million cells/mL (*see* **Note 3**) for 5 days at 37 °C and 5% CO_2 in a cell culture incubator. In parallel, monocytes are resuspended in human serum-supplemented RPMI medium containing CSF-1 (200 ng/mL) in 24-well plates at 250.000 cells per well. On day 3, fresh CSF-1 is added to the culture medium at a final concentration of 200 ng/mL. Differentiation is allowed to proceed until the microscopy experiment is conducted on day 5 (*see* **Note 4**).
3. Prior to harvesting cells on the day of the microscopy experiment (day 5), the FulTrac micro-inserts and 24-well μ-plates, in which the cells are added to for the microscopic assessment, have to be prepared. One micro-insert harbors four micro-wells and therefore is suited for four experimental conditions (Fig. 1a, b). Put as many micro-inserts onto the 24-well plate as experimental conditions you want to include. The induction of pyroptosis and apoptosis in CD4 T cells and MDMs from the same donor will serve as an exemplary experiment throughout

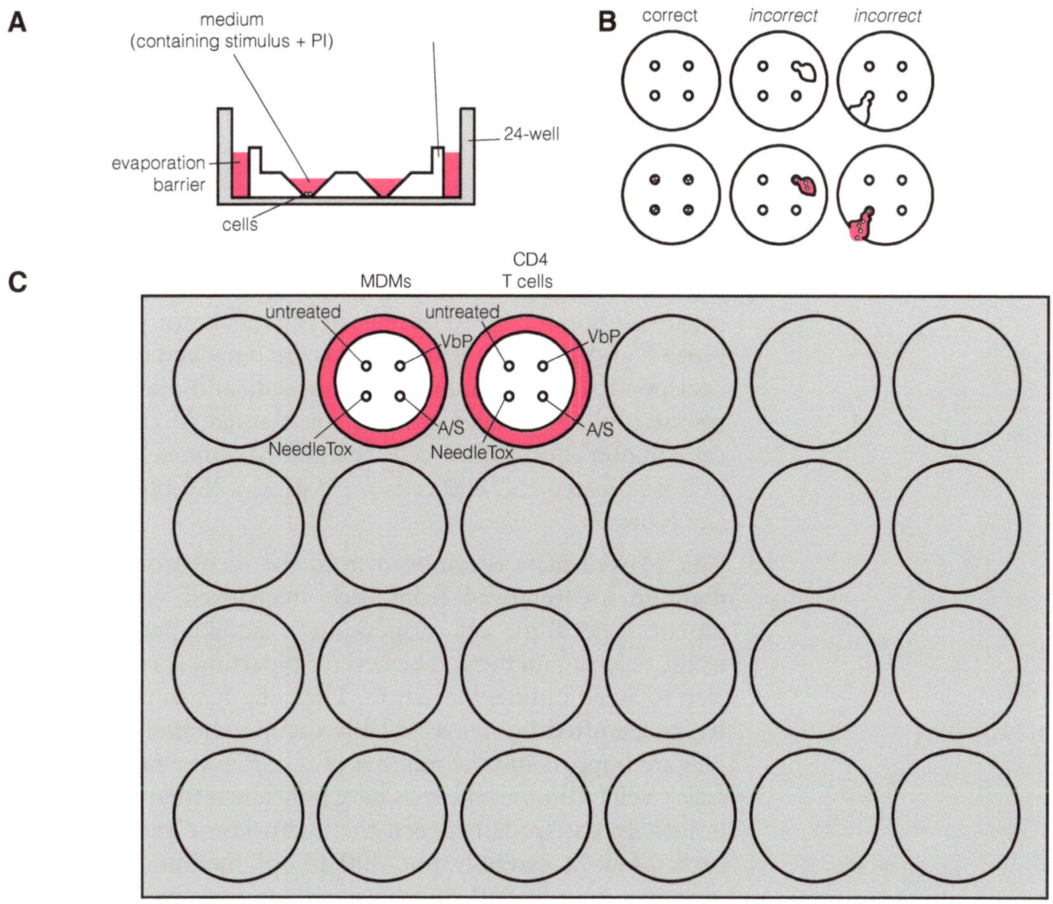

Fig. 1 Preparing the micro-inserts for the live cell imaging of T-cell pyroptosis. (**a**) Schematic side view of the micro-insert after placement in the 24-well. The micro-well is filled with the cells, the stimulus, and the fluorescent dye propidium iodide (PI) to a total volume of 10 μl. The surrounding part of the 24-well is filled with medium as an evaporation barrier. (**b**) When the micro-insert is correctly attached to the surface of the 24-well, cells remain within the field of view (left). Entrapment of air between the micro-insert and the bottom surface of the 24-well will lead to leakage of medium. The cells will move towards the leakage during the course of the live cell imaging experiment and are lost from the field of view (middle and right). (**c**) In order to study four treatments in two cell types, micro-inserts are placed in two of the 24 wells available on the 24-well μ-plate

this protocol. To this end, the two cell types are subjected to four different treatments which warrants the use of two micro-inserts harboring a total of eight micro-wells (Fig. 1c). Putting the micro-inserts in a way that they are evenly attached to the surface of the 24-well and no liquid or cells can leak is critical (*see* **Note 5** and Fig. 1b). In order to prepare the wells of the micro-insert, these have to be filled preliminarily with 10-μl medium and briefly centrifuged at 1500 × *g* for 1 min. Then check via a benchtop transmitted-light inverted microscope whether any air bubble got trapped at the bottom of the well (*see* **Note 6** on how to remove air bubbles) and put the 24-well plate containing the micro-inserts into the incubator at 37 °C and 5% CO_2.

4. After having prepared the micro-inserts, T cells are counted, and the viability is assessed via trypan blue exclusion in a Neubauer chamber or if available an automated cell counter. MDMs that by now have become adherent are harvested from the 24-well plates by removing the culture medium, washing the cells carefully with PBS and adding Accutase cell detachment solution for 30 to 60 min in the incubator at 37 °C (*see* **Note 7** and **Note 8**). Then MDMs are detached by pipetting, resuspended in medium, and counted, and the viability is assessed in a Neubauer chamber or if available an automated cell counter. For both T cells and MDMs, the cell concentration is adjusted to 2000 cells per 8 μL (= 250,000 cells/mL) (*see* **Note 9**).
5. The 24-well plate is removed from the incubator, the 10-μL medium are removed from each micro-well with a 10-μL pipette, and 8-μL cell suspension is added per micro-well using reverse pipetting. The reverse pipetting is critical at this step to avoid bubble formation. Then check again via a benchtop transmitted-light inverted microscope whether any air bubbles are trapped and the number of cells is in the range of 2000 cells/well (although the cells have not fully settled down at this point). In the space between the micro-insert and the wall of each 24-well, carefully put 300 μL of medium or PBS by pipetting twice 150 μL which will reduce evaporation throughout the microscopy experiment (Fig. 1b). Then put the 24-well plate back into the incubator (37 °C and 5% CO_2) and allow the cells to settle down for 1 h (*see* **Note 10**).
6. In the meantime, start the microscope and allow the cell culture chamber to warm up to 37 °C and the CO_2 concentration to equilibrate at 5%.
7. Stimuli are added in 2-μL to 8-μL cell suspension to a final volume of 10 μl. In the exemplary experiment (Fig. 1a), Val-boroPro serves to induce CARD8-dependent pyroptosis in T cells and MDMs, NeedleTox is a rapid inducer of pyroptosis in MDMs but not T cells, and the combination of the BCL2-inhibitor ABT737 combined with the MCL1-inhibitor S63845 (A/S) is used to induce intrinsic apoptosis (*see* **Note 11**). The diverse stimuli are prepared as a 5× working solution in medium that also contains propidium iodide (PI) at a 5× dilution (800 ng/ml) compared to the final concentration (1: 50; 160 ng/ml). PI will serve to monitor cell membrane integrity in the course of the experiment. Prepare a 5× dilution of the diverse stimuli in the PI-containing medium as follows:

Prepare a solution containing 20-μM VbP (final concentration 4 μM, stock concentration 10 mM), a solution containing 1250-ng/ml protective antigen (final concentration 250 ng/ml, stock concentration 1 mg/ml) combined with 125-ng/ml LFn-YcSF (final concentration 25 ng/ml, stock concentration 5 mg/ml), and a solution containing ABT737 (final concentration 1 μM, stock concentration 10 mM) and S63845 (final concentration 1 μM, stock concentration 10 mM). Carefully add the 2 μl of the 5× stimulus and PI-containing working solution to the respective micro-well in order to start stimulation. Add a 5× PI-solution without a stimulus as the "untreated condition." Note the time at which the stimuli were added (*see* **Note 12**).

8. Start the imaging. A Leica DMi8-inverted microscope equipped with a HC PL FLUOTAR L 25×/0.80 IMM objective that uses water for immersion is used for imaging, but other platforms might be equally suitable. First, the 24-well cell culture plate has to be carefully put onto the automatic stage within the incubator of the microscope, and the plate must be covered with the lid that provides the environmental temperature (37 °C) and CO_2 concentration (5%). We chose the transmission light channel as well as the DsRed filter cube (excitation 546/11, dichroic mirror 560, emission 605/75) for the detection of the PI signal. Exposure time has to be adjusted individually. Usually some dead, PI-positive cells are already present at the start of the experiment and can serve to set the exposure time. We employ the so-called "mark and find" mode to identify and mark the different positions on the automated microscope stage. Each micro-well represents one position at which the microscope acquires an image. Using the 25× objective will lead to the micro-well be fully tracked within the image detail of one acquisition. Usually, we take a picture every 2 min, but when imaging more conditions, the acquisition interval might need to be extended to give the microscope time to acquire every position. Allow the experiment to proceed for 16 h (*see* **Note 13**). After having stopped, the acquisition representative images from individual time points or entire movies can be exported and subjected to automated quantification of PI-positive cells. Representative images of individual cells can be acquired at a higher magnification with a 63× objective at the end of the experiment. An example data set and various ways to display the data are shown in Fig. 2. Micro-inserts can be washed and stored for reuse (*see* **Note 14**).
9. For quantification of PI-positive cells, a custom-written *Definiens XD 2.0* script is used (available at https://github.com/GeneCenterMunich/Image-Segmentation-T-cells). The

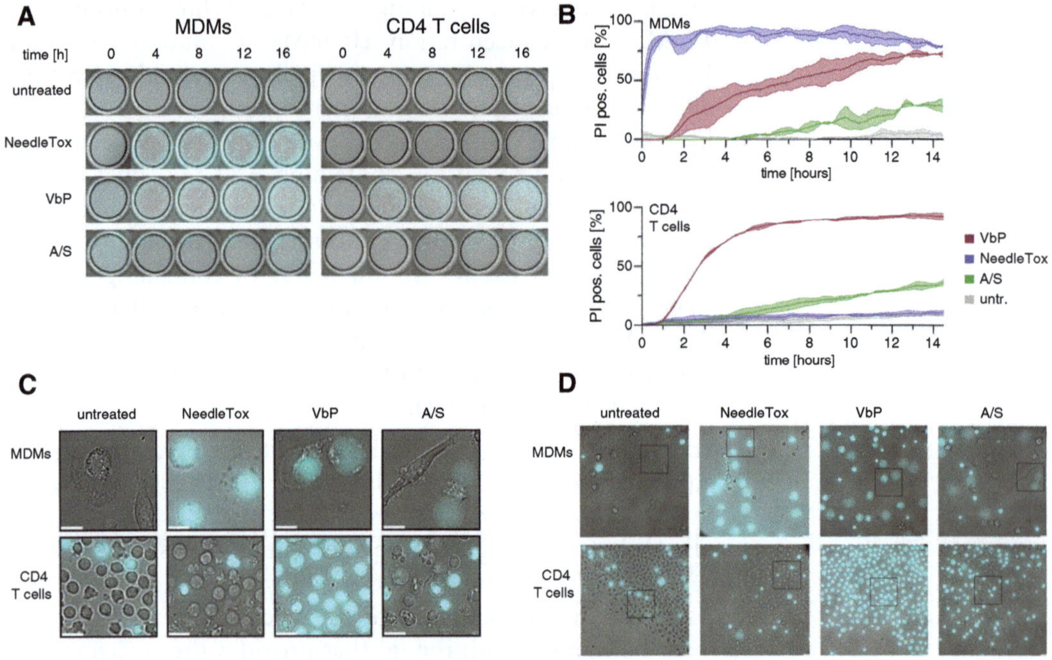

Fig. 2 T cells and monocyte-derived macrophages display morphologic features of pyroptosis. The figure was reproduced from data published in [11]. MDMs and T cells were stimulated with the indicated treatments with Val-boroPro (VbP) serving as an activator of the CARD8 inflammasome, NeedleTox as a stimulus activating the NAIP-NLRC4 inflammasome in MDMs but not CD4 T cells, and the combination of the BCL2-inhibitor ABT737 and the MCL1-inhibitor S63845 (A/S) to induce apoptosis. (**a**) A series of representative images from the indicated time points is shown. (**b**) The fraction of propidium iodide (PI) positive cells was quantified. The quantification reveals that NeedleTox induces a very rapid PI uptake in all MDMs but not CD4 T-cell characteristic of pyroptosis. VbP also induces a relatively fast PI uptake in both MDMs and CD4 T cells compatible with pyroptosis, while the induction of apoptosis through A/S leads to a delayed uptake of PI in only a fraction of cells. (**c**) After 16 h, which was the end of the experiment, representative images were acquired that display cellular ballooning characteristic of pyroptosis as well as the PI-stained nuclei in the VbP-treated T cells and MDMs as well as the NeedleTox-treated MDMs, while apoptotic bodies can be seen in the A/S-treated, PI-negative T cells. (**d**) A lower magnification of the images shown in (**c**) is displayed

image analysis procedure is the following: the contour of the individual cells is determined by using the transmitted light images (Channel1) that was acquired together with the fluorescence signal emitted by the PI-positive T cells (Channel2). In Channel1, the individual cells appear as dark, relatively round objects within a heterogeneous background arising from the inhomogeneous illumination. To facilitate the detection of cells, Channel1 is inverted, and a 3D Gaussian filter with a kernel size of 5 × 5 × 3 pixels is applied, followed by a second 3D Gaussian filter with the same kernel size. Then, the latter image is subtracted from the previous one. This procedure results in a background-subtracted image. Next, a global threshold is applied, and segmentation is carried out using an algorithm implemented in the *Definiens XD 2.0* software platform. Briefly, the used so-called Multi-Threshold Segmentation algorithm splits the image domain and classifies the resulting image objects based on a defined pixel value threshold. This results in segmented patterns classified as "T cells" or "background." The average fluorescence intensity corresponding to the PI level (Channel2) is then computed for each of the resulting "T-cell" objects. This procedure allows extracting the total number of T cells and the number of PI-positive T cells exhibiting a high fluorescence signal in each frame. The fraction of PI-positive cells is then calculated for every image, and the average of PI-positive cells was calculated over a sliding window covering the following 30 min of each image. Note: for quantification of MDMs, this approach is not applicable, as the quantification of the total number of cells by identifying macrophages in an automated fashion is limited due to their heterogeneous morphology. Therefore, the PI signal of each image is normalized to the number of cells, as visually determined at the start of the experiment.

4 Notes

1. To proof a high level of purity of the cell population under study or when a specific T cell subset is of interest, T cells can also be purified via fluorescence-activated cell sorting (FACS). Macrophages differentiated from monocytes (purified through positive selection on CD14 expression) are included as a positive control as this cell type can reliably commit to pyroptosis when exposed to diverse stimuli. Consider that—depending on local regulations—ethical approval has to be obtained from local authorities prior to starting to work with primary human material such as PBMCs.
2. In our hands, thawed T cells are not well suited for studying cell death. A good viability as a starting point is critical. We would refrain from using thawed T cell preparations unless a freeze-thaw protocol is well established and retrieves T cells with a viability greater than 90%. Monitoring viability in T cell cultures is critical.
3. In our hands, culturing primary immune cells in human serum containing cell culture medium works best for both resting T cells and the differentiation of MDMs. Nevertheless, FCS-supplemented medium can also be used given that the overall T cell viability in it is favorable.
4. T cells can be subjected to microscopy right after isolation. As we usually conduct various types of pyroptosis assays once T cells were freshly isolated, it is more convenient to keep the cells

in the incubator until the experiment is conducted. When MDMs are included as a positive control for pyroptosis, to us it is most convenient to allow them to differentiate for 5 days and culture the T cells in parallel. Doing so, one can also isolate T cells on day 1, FACS-purify subpopulations on day 2, and let them rest in IL-7/IL-15 containing medium until the microscopy-based experiment is conducted.

5. In order to place the micro-insert into the 24-well, we use tweezers with which we take up the micro-insert and let it drop onto the bottom surface of the 24-well from a very low distance. Then we press the insert down onto the surface and inspect the 24-well plate from the bottom. Based on the light reflection, no air should be trapped between the silicon surface of the micro-insert and the surface of the well. The micro-wells of the insert should appear as four sharply contoured round dots, and no air must be trapped connecting the micro-wells and the edge of the micro-insert. To remove trapped air, use the tweezers to carefully press onto the micro-insert from the top while looking at it from the bottom. As the micro-insert can be reused (see below) and you usually do not use all 24 wells available on the cell culture plate, do not waste too much time when you do not get the micro-insert tightly closed. Instead, you can put a new micro-insert into another well of the same cell culture plate.
6. The placement and preparation of the micro-inserts are tedious and require some practice. It is critical to avoid air bubbles trapped at the bottom of the micro-insert. Nevertheless, when an air bubble gets trapped, try to remove it by pipetting it with a 10-μl pipette. Trying to disrupt the bubble with a cannula did not prove successful in our hands and carried the risk of damaging the micro-insert or scratch the (glass of the) plate. Prepare additional micro-inserts as a backup when you fail to put the cells without trapping air (compare **Note 5**). The manufacturer's protocol provides additional recommendations regarding the placement and preparation of the micro-insert.
7. Detaching differentiated MDMs from cell culture wells without affecting the viability is sometimes tedious. While minimal amounts of cells but a high viability are required for microscopy-based experiment, MDMs can be differentiated at larger scale in 10-cm tissue culture dishes or even non-tissue culture-treated dishes in which the cells become even less adherent and can therefore be detached more easily.
8. It is sometimes tedious to remove macrophages from the bottom of the culture dish in order to replate them. In our hands, Accutase works best, but also putting them in cold PBS containing 2% EDTA is an option.

9. Given that cell counting is prone to variation, it might be worthwhile titrating once the number of cells that are put per micro-well. Consider that T cells are much smaller but also moving much faster than differentiated MDMs. Cells should be titrated to identify a cell count that allows for the individual cells to be visually discriminated from each other. Generally, a high cell count is preferable as it ascertains a higher validity of the quantification. Therefore, we consider a confluence between 50% and 80% of the micro-well as ideal.
10. It is critical to avoid trapping of air bubbles at the bottom of the micro-insert. Use a 10-μl pipette for pipetting. It is sometimes difficult to put the desired number of cells into the micro-well all the more as automated cell counting and Neubauer chamber cell counting are sometime unreliable. Include a viability dye in your cell counting step, as a low starting viability interferes with the experiment.
11. Depending on the experiment, it might be worthwhile including additional experimental controls. Consider that every condition delays the acquisition time depending on how long the microscope takes to find a position and to acquire images in various channels. When studying T cells of various genotypes, experimental conditions can rapidly exceed what is controllable.

    We would suggest Val-boroPro treatment as bona fide pyroptosis inducer in T cells as well as NeedleTox (LFn-YcsF) for macrophage pyroptosis as these agonists do not require a priming step.
12. Avoid perturbing the T cells that have settled down to the bottom of the micro-well when adding the stimulus as it can take several minutes for them settle down again.
13. When using a 25× objective, the image covers the whole micro-well. Try to center the image around the micro-well. Pyroptosis and apoptosis occur usually relatively fast, and T cells die within few and no more than 24 h. Imaging longer than 16 h can lead to high background cell death due to the unfavorable culture conditions in the micro-inserts.
14. The micro-inserts can be reused. To this end, the inserts are rinsed first with 30 ml ethanol and then with 30 ml PBS in a 50-ml falcon tube. Remove the micro-inserts from the tube by using tweezers and transfer it onto a sterile 10 cm dish. Allow for the micro-inserts to dry in the 10 cm dish with the lid removed on a sterile workbench and store the inserts on another 10 cm dish covered with a lid. Perform the washing of the micro-inserts soon after having finished the microscopy experiment in order to avoid that residual cell culture medium dries out and clots the micro-wells of the insert. Also, the wells on the 24-well plate that were not used during the experiment can be used in a future experiment.

References

1. Spetz J, Presser AG, Sarosiek KA (2019) T cells and regulated cell death: kill or be killed. Int Rev Cell Mol Biol 342:27–71. https://doi.org/10.1016/bs.ircmb.2018.07.004
2. Galluzzi L, Vitale I, Aaronson SA, Abrams JM, Adam D, Agostinis P, Alnemri ES, Altucci L, Amelio I, Andrews DW, Annicchiarico-Petruzzelli M, Antonov AV, Arama E, Baehrecke EH, Barlev NA, Bazan NG, Bernassola F, Bertrand MJM, Bianchi K, Blagosklonny MV, Blomgren K, Borner C, Boya P, Brenner C, Campanella M, Candi E, Carmona-Gutierrez D, Cecconi F, Chan FK, Chandel NS, Cheng EH, Chipuk JE, Cidlowski JA, Ciechanover A, Cohen GM, Conrad M, Cubillos-Ruiz JR, Czabotar PE, D'Angiolella V, Dawson TM, Dawson VL, De Laurenzi V, De Maria R, Debatin KM, DeBerardinis RJ, Deshmukh M, Di Daniele N, Di Virgilio F, Dixit VM, Dixon SJ, Duckett CS, Dynlacht BD, El-Deiry WS, Elrod JW, Fimia GM, Fulda S, Garcia-Saez AJ, Garg AD, Garrido C, Gavathiotis E, Golstein P, Gottlieb E, Green DR, Greene LA, Gronemeyer H, Gross A, Hajnoczky G, Hardwick JM, Harris IS, Hengartner MO, Hetz C, Ichijo H, Jaattela M, Joseph B, Jost PJ, Juin PP, Kaiser WJ, Karin M, Kaufmann T, Kepp O, Kimchi A, Kitsis RN, Klionsky DJ, Knight RA, Kumar S, Lee SW, Lemasters JJ, Levine B, Linkermann A, Lipton SA, Lockshin RA, Lopez-Otin C, Lowe SW, Luedde T, Lugli E, MacFarlane M, Madeo F, Malewicz M, Malorni W, Manic G, Marine JC, Martin SJ, Martinou JC, Medema JP, Mehlen P, Meier P, Melino S, Miao EA, Molkentin JD, Moll UM, Munoz-Pinedo C, Nagata S, Nunez G, Oberst A, Oren M, Overholtzer M, Pagano M, Panaretakis T, Pasparakis M, Penninger JM, Pereira DM, Pervaiz S, Peter ME, Piacentini M, Pinton P, Prehn JHM, Puthalakath H, Rabinovich GA, Rehm M, Rizzuto R, Rodrigues CMP, Rubinsztein DC, Rudel T, Ryan KM, Sayan E, Scorrano L, Shao F, Shi Y, Silke J, Simon HU, Sistigu A, Stockwell BR, Strasser A, Szabadkai G, Tait SWG, Tang D, Tavernarakis N, Thorburn A, Tsujimoto Y, Turk B, Vanden Berghe T, Vandenabeele P, Vander Heiden MG, Villunger A, Virgin HW, Vousden KH, Vucic D, Wagner EF, Walczak H, Wallach D, Wang Y, Wells JA, Wood W, Yuan J, Zakeri Z, Zhivotovsky B, Zitvogel L, Melino G, Kroemer G (2018) Molecular mechanisms of cell death: recommendations of the Nomenclature Committee on Cell Death 2018. Cell Death Differ 25(3):486–541. https://doi.org/10.1038/s41418-017-0012-4
3. Broz P, Dixit VM (2016) Inflammasomes: mechanism of assembly, regulation and signalling. Nat Rev Immunol 16(7):407–420. https://doi.org/10.1038/nri.2016.58
4. Broz P, Pelegrin P, Shao F (2020) The gasdermins, a protein family executing cell death and inflammation. Nat Rev Immunol 20(3):143–157. https://doi.org/10.1038/s41577-019-0228-2
5. Lopes Fischer N, Naseer N, Shin S, Brodsky IE (2020) Effector-triggered immunity and pathogen sensing in metazoans. Nat Microbiol 5(1):14–26. https://doi.org/10.1038/s41564-019-0623-2
6. Davis MA, Fairgrieve MR, Den Hartigh A, Yakovenko O, Duvvuri B, Lood C, Thomas WE, Fink SL, Gale M Jr (2019) Calpain drives pyroptotic vimentin cleavage, intermediate filament loss, and cell rupture that mediates immunostimulation. Proc Natl Acad Sci U S A 116(11):5061–5070. https://doi.org/10.1073/pnas.1818598116
7. Doitsh G, Galloway NL, Geng X, Yang Z, Monroe KM, Zepeda O, Hunt PW, Hatano H, Sowinski S, Munoz-Arias I, Greene WC (2014) Cell death by pyroptosis drives CD4 T-cell depletion in HIV-1 infection. Nature 505(7484):509–514. https://doi.org/10.1038/nature12940
8. Martin BN, Wang C, Zhang CJ, Kang Z, Gulen MF, Zepp JA, Zhao J, Bian G, Do JS, Min B, Pavicic PG Jr, El-Sanadi C, Fox PL, Akitsu A, Iwakura Y, Sarkar A, Wewers MD, Kaiser WJ, Mocarski ES, Rothenberg ME, Hise AG, Dubyak GR, Ransohoff RM, Li X (2016) T cell-intrinsic ASC critically promotes T(H)17-mediated experimental autoimmune encephalomyelitis. Nat Immunol 17(5):583–592. https://doi.org/10.1038/ni.3389
9. Arbore G, West EE, Spolski R, Robertson AA, Klos A, Rheinheimer C, Dutow P, Woodruff TM, Yu ZX, O'Neill LA, Coll RC, Sher A, Leonard WJ, Kohl J, Monk P, Cooper MA, Arno M, Afzali B, Lachmann HJ, Cope AP, Mayer-Barber KD, Kemper C (2016) T helper 1 immunity requires complement-driven NLRP3 inflammasome activity in CD4(+) T cells. Science 352 (6292):aad1210. https://doi.org/10.1126/science.aad1210
10. Johnson DC, Okondo MC, Orth EL, Rao SD, Huang HC, Ball DP, Bachovchin DA (2020) DPP8/9 inhibitors activate the CARD8 inflammasome in resting lymphocytes. Cell

Death Dis 11(8):628. https://doi.org/10.1038/s41419-020-02865-4

11. Linder A, Bauernfried S, Cheng Y, Albanese M, Jung C, Keppler OT, Hornung V (2020) CARD8 inflammasome activation triggers pyroptosis in human T cells. EMBO J 39(19):e105071. https://doi.org/10.15252/embj.2020105071

12. Brosseron F, Marcus K, May C (2015) Isolating peripheral lymphocytes by density gradient centrifugation and magnetic cell sorting. Methods Mol Biol 1295:33–42. https://doi.org/10.1007/978-1-4939-2550-6_3

Index

Thomas A. Kufer and Maria Kaparakis-Liaskos (eds.), *Effector-Triggered Immunity: Methods and Protocols*, Methods in Molecular Biology, vol. 2523, https://doi.org/10.1007/978-1-0716-2449-4,

K

L

M

N

O

P

Q

R

S

T

V

W

MIX
Papier aus verantwortungsvollen Quellen
Paper from responsible sources
FSC® C105338

If you have any concerns about our products,
you can contact us on
ProductSafety@springernature.com

In case Publisher is established outside the EU,
the EU authorized representative is:
Springer Nature Customer Service Center GmbH
Europaplatz 3, 69115 Heidelberg, Germany

Printed by Libri Plureos GmbH
in Hamburg, Germany